Cells in Evolutionary Biology

EVOLUTIONARY CELL BIOLOGY

A Series of Reference Books and Textbooks

PUBLISHED TITLES

**Cells in Evolutionary Biology: Translating Genotypes into Phenotypes –
Past, Present, Future**
Edited by Brian K. Hall and Sally A. Moody

For more information about this series, please visit:
www.crcpress.com/Evolutionary-Cell-Biology/book-series/CRCEVOCELBIO

Cells in Evolutionary Biology

Translating Genotypes into Phenotypes – Past, Present, Future

Edited by
Brian K. Hall
Sally A. Moody

CRC Press
Taylor & Francis Group
Boca Raton London New York

CRC Press is an imprint of the
Taylor & Francis Group, an **informa** business

Cover image: Sequential phases of gene expression result in the cellular processes of condensation (left) and deposition of extracellular matrix (right) to initiate cartilage formation.

CRC Press
Taylor & Francis Group
6000 Broken Sound Parkway NW, Suite 300
Boca Raton, FL 33487-2742

First issued in paperback 2020

ISBN-13: 978-1-4987-8786-4 (hbk)
ISBN-13: 978-0-367-65723-9 (pbk)

Visit the Taylor & Francis Web site at
http://www.taylorandfrancis.com

and the CRC Press Web site at
http://www.crcpress.com

Contents

Series Preface

In recent decades, the central and integrating role of evolution in all of biology was reinforced as the principles of evolutionary biology were integrated into other biological disciplines, such as developmental biology, ecology, and genetics. Major new fields emerged, chief among which are Evolutionary Developmental Biology (or Evo-Devo) and Ecological Evolutionary Developmental Biology (or Eco-Evo-Devo).

Evo-Devo, inspired by the integration of knowledge of change over single lifespans (ontogenetic history) and change over evolutionary time (phylogenetic history) produced a unification of developmental and evolutionary biology that is generating many unanticipated synergies. Molecular biologists routinely employ computational and conceptual tools generated by developmental biologists (who study and compare the development of individuals) and by systematists (who study the evolution of life). Evolutionary biologists routinely use detailed analysis of molecules in experimental systems and in the systematic comparison of organisms. These integrations have shifted paradigms and answered many questions once thought intractable. Although slower to embrace evolution, physiology is increasingly being pursued in an evolutionary context. So too, is cell biology.

Cell Biology is a rich field in biology with a long history. Technology and instrumentation have provided cell biologists the opportunity to make ever more detailed observations of the structure of cells and the processes that occur within and between cells of similar and dissimilar types. In recent years, cell biologists have increasingly asked questions whose answers require insights from evolutionary history. As just one example: How many cell types are there and how did these different cell types evolve? Integrating evolutionary and cellular biology has the potential to generate new theories of cellular function and to create a new field, which we term "*Evolutionary Cell Biology.*"

A major impetus in the development of modern Evo-Devo was a comparison of the evolutionary behavior of cells, evidenced in Stephen J. Gould's 1979 proposal of changes in the timing of the activity of cells in development (heterochrony) as a major force in evolutionary change and in Brian Hall's 1984 elaboration of the relatively small number of mechanisms used by cells in development and in evolution. Given this conceptual basis and the advances in genetic analysis and visualization of cells and their organelles, cell biology is poised to be transformed by embracing the approaches of Evo-Devo as a means of organizing and explaining diverse empirical observations and testing fundamental hypotheses about the cellular basis of life. Importantly, cells provide the link between the genotype and the phenotype, both during development and in evolution. No books that capture this cell focus exist. Hence the proposal for a series of books under the general theme of "Evolutionary Cell Biology," to document, demonstrate and establish a long-sought level in evolutionary biology, viz., the central role played by cellular mechanisms in translating genotypes into phenotypes in all forms of life.

Brian K. Hall

Sally A. Moody

Preface

Cells in Evolutionary Biology: Translating Genotypes into Phenotypes – Past, Present, Future, is the first in a series of books on *Evolutionary Cell Biology*. This book lays out and evaluates how cells have been viewed and have influenced biology, especially evolutionary biology, since 1840 when the Cell Theory was proposed, and how evolutionary theory has influenced the development of cell biology since 1859. All the authors are active researchers and scholars in one or more aspects of cell biology. They represent the best thinking on the topic from the disciplines of philosophy, history of science, developmental and evolutionary biology, and microbiology. An historical approach has been used, spanning 178 years, from the Cell Theory to the present day; no other book has such a wide ambit.

This book begins with how cells were viewed from when the Cell Theory was proposed in 1840 to around 1870 and Darwin's theory of evolution by natural selection and its application to determining relationships between animals and plants. Recognition of the differences between germ, somatic, and stem cells, discussed in Chapter 2 and of cell lineages in ontogeny and phylogeny, discussed in Chapter 3, both had enormous influences on the direction taken by evolutionary biology in the late nineteenth century.

Identification of cells as organisms (protists) led to the recognition of multiple evolutionary routes to multicellularity (Chapter 4), knowledge that is foundational to how we classify life today. Discovery that cells exchanged organelles via symbiosis (the topic of Chapter 5) undercut our organization of life into prokaryotes and eukaryotes, resulted in recognition of multiple (3–6) domains of life, and was fundamental to ongoing re-evaluations of the nature of the Tree of Life. The origination and diversification of cellular signaling centers, discussed in Chapter 6, using vertebrates as the model organisms, facilitated the diversification of tissues, organs, and divergent patterns of morphological evolution.

With increased understanding of cell reproduction, cellular mechanisms were recognized as underpinning tissue and organismal growth that was facilitated by differential mechanisms of growth, tissue-specific patterns of growth, and the evolution of altered timing in ontogeny as important mechanisms of evolutionary change (Chapter 7). Importantly, but still underappreciated, cells form the units that modulate genotype–phenotype interactions, a topic discussed in Chapter 8 in the context of cell behavior as mediating the plasticity that enables change in the phenotype during evolution. The last topic, discussed in Chapter 9, places the behavior of animal and plant cells into the context of developing patterning modules, demonstrating the fundamental evolutionary mechanisms provided by cells in mediating the interactions between genotype, environment, and phenotype in morphological evolution.

By using knowledge from fields as disparate as philosophy, history of science, developmental and evolutionary biology, molecular biology and genetics, *Cells in Evolutionary Biology* evaluates the evolution of cells themselves and the role played by cells as mechanisms of biological change at other levels. Cells provide the link between the genotype and the phenotype in development and in evolution. *Cells in Evolutionary Biology* establish the central role of cellular mechanisms in translating genotypes into phenotypes in all forms of life. To resurrect an old term from development and cytology, there is a *cenotype* between the genotype and the phenotype.

Brian K. Hall

Sally A. Moody

Editors

Brian K. Hall, Ph.D., D.Sc., LL.D.(hc), University Research Professor Emeritus at Dalhousie University in Halifax, NS, Canada, was trained in Australia as an experimental embryologist. His research concentrated on the differentiation of skeletal tissues, especially, how epithelial–mesenchymal signaling initiates osteogenesis and chondrogenesis through the formation of cellular condensations. These studies led him to earlier stages of development and the origin and function of skeletogenic neural crest cells. Comparative studies, using embryos from all five classes of vertebrates provided a strong evolutionary component to his research. These studies, along with analyses of the developmental basis of homology, played significant roles in the establishing of evolutionary developmental biology. He is a Fellow of the Royal Society of Canada, Foreign Fellow of the American Academy of Arts and Science, and recipient of the 2005 Killam Prize in Natural Sciences from the Canada Council for the Arts. He was one of eight individuals awarded the first Kovalevsky Medals in 2001 to recognize the most distinguished scientists of the twentieth century in comparative zoology and evolutionary embryology.

Sally A. Moody, MS, Ph.D., Professor and Chair of the Department of Anatomy and Regenerative Biology at George Washington University, received her Ph.D. in Neuroscience during which she studied motor axon guidance cues in the trigeminal system of the chick embryo. Throughout her career, she has continued her interest in understanding the mechanisms of axon guidance, studying the role of lineage factors in *Xenopus*, extracellular matrix proteins in chick, and genetic mutations in mouse. As a postdoctoral fellow, Sally was introduced to *Xenopus* embryos, which have remained a favorite. She made extensive fate maps of cleavage stage *Xenopus* embryos, identified maternal mRNAs that contribute to neural fate, elucidated proteomic and metabolomic changes that occur within specific lineages during cleavage, and demonstrated lineage influences on the determination of amacrine cell fate in the retina. Currently, her laboratory is studying the gene regulatory network that stabilizes neural fate downstream of neural induction, and identifying novel factors that are required for cranial sensory placode development. Dr. Moody has served on several editorial boards in the fields of neuroscience and developmental biology, on the board of directors of several societies focused on developmental processes, and edited the books: *Cell Fate and Lineage Determination* (Elsevier, 1998) and *Principles of Developmental Genetics*, 2nd ed., (Elsevier, 2014).

Contributors

R. Craig Albertson
Department of Biology
University of Massachusetts at Amherst
Amherst, Massachusetts

Ariane Dröscher
Department of Cultures and
  Civilisations
University of Verona
Verona, Italy

Kathryn D. Kavanagh
Department of Biology
University of Massachusetts
Dartmouth, Massachusetts

Eunsoo Kim
Division of Invertebrate Zoology
American Museum of Natural History
New York, New York

David Kirk
Department of Biology
Washington University in St. Louis
St. Louis, Missouri

Jane Maienschein
Center for Biology and Society
School of Life Sciences
Arizona State University
Tempe, Arizona

and

Marine Biological Laboratory
Woods Hole, Massachusetts

Shinichiro Maruyama
Department of Environmental Life
  Sciences
Division of Ecology and Evolutionary
  Biology
Tohoku University
Sendai, Japan

Vidyanand Nanjundiah
Centre for Human Genetics
Bangalore, India

Stuart A. Newman
Department of Cell Biology and
  Anatomy
New York Medical College
Valhalla, New York

Karl J. Niklas
School of Integrative Plant
  Science
Cornell University
Ithaca, New York

Andrew Reynolds
Department of Philosophy and
  Religious Studies
Cape Breton University
Sydney, Nova Scotia, Canada

Iñaki Ruiz-Trillo
Institut de Biologia Evolutiva
CSIC-Universitat Pompeu Fabra
ICREA
Barcelona, Spain

Richard A. Schneider
Department of Orthopedic Surgery
School of Medicine
University of California San Francisco
San Francisco, California

1 The Role of Cells and Cell Theory in Evolutionary Thought, ca. 1840–1872

Andrew Reynolds

CONTENTS

1.1   INTRODUCTION

The modern student of biology is likely to take for granted that all living things are composed of cells; that cells are the fundamental units of anatomy, physiology, reproduction, and development; and that the evolutionary history of life on earth is characterized by a diversification and specialization of cell types, ranging from primitive and comparatively simple single-celled bacteria to the specialized tissue cells of the multicellular plants and animals—or to use the popular nineteenth century slogan: that evolution proceeds "from monad to man."

Given the ever-expanding warehouse of facts the student of biology must assimilate in order to get up to speed with to the advancing edge of scientific research, this is quite understandable. One consequence, however, is that there is little time for learning about the history of biology, and a student might be forgiven for wondering what purpose there could be in learning about how scientists in the past got things wrong and failed to recognize what we think we know today. But for those actively engaged in research and hoping to move their field forward, an awareness of the history of their subject can help to identify unstated assumptions and provide a valuable range of alternative possibilities as to how to conceptualize and frame their investigations and research questions. Because nature itself does not tell us the terms with which we should describe it, nor how we should understand the relationships and connections between various areas of scientific research and because science is every bit as much a creative activity as it is an analytic exercise, anything that can assist the scientific imagination is to be encouraged. These are just some reasons why scientists might regard the history of science as a topic suitable for more than extra credit alone.[1]

This chapter is concerned with the history of scientific thought about cells,—particularly changes in how the cell has been defined, what have been taken to be its essential properties, and how it stands in relation to larger organisms as a whole (i.e., considerations of anatomy, physiology, and development). In addition, this chapter considers how the cell theory merged with the theory of evolution, following the publication of Darwin's *On the Origin of Species* in 1859. It then became relevant to ask how cells themselves have evolved, both as individual living units and as components of larger social aggregates which form supracellular organisms exhibiting novel features and capacities previously inaccessible to the ancestral cell lineages from which they descended. The discussion is limited roughly to the middle third of the nineteenth century (from 1838 to 1872 or so), ranging from the establishment of what is typically known as "the cell theory" of Schleiden and Schwann to Ernst Haeckel's speculations about the relationship between ontogeny and phylogeny, including an examination of Darwin's thoughts about the cell theory as expressed in his theory of pangenesis of the late 1860s.

While Darwin focused largely on the origin of higher-level taxa such as species, Haeckel and other embryologists paid special attention to the development of individual organisms. As envisioned by Haeckel and his colleagues, comparative embryology would help fill in the gaps in the fossil record of the evolutionary history of the higher taxa, with the premise being that the development of the individual organism from its component cells reveals a brief and shortened version of the evolution of the species to which it belonged—a thesis expressed more memorably by the slogan "ontogeny recapitulates phylogeny." Study of development in representatives of the "lowest" stages in the evolutionary tree (colony-forming protists and sponges for instance) promised to give a glimpse into the evolution of the very first plants and animals, and perhaps explain the basic anatomy and physiology of "higher" organisms such as ourselves.

Attitudes about the nature and the significance of the cell, however, changed considerably over the nineteenth century. Even among those committed to the theory of evolution, it was a valid question to ask whether life arose coincidentally with the first cells or whether life was older than cells?

While initially conceived as a unit of anatomical, physiological, and developmental organization chiefly of significance for understanding current plant and animal life, by mid-nineteenth century the cell came to be regarded as an "elementary organism" in its own right, leading biologists and philosophers to ask whether cells themselves

---

[1] As Thomas Kuhn wrote: "Scientific education makes use of no equivalent of the art museum or the library of classics, and the result is sometimes drastic distortion in the scientist's perception of his discipline's past. More than the practitioners of other creative fields, he comes to see it as leading in a straight line to the discipline's present vantage." Kuhn (2012, 166).

might not be composed of yet smaller submicroscopic units of an even more funda-mental nature and ancient evolutionary history. Having reduced the bodies of plants and animals to living cells, some questioned whether the phenomena associated with life might not be the properties of some smaller unit, or perhaps was inherent in the essentially amorphous and homogeneous chemical substance known as "protoplasm." The cell concept, therefore, faced challenges as the fundamental carrier of life both from below—by "protoplasmists" and proponents of various subcellular entities or molecular structures (e.g., Herbert Spencer's "physiological units," Darwin's "gemmules," or Haeckel's "plastidules")—and from above by "holists" and "organicists" who priori-tized the "organism as a whole" over any of its cellular parts.

By the beginning of the twentieth century, our current view of life as an emergent systems-level property arising from the heterogeneous elements that collectively make up the cell began to crystallize. Aside from vitalists, who insisted life is the result of some special force superadded to the regular material forces of physics and chemistry, life was considered either to be immanent in material particles or to have emerged from complex interactions among a vast number of molecules under favorable conditions.

For nineteenth century biologists cells presented both an opportunity and a chal-lenge: an opportunity to unify the diversity of living organisms under one fundamen-tal form and a challenge to explain how these fundamental living units themselves arose and evolved in the first place. If made out to be too simple (e.g., as homoge-neous and structureless clumps of protoplasm) it is difficult to explain how they manage to carry out all the various vital functions with which they are credited. Similarly, if made out to be too elemental (or "irreducibly complex"), it is difficult to explain how they managed to evolve from any simpler components. In this regard, the emergent systems conception follows a middle path—macromolecular compo-nents of the cell (e.g., enzymes) may be ascribed chemical activity while the cell as a whole system is said to be properly alive. In the words of the pioneering biochemist, Francis Gowland Hopkins (1861–1947): "we cannot, without gross misuse of terms, speak of the cell life as being associated with any one particular type of molecule... 'life,' as we instinctively define it, is a property of the cell as a whole, because it depends on the organization of processes, upon the equilibrium displayed by the totality of the coexisting phases" (Hopkins 1913, 715).[2] The challenge then, as now, was to explain what that organization is and how it comes about.

1.2 CELLS AS ANATOMICAL–PHYSIOLOGICAL–DEVELOPMENTAL UNITS: 1838–1861

The development of the first microscopes in the seventeenth century made it possible to see details of living tissue previously unwitnessed by anatomists and naturalists. In 1665, Robert Hooke described seeing a great number of minute "boxes," "pores," "chambers," or "cells" in sections of cork plant. Motivated as much by specula-tion about the ultimate composition of living bodies as by new optical technology, some natural philosophers sought to resolve organic bodies into more elementary

---

[2] Or as the embryologist E. G. Conklin put it somewhat later, "Life is not found in atoms or molecules or genes as such, but in organization" (Conklin 1940, 18).

structures such as fibers, globules, or cells. Imperfections in the design of microscope lenses that prevented seeing clear and reliable images of the fine anatomical
structures were overcome by the end of the 1830s.

A THE CELL THEORY

For purposes of pedagogical convenience, the brief histories at the beginning of
textbooks typically credit the German biologists Matthias Schleiden (1804–1881)
and Theodor Schwann (1810–1882) alone with creating the cell theory. But they
were neither the first to propose that plants and animals are composed of cells,
nor were their specific ideas about the nature and genesis of cells reflective of
what is today understood to be the cell theory (Sapp 2003; Dröscher 2014). What
Schleiden (a botanist) and Schwann (an animal physiologist) did was to articulate
and to popularize a theory that the cell is the basic unit of life. Schleiden (1838)
argued that all plants are composed of—and more importantly—by cells, which
are the elementary individuals through whose developmental activity the larger
plant body is constructed. Schwann extended this idea to animals in his essay
of 1839, illustrating by means of his own microscopic investigations of animal
development how the various tissues and organs arise from the multiplication,
modification, and in some instances the amalgamation of originally separate and
distinct cells.

Schleiden was of the opinion that new cells arise endogenously from within
existing cells, growing around a preexisting nucleus, the structure first described
by Robert Brown in 1833. The nucleus itself (or *cytoblast* as Schleiden called it), he
believed first arose from granules in a chemical "mother liquid," in a process akin
to the formation of crystals from within a supersaturated fluid medium. Schwann,
on the other hand, maintained that cells grow exogenously from an extracellular
liquid (*cytoblastema*) in the space between cells. Both men were familiar with the
claims of other naturalists that new cells were created by the division of existing cells, but they remained unconvinced of these observations and retained their
belief in the "free formation" of cells from a nutritive liquid. For Schwann, the
attribution of cell formation to physical–chemical forces rather than to the vital
action of preexisting cells satisfied a conviction that a scientific account of organismal development should be consistent with the rest of science and fueled a belief
in spontaneous generation as opposed to special creation by a supernatural agent
(Parnes 2000).

Robert Remak's (1815–1865) study of development in vertebrate embryos in
the 1850s helped to make popular the thesis that new cells arose by cell division.
This thesis became closely associated, however, with Rudolf Virchow (1821–1902),
whose famous dictum "*Omnis cellula e cellula*," states that all cells come from previously existing cells (Virchow 1855). This amendment of the theory of Schleiden
and Schwann established what is typically understood today to be the cell theory:
that all living organisms are composed of one or more cells; that cells are the fundamental living units; and that all cells arise from preexisting cells by binary fission.
Virchow championed this version of the cell theory through his influential book

Cellularpathologie (1858), in which he provided a new account of health and disease firmly rooted in normal and abnormal cell activity.

In the following year, Charles Darwin (1809–1882) published his views on the common evolutionary origins of all living things (Darwin 1859). As the historian Thomas S. Hall explained, Virchow's *"Omnis cellula e cellula"* statement was significant for Darwin's theory because it "supplied the physical basis for that larger continuity of life as a whole which began, according to Darwin, when God first breathed life into an original cell or cells, which has culminated in the variety of forms that inhabit the earth today. The cell, for those who saw it as the irreducible life unit, was thus the basis of the whole history of life" (cited from Hall 1969, 206–207).[3]

B   THE CELL CONCEPT

As the cell theory was becoming better established, researchers were reconsidering the cell concept itself.

Originally used by Hooke to emphasize an empty space or chamber characterized by a solid enveloping wall, investigators were beginning to note that many so-called "cells" lacked any discernible membrane let alone a rigid wall (e.g., the amoeboid "swarmer" cells of fungi and algae, the ova, blood, and even tissue cells of higher animals).[4] Given the vast diversity in cell morphology throughout the organic kingdoms, rejection of an outer wall as an essential characteristic made defining the cell more difficult. Attention turned to the contents of the cell vesicle, to the sticky semifluid substance within. This was variously known as "sarcode," with respect to the infusoria or protozoa, and as "protoplasm" in the case of plants and (rather confusingly) animal embryonal cells. Eventually, it was agreed by most that sarcode and protoplasm were one and the same substance, which provided a means for unifying all the various forms of living beings.

In 1861, Max Schultze proposed a new definition, whereby, a cell was understood to mean essentially "a naked clump of protoplasm containing a nucleus" (Schultze 1861). This protoplasmic cell concept avoided reference to any specific morphological feature aside from a nucleus. In the same year, Ernst Brücke, made popular yet another perspective on cells when he referred to them as "elementary organisms" (Brücke 1861). The suggestion that cells, including those of which human and other animal bodies are composed, are themselves organisms, had obvious resonance with the thesis that all living things share a common evolutionary origin from some more ancient and less complex form of life. Additionally Darwin's younger colleague and disciple T. H. Huxley (1825–1895) had been quite critical of the original cell theory (cf. Huxley 1853), he would go on to

---

[3] It should be noted, however, that in the passage alluded to, Darwin did not speak of cells *per se*, but of life being breathed into "a few forms or into one" (Darwin 1859, 490). It is also worth noting that in the first edition of the *Origin*, Darwin did not mention who or what breathed life into the first form or forms, explicit reference to "the Creator" being added to subsequent editions at the behest of his wife, Emma Darwin.

[4] Cf. Kölliker (1845), Leydig (1857), de Bary (1859).

champion—against vitalists and opponents of evolution—the idea that protoplasm is the "physical basis of life" (Huxley 1868).[5]

1.3   CELLS AS EVOLUTIONARY UNITS: 1844–1868

Having started as a significantly morphological concept, the cell was now increasingly thought of as an elementary organism composed of protoplasm, the fundamental physical–chemical stuff of life. Textbook illustrations of the cell concept from
the mid-nineteenth century on frequently featured amoebae, the supposedly simplest
and most primitive cells and organisms. Often described as mere specks of shape-
less protoplasm, amoebae were notable for their close resemblance to colorless blood
corpuscles and other "amoeboid" cells of humans and other vertebrates.[6] In the context of the Darwinian theory of descent the amoeba exemplified ideas of progressive
transmutation from monad to man and the spontaneous generation of simple life
forms from primitive material conditions under the guidance of natural law.

As the historian Stephen Jacyna (1984) explains, the romantic philosophy
endemic in much of early nineteenth century European and British science, with
its ideas of the unity of life and nature and transmutation of higher forms from
lower, provided fertile soil for the cell theory. This is illustrated by one of the most
sensational examples of evolutionary writing prior to Darwin's *Origin of Species*,
Robert Chambers' *Vestiges of the Natural History of Creation*, published anony-
mously in 1844. Chambers remarked that "It is ascertained that the basis of all
vegetable and animal substances consist of nucleated cells; that is, cells having
granules within them" (Chambers 1994, 170). Chambers also noted the morphological analogy between the reproductive sponge gemmule, the colonial freshwater
algae *Volvox globator* and the early stage mammalian embryo (ibid., 172), and he
discussed evidence for the creation of globules or cells in albumen by means of
electricity (ibid., 173). The "principle of development," Chambers wrote, whereby
"The whole train of animated beings, from the simplest and oldest up to the highest
and most recent, are, then, to be regarded as a series of *advances*...arranged in the
counsels of Divine Wisdom" (ibid., 203); and this he insisted leads us to conclude
that: "The nucleated vesicle, the fundamental form of all organization, we must
regard as the meeting-point between the inorganic and the organic—the end of
the mineral and beginning of the vegetable and animal kingdoms, which thence
start in different directions, but in perfect parallelism and analogy" (Chambers
1994, 204).

When Darwin finally published his evidence for and thoughts on evolution in 1859
he had very little to say about cells other than to note the shared cellular construction
of all living things as evidence for their common descent (Darwin 1859, 484). The
task of elaborating the relevance of the cell theory for the theory of evolution was
enthusiastically taken up by the German zoologist Ernst Haeckel (1834–1919), whose

---

[5] A motivating factor for Huxley's critical attitude toward the Schleiden–Schwann cell theory, as
explained by Richmond (2000), was what he perceived to be its preformationist assumption that
organic form exists already packaged somehow in the cell or the cell nucleus. As an epigeneticist
Huxley believed form gradually develops from a less perfectly structured material.

[6] Reynolds (2008).

specialties were in microscopic single-celled organisms, marine invertebrates, and comparative embryology.[7]

Haeckel was an early proponent of the thesis promoted by Carl von Siebold (1804–1885) that the microscopic creatures known as infusoria or protozoa are unicellular, the entire body consisting of a single cell despite the internal complexity of the paramecia and other ciliates.[8] To the traditional two system classification of plants and animals Haeckel proposed adding a third kingdom comprising all the unicellular forms, that which he called the Protista (Haeckel 1866). Within the Protista Haeckel further distinguished the Monera, a group of microscopic organisms so simple he claimed that they lacked even a nucleus. From a morphological perspective, these supposedly structureless clumps of homogenous protoplasm, typically amoeboid in nature according to Haeckel, had not yet achieved the level of true cells, and for this reason he named them "cytodes" or "organisms without organs" (Haeckel 1870a).

Viewed phylogenetically, cytodes represented the very earliest and most primitive living organisms, which Haeckel believed had formed by spontaneous generation from the chemical elements at the bottom of the ocean.[9] The next stage in evolution would be the differentiation of a portion of the homogenous protoplasm into a nucleus, followed by the creation of simple colonies of these protist cells, and the gradual differentiation of cells into the primary germ layers of the simplest invertebrate marine animals with specialized tissues and organs. Haeckel's "fundamental biogenetic law" declared ontogeny (or the development of the individual animal) to be a brief and condensed recapitulation of the phylogeny of the branch of the evolutionary tree to which it belonged. In this way comparative embryology could be used to reveal the true phylogenetic system of relationships between groups of organisms, and extant organisms of "primitive" organization could serve as evidence of ancient ancestral types. I will return to Haeckel's recapitulation theory later in Section 1.4.

Initially, the cell theory was founded on the idea that cells are the fundamental units of life: that the physiology, reproduction, and development of living organisms is carried out by these ultimate living units. The protoplasm theory involved an attempt to associate life with a simple homogenous substance, but to stop there would be an evasion of the question of how this simple slime-material manages to perform all these feats. This is all the more apparent when we consider the practice of referring to the cell as an elementary organism. To say that multicellular organisms are alive because they consist of more elementary living organisms suggests an explanatory regress. Biologists of the nineteenth century were aware of this and many attempted to explain how the cell or protoplasm was capable of growth, irritability, reproduction, and development by pointing vaguely to its chemical properties, and so ultimately to various molecular and atomic forces.[10] But an alternative

---

[7] Haeckel's contributions to biology were spread over an extensive collection of publications in professional journals and in scientific monographs, but the general details of his views on cells and evolution can be found in his *Generelle Morphologie* (1866) and the *Natürliche Schöpfungsgeschichte* (first edition 1868 followed by subsequent editions, the 11th in 1911).

[8] Von Siebold (1848).

[9] As evidence of this, Haeckel cited the unfortunate specimen *Bathybius haeckelii*, described by Huxley in 1868, but later shown to be an inorganic artifact of the preservation method of samples of deep sea mud. See Rehbock (1975).

[10] See Geison (1969a).

strategy was to hypothesize the existence of some sub-cellular, macromolecular component living within the protoplasmic substance that would be responsible for the physiological activity in question.

A THE SUBCELLULAR MENAGERIE AND EXPLANATIONS OF THE LIFE OF THE CELL

Ernst Brücke's rationale for describing the cell as an elementary organism was not that he considered them to be simple or elementary in structure, indeed just the opposite. What he meant is that the cell is elemental in the same sense as are the chemical elements, viz. that they cannot be reduced to more primitive units while retaining the essential properties by which they are characterized. Brücke highlighted evidence suggesting that the cell must contain a complicated organization of component parts (his term *Werkstücke* has an industrial-machine connotation in German). While he did consider the possibility that this cellular organization might include even smaller and more elementary organisms, he saw no good evidence for that conclusion at that time (Brücke 1861). Others were not so hesitant to develop speculative hypotheses involving the existence of more elementary units or organisms residing within cells. A partial list of this subcellular menagerie would include the following: Karl von Nägeli's "micelles," Herbert Spencer's "physiological units," Charles Darwin's "gemmules," Ernst Haeckel's "plastidules," August Weismann's "biophors," Richard Altmann's "bioblasts," Hugo De Vries's "pangens," Oscar Hertwig's "idioblasts," and Max Verworn's "biogens".

As Hall (1969) explains there have been two basic approaches toward accounting for life: (1) preformationist–vitalist accounts that attribute life's properties to basic living units (e.g., protoplasm, physiological units or micelles), and (2) systems accounts according to which life is a property emergent from the peculiar organization of a complex system of nonliving parts or components. According to the approach taken, life is either immanent (i.e., present at the lowest levels of matter) or emergent. Accordingly, those who speculated about the existence of either visible or invisible subcellular parts may have considered them to be living or merely necessary elements for the cell to be alive. Daniel Nicholson has recently referred to the idea that life must be associated with some ultimate and indivisible vital unit as "biological atomism." "The activity of a living organism" he writes, "is thus conceived as the result of the activities and interactions of its elementary constituents, each of which individually already exhibits all the attributes proper to life" (Nicholson 2010, 203). The cell itself, therefore, either does or does not count as a "biological atom" depending upon how one accounts for its vital properties.

In the following, I restrict my attention to just two of the proposals listed earlier: those of Spencer and Darwin. I choose Darwin for the rather obvious reason that any discussion of cells and evolution ought to include his views on the matter, but also because his opinions on the cell theory have not previously received much attention (though see Müller-Wille 2010). I discuss Spencer's views because they feature rather prominently in Darwin's own thinking about evolution at a microscopic scale.

B Spencer's "Physiological Units" (1867)

The philosopher Herbert Spencer (1820–1903) was an advocate for evolution (or what he called the "development hypothesis") as early as 1852. He was a close acquaintance and correspondent with the likes of T. H. Huxley, J. D. Hooker, and Charles Darwin. In his *Principles of Biology* (1865) Spencer coined the phrase "survival of the fittest" as a less misleading metaphor to replace Darwin's "natural selection." Spencer's lifelong project was to formulate a comprehensive philosophical system deduced from highly abstract first principles that would incorporate physics, biology, psychology, and sociology. *The Principles of Biology* comprised the biological component of Spencer's so-called "synthetic philosophy" and covered the whole range of organic phenomena: from the growth of crystals and the origin of life, to morphological and physiological considerations of living organisms, and the evolution of species diversity. In a chapter on "Waste and Repair" (Chapter 4) Spencer discussed the phenomena of limb regeneration and asexual reproduction by cutting in plants and some lower animals (e.g., begonia and hydra). Rejecting the preformationist hypothesis that the mature parts exist somehow already formed though in miniature within the affected tissue, Spencer insisted, "We have therefore no alternative but to say, that the living particles composing these fragments, have an innate tendency to arrange themselves into the shape of the organism to which they belong. We must infer that a plant or animal of any species, is made up of special units, in all of which there dwells the intrinsic aptitude to aggregate into the form of that species: just as in the atoms of a salt, there dwells the intrinsic aptitude to crystallize in a particular way" (Spencer 1865, 180–181). The intrinsic power on the part of these vital units to assume the correct shape proper to the lost body part or to recreate the whole organism anew Spencer referred to as "polarity." Spencer also extended his discussion of "polarity" to the phenomenon of heredity, whereby, a new organism of the correct shape and form arises from a part (the gamete cells) of a previously existing organism of the same species.

The hypothetical entities responsible for these feats of reproduction and regeneration Spencer called "physiological units" (ibid., 254). These he surmised would be complex macromolecular aggregates distinct from and intermediate between both the *chemical units* of albumin and protein, which are shared alike in the protoplasm of all living bodies, and from the *morphological units* or cells of which living bodies are typically composed. On this latter point, Spencer explained that the cell theory is only approximately true, citing as evidence counter to its universal applicability as follows: (1) the development of fibrous tissue from a "structureless blastema" (ibid., 183) (recall Schwann's idea of free or endogenous cell growth); and (2) the existence of Rhizopods and other amoeboid "specks of protoplasm" which he claims "are not cellular" in organization (ibid.).[11]

In the second volume, Spencer explains the limits of the cell theory with the analogy of a house constructed from clay. While it is true that some homes are constructed

---

[11] Spencer's rationale for denying a cellular organization to these organisms would seem, based on the discussion in the second volume (Spencer 1867, 78), to rely on their lack of any discernible cell wall or membrane. That Spencer still insisted on this feature as a necessary condition for a cell at the time when Schultze's protoplasmic version of the cell concept was catching on is a bit surprising.

from the aggregation of clay bricks (cells in the case of an organism), other simpler homes are made of unmolded clay (protoplasm); nor is every component of a house that is for the most part composed of clay bricks made of bricks. For instance, the chimney pots, drain pipes, and ridge tiles may be formed directly from clay without having first assumed the morphology of a brick, and just so some organic structures may be directly built from unmolded protoplasm without assuming a cellular form (Spencer 1867, 10–11). Huxley's critical review of the cell-theory of 1853 is cited as authoritative support (ibid., 13).

If then cells are not a universal feature of all living organisms and organic form, Spencer reasoned, some other agency or agent must be responsible for the phenomena of growth, development, reproduction, and regeneration. This was his reason for proposing the *physiological units*. By the standards of modern biology, Spencer credited his physiological units with both the heredity function associated with today's genes (nucleic acids) and the constructive developmental ability to create differentiated cells, tissues, and organs that is largely attributed to the activity of proteins and other macromolecules. But like many, at this time, Spencer did not separate the roles of heredity and development—(nor, as we shall see, did Darwin). Spencer proposed that the physiological units must have a plastic nature, thereby, permitting them to be modified by changes in environmental conditions and through the exercise of mature organs, so as to transmit these acquired traits to future generations.

Like Haeckel, Spencer wrote that the theory of evolution suggests that simpler "structureless portions of protoplasm" would have preceded the appearance of cells with nuclei (Spencer 1867, 13) and that the physiological units which comprise protoplasm would have been preceded by yet simpler organic "colloidal" ("jelly-like") molecules, and so on back to the basic elements constitutive in living bodies (carbon, oxygen, hydrogen, and nitrogen) in their simplest and least dynamical or vital arrangements.

C DARWIN'S "GEMMULES" AND THE PANGENESIS HYPOTHESIS (1868)

As the preeminent evolutionary thinker of the nineteenth century, Darwin's thoughts on the cell theory are of special interest. But what we find when we turn to Darwin's statements on the matter of cells and the cell theory may strike the modern reader as a little surprising in two respects. First, for the relative scarcity of discussion in his writings on what one might think is such an important topic for his theory of evolution. Second, for what seems from a modern perspective, according to which students are taught that evolution and the cell theory are the twin crown achievements of nineteenth century biology, to be Darwin's rather tepid endorsement of the latter. Aside from using the term "cell" and cognates (e.g., "cellular") in its anatomical and histological sense in some of his specialist works on barnacles, plants, and worms, Darwin discussed the theory of the cell and its implications for evolution only sparingly. As previously mentioned, *The Origin of Species* does cite the "cellular structure" common to plants and animals as evidence of their shared ancestry from some ancient form of life (Darwin 1859, 484), but the only explicit discussion of the cell theory occurs in *The Variation of Animals and Plants under Domestication* (1st edition 1868, 2nd edition 1875).

This two-volume work was motivated by Darwin's desire to fill in the gap regarding inheritance and variation in the *Origin*'s argument, and it was here that Darwin offered his "provisional hypothesis of pangenesis" featuring the hypothetical subcellular particles he dubbed "gemmules." Like Spencer's hypothesis about physiological units, Darwin's pangenesis was not only an attempt to account for the facts of heredity but of development as well. Through it, Darwin attempted to unify several distinct processes, such as trait inheritance, reproduction (both sexual and asexual), embryonic development, wound repair, and regeneration of lost limbs; all by appeal to the activities of his hypothetical gemmules.[12]

In short, Darwin proposed that throughout the lifetime of a plant or animal, each of its cells casts off small molecular units, or gemmules, which circulate throughout the body and by a mutual attraction or "elective affinity" tend to aggregate especially in the reproductive elements, the gamete cells. Each gemmule was supposed to carry what we would today call "information" about the specialized adult tissue cell from which it originated, and in this way Darwin thought to explain inheritance of innate and acquired characteristics, the occasional reversion to features last seen in a grandparent or more remote generation (on the assumption that gemmules may lie dormant for a generation or more like some plant seeds), wound repair and regeneration of lost limbs, and the development of an adult organism with specialized tissues and organs from an originally undifferentiated zygote. This is but a too brief sketch of what Darwin admitted to be a complicated hypothesis (but not, he remarked (Darwin 1868 II, 402), more complex than the facts they were intended to explain). My chief concern, however, is with Darwin's remarks about the cell theory and his understanding of the relationship between his proposed gemmules and cells. For that reason, I reproduce below two key passages from *Variation of Animals and Plants* so that Darwin's own words will be clearly before us.

Darwin begins his discussion of the cell theory on p. 368 of the second volume of the first edition in a section titled 'The Functional Independence of the Elements or Units of the Body'. Here he cites Virchow and Claude Bernard (1813–1878) as authorities for the judgment that, "Physiologists agree that the whole organism consists of a multitude of elemental parts, which are to a great extent independent of each other" (Darwin 1868 II, 368).[13] Moving downward from larger units to smaller, Darwin writes that each organ enjoys an autonomous life and development within the animal body, while each organ and major system is composed of "an enormous mass of minute centres of action" or cells (ibid., 369). This view of the "independent life of each minute element of the body" was important to Darwin's approach to the problems of inheritance, variation, and reproduction, for it allowed him to propose the independent activity of his gemmule-particles.

---

[12] The term "gemmule" was originally used from the beginning of the nineteenth century to denote small asexual reproductive bodies in some plants and sponges. Today it is typically restricted to the multicellular masses or "buds" found in freshwater sponges. These act like spore sacs to preserve the organism/colony over times of harsh conditions.

[13] Regarding the cell theory, Darwin also cites the 1858 English translation of Hugo von Mohl's "The Vegetable Cell," originally published in German in 1853 (Darwin 1868, II 346, n. 22).

I reproduce now the first passage, in which Darwin discusses the cell theory, with footnote material inserted in square brackets, following the original footnote number for convenience. Darwin wrote:

Whether each of the innumerable autonomous elements of the body is a cell or the modified product of a cell, is a more doubtful question, even if so wide a definition be given to the term, as to include cell-like bodies without walls and without nuclei.[25] [For the most recent classification of cells, see Ernst Häckel's "Generelle Morpholog.," Band ii., 1866, s. 275.] Professor Lionel Beale uses the term "germinal matter" for the contents of cells, taken in this wide acceptation, and he draws a broad distinction between germinal matter and "formed material" or the various products of cells.[26] ["The Structure and Growth of Tissues," 1865, p. 21, & c.]. But the doctrine of *omnis cellula e cellulâ* is admitted for plants and is a widely prevalent belief with respect to animals.[27] (Dr. W. Turner, *"The Present Aspect of Cellular Pathology,"* *"Edinburgh Medical Journal,"* April, 1863.) Thus Virchow, the great supporter of the cellular theory, whilst allowing that difficulties exist, maintains that every atom of tissue is derived from cells, and these from pre-existing cells, and these primarily from the egg, which he regards as a great cell. That cells, still retaining the same nature, increase by self-division or proliferation, is admitted by almost every one. But when an organism undergoes a great change of structure during development, the cells, which at each stage are supposed to be directly derived from previously-existing cells, must likewise be greatly changed in nature; this change is apparently attributed by the supporters of the cellular doctrine to some inherent power which the cells possess, and not to any external agency.

Another school maintains that cells and tissues of all kinds may be formed, independently of pre-existing cells, from plastic lymph or blastema; and this it is thought is well exhibited in the repair of wounds. As I have not especially attended to histology, it would be presumptuous in me to express an opinion on the two opposed doctrines. But everyone appears to admit that the body consists of a multitude of "organic units,"[28] This term is used by Dr. E. Montgomery ("On the Formation of so-called Cells in Animal Bodies," 1867, p. 42), who denies that cells are derived from other cells by a process of growth, but believes that they originate through certain chemical changes, each of which possesses its own proper attributes, and is to a certain extent independent of all others. Hence it will be convenient to use indifferently the terms cells or organic units or simply units. (Darwin 1868 II, 370–371)

These remarks give the impression that Darwin was somewhat agnostic on the question of the "cellular-doctrine," or that perhaps like Spencer he considered it to be of less than universal validity. Note also his remark about "another school of thought" that admits the free formation of new cells and tissues from "plastic lymph or blastema" independent of cell-division; viz. the Schleiden–Schwann theory which analogized the generation of new cells to the process of crystallization. Darwin appears to reserve judgment on whether cells are the causal agents responsible for the formation of differentiated tissue or whether some other force or forces—originating perhaps from the environment external to cells—is responsible for molding homogeneous cells into specialized tissues and organs.

I return to the question of free-cell formation after we consider the second chief passage, in which Darwin explains his "Provisional hypothesis of pangenesis." Some words have been italicized to emphasize the peculiar cast of Darwin's ideas and how they diverged from the generally accepted version of the cell theory of the time. Darwin writes:

> It is *almost* universally admitted that cells, or the units of the body, propagate them-selves by self-division or proliferation, retaining the same nature, and ultimately becoming converted into the various tissues and substances of the body. *But besides this means of increase* I assume that cells, before their conversion into completely passive or "formed material," throw off minute granules or atoms, which circulate freely throughout the system, and when supplied with proper nutriment multiply by self-division, *subsequently becoming developed into cells like those from which they were derived.* These granules for the sake of distinctness may be called cell-gemmules, or, *as the cellular theory is not fully established,* simply gemmules. They are supposed to be transmitted from the parents to the offspring, and are generally developed in the generation which immediately succeeds, but are often transmitted in a dormant state during many generations and are then developed. Their development is supposed to depend on their union with other partially developed cells *or gemmules* which precede them in the regular course of growth…Gemmules are supposed to be thrown off by every cell or unit, not only during the adult state, but during all the stages of develop-ment. Lastly, I assume that the gemmules in their dormant state have a mutual affinity for each other, leading to their aggregation either into buds or into the sexual elements. Hence, speaking strictly, it is not the reproductive elements, nor the buds, which gener-ate new organisms, but the cells themselves throughout the body. These assumptions constitute the provisional hypothesis which I have called Pangenesis. (Darwin 1868 II, 374) (italics added)

Two things are worth noting here. One is the continuation of Darwin's apparent reserva-tions about the universal validity of the cell theory, (specifically, that all organic tissues are composed of cells and that each cell originates from the division of a previous cell). The other is the apparent suggestion that, aside from the widely recognized process of cell proliferation by division, new cells can arise when gemmules accumulate in condi-tions of proper nutriment. When aligned with the statement that gemmules circulate freely throughout the body, this sounds as though Darwin is himself advocating a form of the free-cell formation thesis. If correct, this would explain his cautious remarks regarding the "cellular theory" espoused by the likes of Virchow, Haeckel, and others.

Darwin's caution toward the cell theory would also be explicable if Hughes (1959, 77) were correct that Darwin had conceived his "pangenesis" hypothesis in the 1840s, years before the publication of Virchow's *Cellular Pathology* and at a time when the free-cell formation was widely accepted among British scientists. Geison (1969b), however, argues that while Darwin was certainly interested in the issue of inheritance in the 1840s, he did not work out the specifics of the pangenesis hypothesis until the early 1860s; which means the English translation of Virchow's *Cellular Pathology* (published in 1860) was available to him. Even so, it was not well established that cell-division is the sole means of cell generation until after 1875, and then still the existence of syncytia—(continuous masses of protoplasm with multiple nuclei such as are commonly found in muscle tissue and in the developmental stages of some animals)—continued to be cited as evidence against the cell-doctrine into the 1890s.

In an extended footnote (Darwin 1868, p. 375, n. 29), Darwin compared his "gemmules" to Spencer's "physiological units," which he says "agree with my gemmules" in several respects, including being "the efficient agents in all the forms of reproduction and in the repairs of injuries; they account for inheritance" (though are not brought to bear by Spencer on reversion or atavism to Darwin's surprise). A key difference, Darwin notes, is that several gemmules or a mass of them are required "for the development of each cell or part." This too, therefore, gives the impression that Darwin thought of his gemmules as having the power to create cells independently of the division of a previously existing cell.

This impression is strengthened when Darwin states: "Physiologists maintain, as we have seen, that each cell, though to a large extent dependent on others, is likewise, to a certain extent, independent or autonomous. I go one small step further and assume that each cell casts off a free gemmule, *which is capable of reproducing a similar cell*" (Darwin 1868 II, p. 377) (italics added). The crucial question is, what does Darwin mean exactly by saying a gemmule (or a number of them) is capable of "reproducing" a similar cell? Some light appears to be thrown when he later explains how he conceives of the gemmules as entering into cells and metaphorically "fertilizing" them (ibid., pp. 388, 389). This, when read in tandem with the statement that a cell is requisite for the gemmule's development (ibid., p. 381, and p. 388), would seem to settle the matter against any suggestion of free-cell formation. Even more definitive would appear to be a letter written to Miles Joseph Berkeley on September 7, 1868 in which Darwin emphatically states: "I have never supposed that they [gemmules] were developed into free cells, but that they penetrated other nascent cells and modified their subsequent development." (Darwin Correspondence Project online, letter DCP-LETT-6353).

And yet, in an earlier letter (June 12, 1867) from Darwin to T. H. Huxley explaining that he was unhappy with "pangenesis" as a term for his hypothesis, he states:
Now I want to know whether I could not invent a better word.

Cyttarogenesis, i.e., cell-genesis is more true & expressive but long.—
Atomogenesis, sounds rather better, I think, but an "atom" is an object which cannot be divided; & term might refer to the origin of atom of inorganic matter. (Darwin Correspondence Project on-line, letter DCP-LETT-5568)[14]

It seems the key point is that, Darwin was not really thinking of the genesis of a *de novo* cell, but the genesis of a *differentiated* cell from one in a previous state of less differentiation, what Darwin describes multiple times as a "nascent" cell.

If Darwin seems to have been a bit unclear in his mind, or at least in his language, in the period 1867–1868, that was no longer the case by 1875 when the second edition of *Variation* appeared. Several important changes were made in this revised edition. For instance, Darwin revised the line, "It is almost universally admitted that cells…propagate themselves by self-division…" from the 1st edition, to "It is universally admitted…" (Darwin 1875, 369); and the line introducing the gemmules,

---

[14] "Cyttarogenesis" had been suggested by his son George Howard Darwin (1845–1912) who was at the time a student at Cambridge (cf. DCP-LETT-5561, June 3, 1867). I am most grateful to Rosemary Clarkson of the Darwin Correspondence Project at the Cambridge University Library for answering my inquiry about some of Darwin's letters.

which originally seemed to express some tentativeness about the cell theory ("These granules for the sake of distinctness may be called cell-gemmules, or, as the cellular theory is not fully established, simply gemmules") was simplified to "These granules may be called gemmules" (Darwin 1875, 369). Moreover, he now states clearly that:

> It has also been assumed that the development of each gemmule depends on its union with another cell or unit which has just commenced its development, and which precede it in due order of growth…As the tissues of plants are formed, as far is known, only by the proliferation of preexisting cells, we must conclude that the gemmules derived from the foreign pollen do not become developed into new and separate cells, but penetrate and modify the nascent cells of the mother plant…In this case and in all others the proper gemmules must combine in due order with preexisting nascent cells, owing to their elective affinities. (1875 II, 375)

Darwin, like other biologists of his time, believed that the tasks of heredity—of transmitting what we would today call the genetic "information" for form and function—and of development were carried out by one and the same agent. For Darwin, gemmules; for Spencer, physiological units; for Haeckel (1870b), plastidules, and so on.[15] Today we separate the genetic function, performed by the nucleic acids (DNA, RNA), from the task of development, whereby, specialized tissues, organs, and organisms are constructed by cells themselves, largely through the activity of their protein "machinery." Darwin, however, wanted a hypothesis to unite the phenomena of development, regeneration, variation, and inheritance because he believed the capacity of a hypothesis to pull off what William Whewell called a "consilience of inductions" to be a sign of its verisimilitude.[16]

In general, the cell theory was important for Darwin's approach to the problem of inheritance and development because it encouraged him to think of the life of plants and animals as involving a hierarchy of distributed and relatively autonomous agents (Müller-Wille 2010). This is aptly illustrated in the memorable passage from the concluding section on pangenesis where Darwin writes: "We cannot fathom the marvelous complexity of an organic being, but on the hypothesis here advanced this complexity is much increased. Each living creature must be looked at as a microcosm—a little universe, formed of a host of self-propagating organisms, inconceivably minute and as numerous as the stars in heaven" (Darwin 1868 II, 404).[17]

---

[15] Haeckel's "plastidules" were hypothetical molecular units comprising the "plasson" or protoplasm of cells and cytodes, which according to his hypothesis were capable of transferring specific frequencies of molecular vibration from one generation to the next and so transmitting hereditary information.

[16] Darwin (1868 II, 357) cites Whewell in defense of the use of hypotheses in the section devoted to pangenesis, and discusses the consilience of inductions (though not using that specific term) in regards to natural selection in the first volume (Darwin 1868 I, 8f). See Ruse (2000) for discussion of Darwin's ideas about the evidentiary aspects of hypothesis and the influence of Whewell and Sir John Herschel on Darwin's understanding of the methodology and philosophy of science.

[17] This passage, incidentally, led Hall (1969, 321–322) to conclude that gemmules were in Darwin's mind "constitutive as well as genetic," that is, that they are supposed to be the true units of life, more fundamental even than cells. But that would assume, and I think incorrectly for reasons outlined in this paper, that Darwin intended them to be capable of free-cell formation. Darwin's gemmules, I believe, were intended to be active without being fully living. The discussion and title of a paper like (Ryder 1879), however, suggests that confusion remained in Darwin's time about whether or not gemmules were intended to be autonomous of cells.

This view of the organism as a complex and compound arrangement would gradually be extended to each individual cell by those like the cytologist Oscar Hertwig (1849–1922) who, echoing Darwin's language, declared that the cell itself "is a marvellously complicated organism, a small universe, construction of which we can only laboriously penetrate by means of microscopical, chemicophysical, and experimental methods of inquiry" (Hertwig 1895, ix). Such sentiments, extending back to Brücke's essay of 1861, should give the lie to the claims of Intelligent Design proponents that the cell was regarded by Darwin and his contemporaries as a simple homogenous lump of protoplasm. One needs to read beyond the popular pronouncements of those like Huxley and Haeckel made for nonscientific audiences to learn what biologists of the time actually thought.

1.4   CELLS AS UNITS OF ONTOGENY AND PHYLOGENY: RECAPITULATION AND HAECKEL'S GASTRAEA THEORY (CA. 1872)

Despite this turn toward speculation about ultramicroscopic units, not all attention was concentrated at the subcellular level. In fact, the efforts of embryologists of the nineteenth century largely remained focused on the gross cellular events leading to the formation of specialized tissues and other structures pertinent to the development of organs and organisms as a whole. While scientists could only speculate about what was going on inside cells, their movements and transformations into differentiated tissues and organs could be observed with the microscope. This was most easily done with frogs, chickens, and a range of sea creatures with oviparous development. Embryologists of the early nineteenth century such as Christian Pander (1794–1865) and Karl Ernst von Baer (1792–1876) had resolved the early vertebrate embryo into three primordial "germ layers" from which the major organs and body parts gradually develop. These eventually came to be known as the ectoderm (outer layer), mesoderm (middle layer), and endoderm (inner layer). Extension of the germ layers to the invertebrates soon followed, with Martin Rathke (1793–1860) exhibiting in 1825 their presence in the development of the crayfish, T. H. Huxley establishing in 1849 that the two layers of tissues in adult jellyfish are homologous to the ectoderm and endoderm layers of the vertebrate embryo, and in the 1860s Alexander Kovalevsky (1840–1901) provided evidence that the germ layers are present throughout the invertebrates. This germ layer theory played a significant role in the attempts to establish a rational system of animal taxonomy.

Another idea popular among early nineteenth century thinkers, such as Johann Friedrich Meckel (1781–1833) and Étienne Serrès (1786–1868), was that the sequence of events exhibited in the development of one of the "higher" animals paralleled the series of beings represented by the *scala naturae*, from simplest to most perfect. In the words of Serrès: "The development of the individual organism obeys the same laws as the development of the whole animal series; that is to say, the higher animal, in its gradual evolution, essentially passes through the permanent organic stages which lie below it."[18]

---

[18] Quoted in Coleman (1977, 50).

This thesis of recapitulation was also promoted by leading scientists such as Louis Agassiz (1807–1873) and Richard Owen (1804–1892), who though rejecting the idea of evolution or transmutation from one ideal form to another, believed in a threefold parallelism between the great chain of being, embryogenesis, and the fossil record. The parallel in these distinct series was for Agassiz and Owen signs of a unified and specially designed, divine plan. Darwin himself helped to show how developmental history could be used to uncover the natural relationship between barnacles and the other crustacea (Darwin 1851–1854).[19] Darwin used homologies in developmental structure as a guide to taxonomic relationships among barnacle and other species, and in the *Origin* presented the evidence for his contention that "community in embryonic structure reveals community of descent" (Darwin 1859, 449).

Darwin had also written in the *Origin* that "Embryology rises greatly in interest when we thus look at the embryo as a picture, more or less obscured, of the common parent-form of each great class of animals" (ibid., 450). This inspired the German naturalist Fritz Müller (1821–1897) to investigate further the similarities in developmental structure and processes among the crustacea. In his 1864 collection of essays *Für Darwin*, Müller revolutionized the thesis of recapitulation, arguing that individual development proceeds through a series of stages laid down by earlier ancestors, which are more or less faithful to the original depending upon the degree to which changing environmental conditions have placed a selective pressure on larval or embryonal forms to adapt and change in order to survive (Müller 1864).

This evolutionary version of the recapitulation doctrine was enthusiastically taken up by Haeckel who declared it the "fundamental biogenetic law": that ontogeny recapitulates phylogeny.[20] In this way, in the absence of a clear and complete paleontological fossil record, comparative embryology could be used to fill in the gaps and to reveal the true phylogenetic relationships between species. Moreover, extant organisms of "primitive" organization—those whose stage of development had failed to progress any higher in the evolutionary tree—provided additional evidence of ancient ancestral form.

In a series of publications between 1866 and 1871, the Russian biologist, Alexander Kovalevsky, used embryological development to reveal the close relationship between what was considered to be the most primitive vertebrate, the lancet or Amphioxus, and the larval tadpole of the invertebrate ascidian or tunicate. Like Amphioxus and other vertebrates, the ascidia develop a notochord during the larval stage (only to lose it in adulthood), in addition to sharing other features of vertebrate development. This suggested the long-sought bridge between invertebrates and vertebrates, and on these grounds, Haeckel established the Chordata as a new phylum in 1870. This was the context in which Haeckel decided to investigate development in the calcareous sponges (Haeckel 1872), which convinced him that he had found a deep ontogenetic homology uniting all the tissue-forming animals or Metazoa. This, Haeckel believed allowed him to look deep into the phylogenetic tree and to reconstruct the very first multicellular animal with differentiated tissues.

---

[19] See Richmond (2007).
[20] Haeckel discussed the recapitulation thesis in (Haeckel 1866), but it was not until Haeckel (1870c) that he began calling it the *biogenetische Grundgesetz*. See Churchill (2007).

Haeckel had originally considered sponges to be colonies of single-celled protozoa, not true multicellular animals. But after studying their development more
closely, he reversed this opinion and came to regard sponges as the most primitive
and earliest forms of an animal with differentiated tissues. The simplest form of a
sponge, he believed, had a hollow cup-shaped body composed of two cell layers
which he argued were homologous in form and development to the ectoderm and
endoderm layers found in all animals. The cells of the two layers were differentiated, those of the outer layer bearing flagella while those lining the inner cavity
did not. From his observations of development in these sponges, Haeckel concluded
that this two-layered construction resulted when the blastula stage (a spherical and
hollow ball of flagellated cells) began to push in at one end to create an inner secondary endodermal layer from the former single layer of ectoderm. Once situated
inside the cup-shaped arrangement these inner cells withdrew their flagella and further differentiated themselves as to color, shape, and size. Hence by invagination of
the single-layered blastula, a two-layered gastrula is created. In many other animal
phyla, gastrulation is followed by the emergence of a third mesoderm layer and more
specialized structures, and organs develop from all three. Haeckel interpreted the
morphological condition of these mature sponges as being equivalent to the gastrula
stage in animal development. They produced no further structures or tissues other
than the hole created at the site of invagination (the blastopore), which he suggested
represented a primitive rudiment of a mouth (the *prostoma*), and the internal cavity
which he considered to be a primitive but specialized gut rudiment (the *progastra*).[21]

This, in Haeckel's opinion, was key evidence that the sponges belonged to the
other tissue-forming animals. More importantly, it suggested to him, in accordance
with the fundamental biogenetic law, that the tissue-forming animals or Metazoa
were monophyletic. The common ancestor from which all animals had evolved,
Haeckel inferred, had the permanent morphology of a gastrula, and so he named it
the *Gastraea* (Haeckel 1872, 1874a, b). Haeckel also referred to the gastraea as the
Urdarmthiere (German for "primitive gut animal"). The gastraea he proposed had
evolved from a *Blastaea*, a spherical and hollow colony of ciliated cells that would
have been arranged as a single layer without differentiated tissues. The blastaea, in
turn, would have emerged from a heap of homogenous amoeboid cells (a *Moraea* or
Synamoebium) that had failed to separate upon repeated division of a single amoeba
(*Cytaea*), the original *Urstamm-Zelle* or "stem cell" (see Chapter 2 by Dröscher
this volume). The nucleated unicellular amoebae had, in turn, evolved from a still
more primitive non-nucleated moneron similar to the earlier-mentioned *Bathybius*.[22]
This phylogenetic series was paralleled (and therefore could be witnessed still today)
by the ontogenetic series of developmental stages: monerula, ovula, morula, blastula,
gastrula (Figure 1.1).

---

[21] See Leys and Eerkes-Medrano (2005) for a modern assessment of Haeckel's account of sponge development. See Hopwood (2015) for the most recent and extensive account of Haeckel's attempts to illustrate the theory of recapitulation with images of vertebrate embryos.

[22] See Reynolds and Hülsmann (2008) for discussion of Haeckel's attempts to identify a living representative of the blastaea ancestor. Like many of his time, Haeckel considered the amorphous amoebae to
be of more primitive ancestry than the more morphologically constant bacteria.

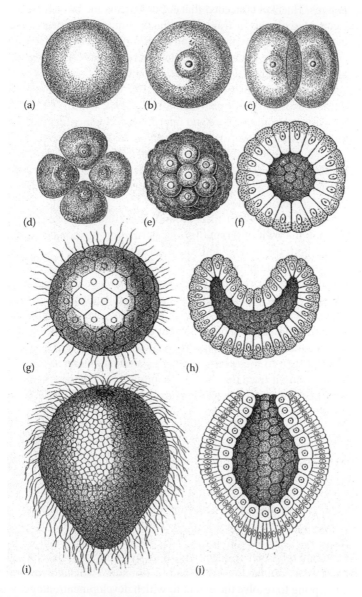

FIGURE 1.1 Early stages of development in the coral *Monoxenia Darwinii*. (a) Monerula, (b) cytula-ovula, (c–e) morula, (f, g) blastula, (h–j) gastrula. (From Haeckel, E., *Natürliche Schöpfungsgeschichte*, 7th ed., 448, 1879. With permission.)

 With the Gastraea theory Haeckel sought to furnish the main trunk of the animal tree: from gastraea it branched threefold to the sponges, Coelenterata (corals, anemones, ascidians, medusae) and the worms upward to all the rest. Gastrulation, in fact, looks quite different across the various animal phyla (Oppenheimer 1967, Hall 1998), and in the Vertebrata, it has very little resemblance with gastrulation in

Haeckel's sponges. Haeckel conceded that *Amphioxus* is the only vertebrate to display what he considered to be gastrulation in the original form. However, rather than conclude that this falsified his recapitulation theory, Haeckel explained (away) these divergences in the evidence as arising from later adaptations (*cenogenesis*) that had "falsified" or altered the original (*palingenetic*) pattern.

Despite the difficulty of discerning palingenetic from cenogenetic patterns in development, the Gastraea theory provided a framework within which and against which other researchers could explore and interpret developmental patterns and phylogenetic relations among the various animal phyla. E. Ray Lankester (1847–1929), for instance, in the 1870s argued that the first diploblastic (or two-layered) animals developed, not through a cup-shaped gastrula formed by invagination of the blastula stage, but by delamination of an original single germ layer that formed a flatworm-like planula, as is common to the coelenterates.

Elie Metchnikoff (1845–1916) also around this time proposed an alternative model, based upon his own observations of gastrulation in sponges, hydroids, and medusa, whereby both the middle and inner germ layers develop secondarily from ectoderm cells that migrate into and fill the hollow blastocoel cavity. He suggested that the ancestral metazoan, which he called a *parenchymella* (or later *phagocytella*), likely resembled a colony of externally flagellated and internally amoeboid cells capable of intracellular digestion, rather than Haeckel's gastraea which had specialized digestion extracellularly to a primitive gut cavity.[23]

Gradually embryologists shucked the subservience of embryology to the interests of phylogeny and began to focus on ontogeny as a phenomenon worthy of study in its own right. Releasing it from the grip of recapitulationist theory meant that explanations of ontogenetic events were no longer to be sought in the deep (and somewhat mysterious) forces of phylogenetic history, but in more immediate and causally mechanistic processes. The pursuit of these material causes required greater attention to the minute details of how individual cells divide, move, transform, and interact with one another in the developing embryo. These details were collected through cell lineage studies that traced the emergence of various tissues and organs back to particular early cleavage cells and ultimately to specific regions of the original fertilized ovum. Embryologists such as Charles Otis Whitman, Edwin G. Conklin, Edmund B. Wilson and others illustrated the results of these cell lineage studies by means of "fate maps" (see Maienschein (1978) and in this volume, in Chapter 3).

It would be sometime before questions about evolution and development would again be entwined as intimately as they had been under Haeckel's recapitulationist approach. Attempting to resolve the extent to which development/ontogeny is caused by evolution/phylogeny and vice versa became the concern of evolutionary developmental biology (Evo-Devo) from about the 1970s on.[24] From a modern perspective, Haeckel might perhaps be regarded as having attempted to account for cell behavior by appeal to evolutionary history imprinting what some biologists referred to as a "program" on cell development. Explaining how that program works and

---

[23] Tauber and Chernyak (1991), 53–67.

[24] For the complicated histories of embryology, developmental biology, evolutionary biology, evolutionary genetics and their relation to evolutionary developmental biology see Laubichler and Maienschein (2007).

gets "encoded" has been a major problem for cell and molecular biology, but more recently those engaged in evolutionary developmental biology have shifted their conceptual focus to talk of plasticity, epigenetics, emergent properties, cell–cell communication, and feedback within signaling pathways and networks. In the process, those elementary organisms known as cells have also been subjected to evolutionary analysis. The rise of the molecular revolution in biology opened the way for discussion of "molecular evolution," and cells themselves can now be taxonomized on the basis of protein (amino acid) sequence and DNA homologies for instance. Cells have truly emerged as complex and organized systems of diverse parts and processes.

1.5 CONCLUSION

Two great problems occupied nineteenth century biology. One concerned the nature of life: What *is* it? And how do living organisms "work?" The other concerned the different *types* of living things on earth: How did all the different species of living organisms come to be? Darwin showed the way toward answering the second question with the theory of evolution, albeit leaving many questions about details and mechanisms unresolved. Identification of the cell as the fundamental unit of life provided an answer to the first problem, but it was not an absolute stopping point, for the existence and nature of cells also have to be explained.

While Darwin himself was chiefly concerned with the question of how species evolve, many questions remained as to whether and how cells evolve. What are the relevant causal agents of development and evolution? Does the answer lie in adult organisms, embryos, germ layers, cells, physiological units, gemmules, proteins, nucleic acids, all of the above? Do cells cause other cells to form or do some more fundamental unit, something like Darwin's gemmules perhaps, cause cells to form and to differentiate? If phylogeny is the cause of ontogeny, in some sense, then what causal agency do cells have? How do cells "record" past phylogenetic history (palingenesis) and how do they manage to introduce novelty (cenogenesis) into the developmental plans that produce adaptive changes in organisms and species?

The rest of the chapters in this volume discuss the challenges and successes of the efforts to answer these questions, by differentiating between germ and somatic cell lines, by careful observation of the differential fates of embryo cells, by investigating what if any causal agent is wielded by genes (the conceptual offspring of Spencer's physiological units and Darwin's gemmules), by noting the existence and effects of intimate cell-within-cell symbioses, and by further dissecting the complex cell-system into signaling and regulative pathways and networks, by means of which these elementary organisms are able to communicate with, influence, and be influenced by one another and their external environments.

REFERENCES

Brown, R. 1833. On the organs and mode of fecundation in orchideae and asclepiadeae. *Trans Linn Soc Lond* 16: 685–745.
Brücke, E. von 1861. Die Elementarorganismen. *Sitzungsberichte der Mathematisch-Naturwissenschaftlichen Classe der Kaiserlichen Akademie der Wissenschaften* 44: 381–406.

Chambers, R. 1994 (1844). *Vestiges of the Natural History of Creation and Other Evolutionary Writings*. Edited with a new introduction by J. A. Secord. Chicago, IL: University of Chicago Press.

Churchill, F. B. 2007. Living with the biogenetic law: A reappraisal. In *From Embryology to Evo-Devo: A History of Developmental Evolution* (Eds.) M. D. Laubichler and J. Maienschein, pp. 37–81. Cambridge, MA: MIT Press.

Coleman W. 1977. *Biology in the Nineteenth Century: Problems of Form, Function, and Transformation*. Cambridge, UK: Cambridge University Press.

Conklin, E. G. 1940. Cell and protoplasm concepts: Historical account. In *Cell and Protoplasm*: Publication of the American Association for the Advancement of Science, No. 14 (Ed.) F. R. Moulton, pp. 6–19. Washington, DC: The Science Press.

Darwin, C. 1851. *Living Cirripedia, A Monograph on the Sub-Class Cirripedia, with Figures of All the Species; or, Pedunculated Cirripedes*. London, UK: The Ray Society.

Darwin, C. 1854. *Living Cirripedia, The Balanidae, (or sessile cirripedes); The Verricudae*. London, UK: The Ray Society.

Darwin, C. 1859. *On the Origin of Species by Means of Natural Selection or the Preservation of Favoured Races in the Struggle for Life*. London, UK: Charles Murray.

Darwin, C. 1868. *The Variation of Animals and Plants under Domestication*, 2 Vols. London, UK: John Murray.

Darwin, C. 1875. *The Variation of Animals and Plants under Domestication*, 2 Vols. 2nd ed. revised. London, UK: John Murray.

de Bary, H. A. 1859. Die Mycetezoen. Ein beitrag zur kenntniss der niedersten thiere. *Z wiss Zool* 10: 88–175.

Dröscher, A. 2014. History of cell biology. In *Encyclopedia of Life Sciences (ELS)*. Chichester: John Wiley & Sons. (doi:10.1002/9780470015902.a0021786.pub2. Accessed June 17, 2014.)

Geison, G. L. 1969a. The protoplasmic theory of life and the vitalist-mechanist debate. *Isis* 60: 273–292.

Geison, G. L. 1969b. Darwin and heredity: The evolution of his hypothesis of pangenesis. *J Hist Med Allied Sci* 4: 375–411.

Haeckel, E. 1866. *Die Generelle Morphologie der Organismen. Allgemeine Grundzüge der Organischen Formen-Wissenschaft, mechanisch begründet durch die von Charles Darwins reformierte Deszendenz-Theorie*, 2 Vols. Berlin, Germany: G. Reimer.

Haeckel, E. 1868. *Natürliche Schöpfungsgeschichte. Gemeinverständliche wissenschaftliche Vorträge über die Entwicklungslehre im Allgemeinen und diejenige von Darwin, Goethe und Lamarck im Besonderen, über die Anwendung derselben auf den Ursprung des Menschen und andere damit zussamenhängende Grundfragen der Naturwissenschaften*, 2 Vols. Berlin, Germany: G. Reimer.

Haeckel, E. 1870a. *Studien über Moneren und andere Protisten*. Leipzig, Germany: Engelmann.

Haeckel, E. 1870b. Beiträge zur plastidientheorie. *Jenaische Zeitschrift für Medicin und Naturwissenschaft* 5: 499–519.

Haeckel, E. 1870c. *Natürliche Schöpfungsgeschichte. Gemeinverständliche wissenschaftliche Vorträge über die Entwicklungslehre im Allgemeinen und diejenige von Darwin, Goethe und Lamarck im Besonderen, über die Anwendung derselben auf den Ursprung des Menschen und andere damit zussamenhängende Grundfragen der Naturwissenschaften*, 2 Vols. 3rd edition. Berlin, Germany: G. Reimer.

Haeckel, E. 1872. *Die Kalkschwämme (Calcispongae). Eine Monographie. I: Biologie der Kalkschwämme (Calcispongae oder Grantien). II: System der Kalkschwämme. III: Atlas der Kalkschwämme*. Berlin, Germany: G. Reimer.

Haeckel E. 1874a. Die Gastraea-Theorie, die phylogenetische classification des thierreichs und die homologie der keimblätter. *Jena Z Naturwiss* 8: 1–55.

Haeckel, E. 1874b. *Anthropogenieoder Entwicklungsgeschichte des Menschen. Gemeinverständliche Vorträge über die Grundzüge der menschlichen Keimes- und Stammes-Geschichte.* Leipzig, Germany: Wilhelm Engelmann.

Haeckel, E. 1879. *Natürliche Schöpfungsgeschichte,* (7th ed.), Berlin, Germany: G. Reimer, p. 448.

Hall, B. K. 1998. Germ layers and the germ-layer theory revisited: Primary and secondary germ layers, neural crest as a fourth germ layer, homology, and demise of the germ-layer theory. *Evolutionary Biology,* 30 Vol (Eds.) M. K. Hecht, R. J. MacIntyre, and M. T. Clegg, pp. 121–186. New York: Plenum Press.

Hall, T. S. 1969. *Ideas of Life and Matter: Studies in the History of General Physiology 600 B.C.–1900 A.D. 2 Vol. From the Enlightenment to the End of the Nineteenth Century.* Chicago, IL: University of Chicago Press.

Hertwig, O. 1895. *The Cell: Outlines of General Anatomy and Physiology.* Translated by M. Cambell and H. Johnstone Campbell (Eds.). London, UK: Swan Sonnenschein.

Hooke, R. 1665. *Micrographia: Some Physiological Descriptions of Minute Bodies made by Magnifying Glasses with Observations and Inquiries Thereupon.* London, UK: Jo. Martyn and Ja. Allestry, Printers to the Royal Society.

Hopkins, F. G. 1913. The dynamic side of biochemistry. Reports of the British Association for the Advancement of Science, pp. 652–668.

Hopwood, N. 2015. *Haeckel's Embryos: Images, Evolution, and Fraud.* Chicago, IL: University of Chicago Press.

Hughes, A. F. W. 1959. *A History of Cytology.* London and New York: Abelard-Schuman.

Huxley, T. H. 1853. The cell theory. *British and Foreign Medico-Chirurgical Review* 12: 285–314.

Huxley, T. H. 1968 (1868). On the physical basis of life. *Collected Essays,* 1 Vol., pp. 130–165. New York: Greenwood Press.

Jacyna, L. S. 1984. The romantic programme and the reception of cell theory in Britain. *J Hist Biol* 17(1): 13–48.

Kölliker, R. von 1845. Die Lehre von der thiereschen Zelle. *Z Wiss Bot* 2: 46–102.

Kuhn, T. S. 2012. *The Structure of Scientific Revolutions: 4th Edition with an Introductory Essay by Ian Hacking.* Chicago: University of Chicago Press.

Laubichler, M. D., and J. Maienschein (Eds.). 2007. *From Embryology to Evo-Devo: A History of Developmental Biology.* Cambridge, MA: MIT Press.

Leydig, F. 1857. *Lehrbuch der Histologie des Menschen und der Thiere.* Frankfurt, Germany: Meidinger Sohn.

Leys, S. P. and Eerkes-Medrano, D. 2005. Gastrulation in calcareous sponges: In search of Haeckel's Gastraea. *Integr Comp Biol* 45: 342–351.

Maienschein, J. 1978. Cell lineage, ancestral reminiscence, and the biogenetic law. *J Hist Biol* 11(1): 129–158.

Müller, F. 1864. *Für Darwin.* Leipzig: Engelmann.

Müller-Wille, S. 2010. Cell theory, Specificity, and reproduction, 1837–1870. *Stud Hist Philos Biol Biomed Sci* 41(3): 225–231.

Nicholson, D. 2014. The return of the organism as a fundamental explanatory concept in biology. *Phil Comp* 9: 347–359.

Oppenheimer, J. M. 1967. The non-specificity of the germ-layers. *Essays in the History of Embryology and Biology* (Ed.) J. M. Oppenheimer, pp. 256–294. Cambridge, MA: MIT Press.

Parnes, O. 2000. The envisioning of cells. *Science in Context* 13(1): 71–92.

Rehbock, P. F. 1975. Huxley, Haeckel, and the oceanographers: The case of Bathybius haeckelii. *Isis* 66(4): 504–533.

Remak, R. 1855. *Untersuchungen über die Entwickelung der Wirbelthiere.* Berlin, Germany: G. Reimer.

Reynolds, A. S. 2008. Amoebae as exemplary cells: The protean nature of an elementary organism. *J Hist Biol* 41: 307–337.

Reynolds, A. S. and Hülsmann, N. 2008. Ernst Haeckel's discovery of Magosphaera planula: A vestige of metazoan origins? *Hist Philos Life Sci* 30:339–386.

Richmond, M. L. 2000. TH Huxley's criticism of German cell theory: An epigenetic and physiological interpretation of cell structure. *J Hist Biol* 33(2): 247–289.

Richmond, M. L. 2007. Darwin's study of the Cirripedia. *The Complete Work of Charles Darwin Online* (Ed.) John van Wyhe. (http://darwin-online.org.uk/EditorialIntroductions/Richmond_cirripedia.html. Accessed July 26, 2016.)

Ruse, M. 2000. Darwin and the philosophers: Epistemological factors in the development and the reception of the theory of *The Origin of Species. Biology and Epistemology*, (Eds.) R. Creath and J. Maienschein, pp. 3–26. Cambridge, UK: Cambridge University Press.

Ryder, J. A. 1879. The Gemmule versus the Plastidule as the ultimate physical unit of living matter. *Am Nat* 13(1): 12–20.

Sapp, J. 2003. *Genesis: The Evolution of Biology*. New York: Oxford University Press.

Schleiden, M. 1847 (1838). *Contributions to Phytogenesis*. Translation by H. Smith. London, UK: Sydenham Society.

Schultze, M. 1987 (1861). Über Muskelkörpchen und das was man eine Zelle zu nennen habe. *Archiv Anatomische Physiologische wissenschaftlich Medicin*, S.1-27: Reprinted in *Klassische Schriften zur Zellenlehre von Matthias Jacob Schleiden, Theodor Schwann, Max Schultze*, eingeleitet und bearbeitet von I. Jahn, Leipzig, Germany: Akademische Verlagsgesellschaft, Geest & Portig K.-G.

Schwann, T. (1839). *Microscopical Researches into the Accordance in the Structure and Growth of Animals and Plants*. Translation by H. Smith. London, UK: Sydenham Society.

Siebold, C. von 1848. Über einzellige pflanzen und tiere. *Z wiss Zool* 1: 270–294. (Reprinted in English translation in *Q J Microsc Sci* 1(2): 111–121; 195–206, 1853.)

Spencer, H. 1865–1867. *Principles of Biology*, (Vol. I 1865; Vol. II 1867). London, UK: Williams & Norgate.

Tauber, A. I. and L. Chernyak. 1991. *Metchnikoff and the Origins of Immunology: From Metaphor to Theory*. New York: Oxford University Press.

Virchow, R. 1855. Cellular-pathologie. *Virchows Arch* 8(1): 3–39.

2 Germ Cells and Somatic Cells in Evolutionary Biology
August Weismann's Germ Plasm Theory

Ariane Dröscher

CONTENTS

2.1 INTRODUCTION

In the late nineteenth and early twentieth centuries, germ cells acquired a central position in the evolutionary debate. This holds true principally for the underlying conceptual implications of the germ plasm theory rather than for germ cells proper, which, on the contrary, soon faded away from the mainstream research focus. The main theoretical value consisted in opening a new approach to the investigation of the mechanisms of inheritance and development, in linking this approach with the issues of the contemporary evolutionary debate, and in becoming a decisive argument in the battle between neo-Lamarckians and neo-Darwinians. The former defended the possibility of inheritance of acquired characters (soft inheritance), the latter maintained that natural selection was sufficient to explain evolutionary change.

The scholar mostly coupled with the germ plasm theory is August Weismann (1834–1914). By the turn of the twentieth century, his assumption of particulate inheritance, linked to a specific substance, the "germ plasm," which is separate from the plasm of the somatic cells, but continuous with the germ plasms of the preceding and the following generations, due to germinal selection, provided the framework to revisit the fundaments of Darwinism. In 1986 Ernst Mayr (1904–2005), one of the protagonists of the evolutionary synthesis wrote:

> August Weismann is one of the towering figures in the history of evolutionary biology. If we ask who in the nineteenth century after Darwin had the greatest impact on evolutionary theory, the unequivocal answer must be Weismann. It was he who was responsible for what Romanes later called neo-Darwinism. (Mayr 1985, p. 295)

In the first decade of the twentieth century, Weismann was praised mostly for his pioneering concepts in heredity research. It is to Mayr's credit to have emphasized the importance of Weismann's contributions to germinal selection as a form of natural selection. Historian Jean Gayon agreed that the reconciliation of the theory of natural selection with the contemporary achievements of heredity research was the major challenge faced by Darwinism during its first 60 years and that this finally resulted in the modern synthesis (Gayon 1998). Linking Darwinism and heredity, Weismann was supposed to have established the firm bases for definitely refuting all kinds of soft inheritance and neo-Lamarckism.

Whatever the *a posteriori* judgment about the implications and results of the germ plasm theory may be, it must not be forgotten that the scholars involved in the Neo-Darwinian debate, developed their conceptions for reasons which often differed and aimed at different audiences than Weismann's. Recent historical analyses have, in fact, demonstrated that during the ongoing discussion, single aspects of his works were extrapolated, reinterpreted, and used for proper argumentation. Mayr, for instance, was anything but disinterested in the neo-Darwinian debate and so he adopted Weismann's ideas to support his own vision. However, the relationship between Weismann, Weismannism, Darwin, and Darwinism is not that simple, as demonstrated by the opinion of the evolutionary biologist George John Romanes (1848–1894), a friend of Darwin and creator of the terms "neo-Darwinism" and "Weismannism." In 1889, in a letter to his colleague Edward Bagnall Poulton (1856–1943) Romanes expressed his intimate conviction that "had Darwin lived till now, he would almost certainly have been opposed to Weismann. This is not a thing I should like to say in public, but one that I should like to feel practically assured about in my own mind" (Romanes 2011, p. 230).

The meaning of "Weismannism" is hence complex and variegated, differing according to the author and his context. Romanes' original definition explicitly excluded "any reference to the important question with which the name of Weismann has been mainly associated—that is, the inheritance or noninheritance of acquired characters" (Romanes 1893, p. vii). Yet, very persistently, Weismannism continued to be associated with "the idea of inviolate, eternal germ plasm, uninfluenced by the life history of the carrier organism" (Burnham 1972, p. 326) and hence, according to Weismannists, the factual impossibility that acquired modifications could be inherited. Later, some researchers extended the meaning of the term to apply to one of the (controversially debated) tenets of classical molecular biology i.e. the central dogma: "In the development of an individual, DNA causes the production both of DNA (genetic material) and of protein (somatic material). The reverse process never occurs. Protein is never a cause of DNA" (MacLaurin 1998, p. 37). Molecular Weismannism has thus undergone the remarkable conceptual shift to mean the unidirectional flow of molecular information as basic to evolutionary theory.

Recent epigenetics and the revival of less strict views on soft inheritance has brought along a reconsideration of Weismann and Weismannism. Whereas the latter has lost a good part of its explanatory value, rereading Weismann's original writings shed new light on his thinking. While Mayr even advanced some sort of apologies for Weismann's juvenile "errors and failures" (Mayr 1985, p. 295), in the past decades, just these "errors" became main points of interest. The works of Griesemer (2005), Wimsatt and Griesemer (2007), Winther (2001), Stamhuis (2003), Stanford (2005), Weissmann (2010), Dröscher (2008; 2014; 2015), and above all Churchill (1968; 1985; 1986; 1999; 2015), have strongly contributed to the recent renaissance of Weismann studies. It became evident that in the course of the twentieth century his concepts had been straightened and purged of all contingencies. On the other hand, the historical studies emphasized that Weismann did not stand alone. He was neither the first nor the only biologist of his time to propose a theory aimed at connecting concepts of cytology, inheritance, development, and evolution. Rather, concepts of this kind originated and grew on a fertile ground. Far from being able to reconstruct the whole richness of the debate, three theories deserve special attention: (1) Darwin's pangenesis theory, (2) Richard Owen's theory of the continuity of the germ-mass, and (3) Francis Galton's concept of stirps.

This historical review first illustrates some main ideas on the relationship between inheritance and evolution circulating in the second half of the nineteenth century. The third and fourth part concentrates on Weismann and his germ plasm theory, while the last part treats the reception and transformation of Weismann's theories.

2.2 CELLS, EVOLUTION, AND HEREDITY IN THE MID-NINETEENTH CENTURY

Concepts of cells and of evolution were linked much earlier than it is generally thought. Especially, in the German-speaking world, an increasing number of attempts to establish common laws of vital organization and functioning characterized the first decades of the nineteenth century. Rejecting the older concepts of superior and immaterial guiding principles, forces, Bauplans (architectural body plans), or similar basic assumptions of idealist morphology and Romantic science, the emerging cell theory provided several young naturalists, especially botanists, with a valuable new way to understand the tie that links together the elements of the "chain of being" and guarantees the unity of living matter. This unity comprised a single organism and its developmental stages, as well as the entire organismal world and its transformations or metamorphoses. In the words of the Italian botanist Giuseppe Meneghini (1811–1889): "All plants, as many as there are, according to us, can be reduced to a vegetal vesicle, and the various phases of its development all swallowed within few expressions, the

sixty thousand species counted by the botanists." (Meneghini 1838, p. 48, my translation). Ten years later, Matthias Schleiden (1804–1881) repeated in a very similar way:

> "Then, if the first cell be given, the foregoing points out how the whole wealth of
> the vegetable kingdom may have been formed by a gradual passage from it through
> varieties, sub-species and species, and thus onward, beginning anew from each
> species—in a space of time, indeed, of which we have no conception." (Schleiden
> 1848, p. 291)

These conceptions did not need any theory of heredity. Many pioneers of cell theory
conceived of organic nature as an entity in a permanent process of progressive transformation. These changes concerned individual development (ontogenesis), as well
as species development (phylogenesis) and often included the border between the
living and the nonliving world. The cell represented the central junction of all these
material and historical metamorphoses. The cell transformed the inorganic chemical
elements of the soil and the air into the organic constituents of the organism (e.g.,
Endlicher and Unger 1843: 17–29). The cell performed the basic physical and chemical processes of the body. The cell accomplished the transformations from the egg
to the adult organism. Finally, the cell represented the unifying principle between
"lower" (one-celled) and "higher" (multicellular) life forms (Dröscher 2016).

It was an approach which to some extent opposed Darwin's, notwithstanding his
long-term interests in generation, zoophytes, and monads. The young generation of
botanists who in the 1820s and 1830s broke with pure taxonomy laid their emphasis
on structure and its changes through chemical and physiological activity. They considered cells and their internal physical–chemical processes as the driving force of
the progressively complex transformation of matter, while Darwin's starting point
was that of a more classical natural historian and taxonomist who focused on the
macrolevel, that is, populations of higher organisms and their complex characteristics and variations (Charpa 2010). From this point of view, it is easy to understand
why there was no immediate merging of cell theory and Darwinism. For Darwin,
cells and their activities were important aspects, which demanded consideration, but
their relevance was secondary.

Moreover, over the course of the following decades, the rather vague notions of
organic continuity, generation, and progression were no longer sufficient to provide
a valid framework for cell research and organic evolution. They had to give way
to more concrete models for the mechanisms of inheritance. In fact, in the second
half of the century at least five different theories of some kind of material carriers
responsible for ontogeny and for the transmission of characters from one generation
to the next, found a broader audience (Churchill 1987). Darwin's and Weismann's
theories were but two of them. These mechanisms had to take into consideration the
new findings of cell research. Moreover, any new conception of inheritance had also
to account for the mechanisms of natural selection and/or the inheritance of acquired
characters. Finally, these conceptions had to be able to provide a plausible framework to explain either the gradual steps or the sudden drastic changes (saltations) of
evolution.

A CHARLES DARWIN

Researchers like Charles Darwin (1809–1882) felt the need for a better model to understand the continuities and discontinuities of generation; that is, a concept of heredity able to provide a plausible mechanism of the transmission and selection of characters. According to Romanes (1893, p. viii), Darwin meditated on several models, including ideas similar to Weismann's later theory of a continuity of germ plasm. Probably he was still too much rooted in early the nineteenth-century biology to feel comfortable with such a radical theory. Phillip Sloan argued several years ago (Sloan 1986, pp. 373–393) that the young Darwin, notwithstanding all differences, owed much to the botanists, in particular, Robert Brown (1773–1858), Alexandre Brongniart (1770–1847), and Augustin Pyrame de Candolle (1778–1841). Since 1831, through his teacher John Stevens Henslow (1796–1861), Darwin was persistently concerned with granular matter and the corresponding discussions about their vitality, autonomy, and reproduction. During his microscopical studies of the "fine granular pulpy matter" of live arrow worms (chaetognaths) on board the *Beagle*, Darwin transferred these botanical ideas to the level of lower animals and often called the observed granules "gemmules" (ibid., p. 388).

Of particularly importance were Darwin's studies on "zoophytes," organisms like corals and sponges, which at that time were supposed to stand between the animal and the plant kingdom and which produce a great quantity of gemmules to reproduce asexually. In 1837, Darwin set down in his Red Notebook:

> Propagation. whether ordinary. hermaphrodite. or by cutting an animal in two. (gemmiparous. by nature or accident). we see an individual divided either at one moment or through lapse of ages. —Therefore we are not so much surprised at seeing Zoophite producing distinct animals. still partly united. & egg[s?] which become quite separate. —Considering all individuals of all species. as *[each]* one individual *[divided]* by different methods, associated life only adds one other method where the division is not perfect. (quoted from Herbert 1980, p. 67)

Darwin thus connected his conception of zoophyte generation with his thoughts about the relationship of individuals within a species. This enabled him to explain the dynamics of species, starting from primordial granules (Sloan 1986, p. 431).

In these years, Darwin became convinced that the ultimate entities of life were neither the cell nor infusoria nor monads, as some suggested, but these are still smaller granular particles. All life forms reproduced through them. Around 1837, he wrote another significant note in his Red Notebook: "The living atoms having definite existence, those that have undergone the greatest number of changes towards perfection (namely mammalia) must have a shorter duration, than the more constant: This view supposes the simplest infusoria same since commencement of world" (quoted from Herbert 1980, p. 32). The underlying impetus was hence some sort of vital force, which acted on these atoms and induced them to progress.

About 30 years later, Darwin published his "provisional hypothesis of pangenesis" as a separate chapter in the second volume of *The Variation of Animals and Plants under Domestication* (1868) (for a more detailed account Chapter 1 by Reynolds in this volume). In the second edition (Darwin 1875), the text had undergone a considerable modification to become more compelling. The pangenesis theory played an important role in corroborating his evolutionary theory. As is well known, Darwin assumed a gradual process of evolution, rejecting the idea of sudden great changes championed by the supporters of saltationism. Darwin's insistence on gradualism implied the inheritance of small variations or "many slight differences which appear in the offspring from the same parents" (Darwin 1859, p. 45). Therefore, it was much more plausible to embrace a particulate theory of inheritance and to assume carriers, which represent single characters and not the entire organism. Resuming the above-mentioned ideas about the reproduction of lower plants and his work on zoophytes, he imagined that gemmules played that role in the inheritance of more complex organisms, too. According to him, cells repeatedly throw off gemmules, which then freely float throughout the body. A set of gemmules of all cells was collected in the reproductive organs and thus transmitted to the offspring. In the embryo, each gemmule then developed into a cell.

A second aspect that convinced Darwin to develop his pangenesis theory was that it displayed a mechanism, which explained the inheritance of acquired characters. In the subchapter *The Functional Independence of the Elements or Units of the Body*, Darwin declared that: "everyone appears to admit that the body consists of a multitude of 'organic units,' each of which possesses its own proper attributes, and is to a certain extent independent of all others" (Darwin 1868, II, pp. 370–371). He quoted Claude Bernard (1813–1878), Rudolf Virchow (1821–1902), and James Paget (1814–1899) as heralds of the autonomy of cells, but the most lasting influence on his conception of the role of cells in reproduction and variability seems to have been the physician Lionel Beale (1828–1906). Darwin adopted his use of the term "germinal matter" for the contents of cells, and for drawing a broad distinction between germinal matter and "formed matter" or the various products of the cell (ibid., p. 370). He then explained that according to Beale, nerve fibers:

> "are renovated exclusively by the conversion of fresh germinal matter (that is the so-called nuclei) into 'formed material'. However this may be, it appears probable that all external agencies, such as changed nutrition, increased use or disuse, &c., which induced any permanent modification in a structure, would at the same time or previously act on the cells, nuclei, germinal or formative matter, from which the structures in question were developed, and consequently would act on the gemmules or cast-off atoms." (ibid. p. 382)

Being thrown off during the whole lifespan of an organism, many gemmules represented characters of different stages of life. Thus, varieties of the same character were inherited:

> The retention of free and undeveloped gemmules in the same body from early youth to old age may appear improbable, but we should remember that many rudimentary and useless organs are transmitted and have been transmitted during an indefinite number of generations. We shall presently see how well the long continued transmission of undeveloped gemmules explains many facts. (ibid. p. 378)

Darwin was well aware of the highly speculative nature of his hypothesis and therefore expended considerable energy to render plausible to his readers the existence of myriads of minuscule gemmules as well as their capacity to divide. In the second edition, he emphasized with even more vigor:

"But I have further to assume that the gemmules in their undeveloped state are capable of largely multiplying themselves by self-division, like independent organisms. Delpino insists that to 'admit of multiplication by fissiparity in corpuscles, analogous to seeds or buds [...] is repugnant to all analogy.' But this seems a strange objection, as Thuret has seen the zoospore of an alga divide itself, and each half germinated. Haeckel divided the segmented ovum of a siphonophora into many pieces, and these were developed. Nor does the extreme minuteness of the gemmules, which can hardly differ much in nature from the lowest and simplest organisms, render it improbable that they should grow and multiply. A great authority, Dr. Beale, says 'that minute yeast cells are capable of throwing off buds or gemmules, much less than the 1/100000 of an inch in diameter;' and these he thinks are 'capable of subdivision practically ad infinitum'." (Darwin 1875, II, p. 372)

His pangenesis theory aimed at explaining variability, as well as constancy, and based on his assumption of the inheritance of characters modified by use or disuse. This Lamarckian aspect is probably one of the reasons why later Darwinists often considered this theory "an error." Yet for Darwin, it was essential to conceive a mechanism able to explain the transmission of the characters of all single features of the body. The pangenesis hypothesis enabled Darwin to understand many frequent and rare phenomena of inheritance, such as the transmission of unchanged characters through many generations, the mixture of paternal and maternal traits, but also the dominance of certain characters, their reappearance after many generations of nonexpression (through the reactivation of dormant gemmules), and, most importantly for Darwin, the emergence of small new variations (Deichmann 2010, pp. 94–95).

Far from attributing the whole credit for the profound changes in nineteenth-century biology to Darwin alone, it is evident that his works put several issues definitely onto the general research agenda, regardless of whether one agreed with him or not.

B RICHARD OWEN

Weismann's theory was inspired by several questions circulating in the mid-nineteenth century, regarding embryology, cytology, and Darwinism. An essential aspect that was missing in Darwin's thoughts, judging from a Weismannian point of view, was the role of germ cells. Darwin did not need to consider them because, for him, it was not the reproductive organs that generated new organisms, but the sum of the somatic cells via the gemmules. Weismann significantly contributed to the shift of attention toward germ cells. Yet it was not completely original. Seven years after the appearance of *Die Continuität des Keimplasmas* (1885), Weismann rectified in a short historical survey that he, at last, had come to know that other scholars had developed similar ideas before him. He cited Richard Owen (1804–1882), Francis Galton (1822–1911), Gustav Jäger (1832–1917), August Rauber (1841–1917), and Moritz Nussbaum (1850–1915). He concluded, however, that all preceding authors had remained without influence (Weismann 1893, pp. 198–202). Independently from claims of priority or major importance, it is revealing that similar ideas came up at about the same time.

Not surprisingly, Darwin's pangenesis theory had found particularly fertile ground in the Anglo–Saxon world. Richard Owen, Darwin's former friend and later rival and critic, was probably the first to assume the existence of a germ line. Like Darwin, he was not a cell researcher, but concentrated on comparative anatomy and palaeontology (Rupke 2009), but, like Darwin, he understood the necessity to go down to the cellular level in order to explain phenomena at the macrolevel. Owen's ideas are not tenable from today's point of view; however, they well illustrate the approach, as well as the conceptual difficulties of this generation of biologists, when trying to understand the mechanisms of inheritance.

During his early works on invertebrates, Owen investigated the embryology of aphids. He noticed in their larvae a little mass of cells distinct from the rest of the body cells. He called this clump the "germ mass," because he conceived of them as being the "progeny of the primary impregnated germ-cell" (Owen 1849, p. 70). During the development of the organism, this cellular mass was not consumed but preserved for the production of the offspring. In his book on parthenogenesis, Owen interpreted this finding within the framework of Müllerian Romantic science, Aristotelian orthogenetic "organizing forces" and a developmental concept of evolution:

> "[In] proportion to the number of generations of germ-cells, with the concomitant dilution of the spermatic force, and in the ratio of the degree and extent of the conversion of these cells into the tissues and organs of the animal is the perfection of the individual, and the diminution of its power of propagating without the reception of fresh spermatic force. In the vertebrate animal the whole of this force originally diffused amongst the cells or nuclei of the germ-mass is exhausted in the development of the tissues and organs of the individual, in the mysterious renovation of the spermatic power in the male by a special organ, and in the development of ova or cells prepared fit for its reception in the female." (Owen 1849, p. 69)

Owen did not intend to become a philosopher of inheritance, nor did he possess profound cytological knowledge. Moreover, he was an opponent of natural selection. The theoretical importance of the above-mentioned passage was that it could be read as a statement about the independence and continuity of the germ cells, and hence as an argument against the transmission of any modification of the body cells of an organism into the next generation.

C FRANCIS GALTON

Another supposed forerunner of Weismann's germ plasm theory was Francis Galton. Romanes even considered his theory as "virtually identical," anticipating that of Weismann by some 10 years (Romanes 1893, p. 59). In fact, Galton had already developed his theory in the 1870s. He was neither an anatomist, like Owen, nor an embryologist or a cytologist, but an anthropologist engaged with questions of human heredity. He was one of the first to pick up and test Darwin's gemmules theory because it promised to provide a model for the underlying mechanisms of transmission. Yet, his blood transfusion experiments with rabbits failed to provide the hoped-for confirmation of pangenesis (Galton 1871). Assuming that the gemmules circulated in the blood, he hoped to obtain transmission of at least some characters from the blood donor to the blood recipient rabbit, but no characters were transmitted. He, therefore, set out to develop his own theory (Bulmer 1999).

Galton supposed that the fertilized egg contained the "stirp," that is, a complete set of the "germs," as he called them, and all their varieties. The germs represented individual characters. Some of them—the "patent" germs—developed into cells, while others—the "latent" ones—did not express themselves in this organism. A selection of the latent germs was passed over to the offspring where again they could become patent and develop or remain latent and be selected for inheritance. The transmission to the next generation took place through the germ cells, directly formed by the latent germs. Like Owen's theory, Galton's theory could be understood as largely excluding the possibility of the inheritance of acquired characters. However, Galton, like Darwin and later Weismann, did not completely reject soft inheritance. Therefore, he was compelled to add an auxiliary theory, which further complicated his concept: Cells may throw out some germs, which represent the state of the cell at this moment. These germs occasionally may find their way to the germ cells and then be transmitted.

After this publication, Galton soon returned to focus on the statistical aspects of cross-generational transmission. He was rather innovative in this regard, but he cared little about the embryological or cellular aspects. His concept of cell formation, in fact, recalls Schleiden's theory. Similar to the *Zellenlehre*, formulated more than 30 years before, for Galton, the germs grew, aggregated, and transformed into new cells.

2.3 AUGUST WEISMANN AND THE PREMISES OF HIS THEORY

August Weismann's entrance into the world of science was not straightforward (Churchill 2015). In his youth, he collected plants and insects, above all butterflies, but since the natural sciences still were not institutionalized nor offered any professional future, from 1852 to 1857, he studied medicine in Göttingen. Attracted by the lessons of chemist Friedrich Wöhler (1800–1882), his first research steps went into this direction, but he soon renounced chemistry. Upon graduation, he passed one year at the city hospital in Rostock, attended some lectures at the University of Vienna, and worked as a practicing physician in Frankfurt. Not happy with this position, Weismann enrolled as a military physician in the Italo–Franco–Austrian war of 1859, and then visited several North Italian cities, Paris and zoologist Rudolf Leuckart (1822–1898) in Gießen. The following two years, he passed at the castle of Schaumburg as a personal physician of the banished Grand Duke Stephan of Austria. Here he had plenty of time to study the histology of muscle fibers and the metamorphosis of dipteran insects. The ducal librarian advised him to read Charles Darwin's *On the Origin of Species*. As a consequence, in 1863 Weismann definitively decided to become a scientist, wrote his *Habilitationsschrift* and started teaching as *Privatdozent* at the small University of Freiburg im Breisgau. In 1867, he was appointed as an associate professor of zoology. In 1873, he became a full professor, and in 1886, the director of a new institute. Though receiving numerous calls to other prestigious chairs, Weismann remained at Freiburg until his retirement in 1912.

As we have seen, Weismann's ideas about development, inheritance, and evolution were not completely new. However, even for those scholars who may be considered as his precursors, the theories of particulate inheritance and the role of germ cells had only secondary importance, auxiliary for underpinning their proper research thesis. With Weismann, these issues definitively underwent a leap in quality. Germ cells were at the center of his elaborate theory. Moreover, with respect to Darwin, Owen, and Galton, he was the best informed about the most recent developments in cytology.

Since the 1840s cell research had undergone a considerable quantitative and qualitative increase (Chapter 1 by Reynolds in this volume). Numerous new findings and conceptions had been made and discussed. Some were of particular importance for Weismann. One of them regarded germ cells. In 1870, Wilhelm von Waldeyer-Hartz (1836–1921) published a path-breaking report about the formation of germ cells in female mammals, demonstrating for the first time their origin from the ovarian surface epithelium (Waldeyer 1870, p. 43). This authoritative and widely confirmed view entered nearly all textbooks of embryology of the next decades. In Weismannian's later debate, it was pointed out that germ cells originate from somatic cells. A few years later, however, Alexander Goette (1840–1922) and Francis Balfour (1851–1882) failed to confirm in salamanders and in the catshark *Scyllium canicula*, a genetic relationship of germ cells with any particular tissue (Goette 1874–1875; Balfour 1878). In 1880, Moritz Nussbaum described in frogs and bony fish early segregation of the so-called primordial germ cells. Although he was not able to observe the events in early cleavage, he was convinced that he had detected during embryogenesis a cell-mass (*Zellhaufen*), which was always distinguishable from the somatic cells through their possession of yolk platelets (Nussbaum 1880, 3). He called them sex cells (*Geschlechtszellen*), because, at a certain point, they migrated and penetrated the layer of the genital glands, where they formed the progeny of the sexual cells. Linking his histological observations with mid-nineteenth century evolutionary theory, he conceived the sex cells as undifferentiated and ancestral. They were completely independent of the somatic cells, and formed a direct continuity within the organism and between the *Geschlechtszellen* of succeeding generations:

"The cleaved egg thus divides into the cellular material of the individual and into the cells made for the preservation of the species. In both parts, the multiplication of cells proceeds continuously, save that in the body of the individual happens also the division of labour, whereas in its sex cells only a simple additional division takes place. Both groups of cells and their offspring definitely multiply independently of one another, so that the sex cells do not participate in the construction of the tissues of the individual, and not a single sperm or egg cell arises out of the cellular material of the individual." (Nussbaum 1880, p. 112, my translation)

From 1877 onward, Weismann began to study the formation of germ cells in hydrozoans. The years 1881–1882 marked a watershed in his conceptualization of germ cells (Churchill 2015, pp. 161–164), and in 1883 he published his monograph *Entstehung der Sexualzellen bei den Hydromedusen* (The Origin of the Sex Cells in Hydromedusae). His researches on the embryology of different species of hydromedusae did not show the origin of germ cells because he was not successful in following them back to the youngest stages. Yet, he claimed that they were initially merely indistinguishable from the surrounding cells. At a certain stage, however, he noted them and described, independently from Nikolaus Kleinenberg (1842–1897), in detail the amoeboid movements of the *Urkeimzellen* (primordial germ cells) from the coenosac, along with the hydranth, and toward the gonophores. Now, germ cell migration became a central focus of his studies.

In many hydromedusae, Weismann was able to establish a direct lineage of germ cells, which went through all stages from the fertilized egg cell to the adult sex cells and remained always distinct from the somatic cells. He thus asserted in his famous theoretical treatise *Die Continuität des Keimplasmas als Grundlage einer Theorie der Vererbung* (1885) a continuity of the germ plasm, handed over from one germ cell to the next, during the development of an organism (intragenerational germline), and stated that only this germ plasm was transmitted to the successive generation (intergenerational germline). Later, he enforced this point arguing that:

> "the existence in the germ-cell of a reproductive substance, the *germ-plasm*, which cannot be formed spontaneously, but is always passed on from the germ-cell in which an organism originates in direct *continuity* to the germ-cells of the succeeding generations. [...] The germ-cells alone transmit the reproductive substance or germ-plasm in uninterrupted succession from one generation to the next, while the body (soma) which bears and nourishes the germ-cells, is, in a certain sense, only an outgrowth from one of them." (Weismann [1892] 1893, p. 9)

Weismann thus formulated the first main point of his theory: the neat separation between two cell-lines, the somatic one, which forms the body, but has no influence on the constitution of the germ cells, and the germ track, which, being potentially immortal, preserves the original set of hereditary carriers and forms a continuity with the germ plasm of the preceding and the following generations. In this and in his later publications, he cited Nussbaum and other authors working on germ cells. However, it seems that after having become convinced of this point, he paid little attention to the further development of these studies, probably because they treated germ cells, and hence embryology, and not the germ plasm, as fundamental for his theory of inheritance.

A second important aspect of the germ plasm theory regarded general cytology. The morphological cell studies of the second half of the nineteenth century revealed, on the one hand, an ever greater number of distinct intracellular structures, and, on the other hand, an increasing number of different cell types, each presenting a significantly different internal organization. It became unfeasible to speak of *the* cell and generically assigning it the explanation of all secrets of life. Of all endocellular structures, the nucleus received special attention when it proved to possess the capacity of division and itself a distinct internal organization. In 1873, Anton Schneider (1831–1890) clearly described chromosomes. In the mid-1870s, the elucidation of the process of mitosis found its first climax in the works of Eduard Strasburger (1844–1912) and Walther Flemming (1843–1905), and in 1883, Édouard van Beneden (1846–1910) gave a decisive contribution to enlighten the behavior of chromosomes during meiosis.

Weismann was particularly interested in the cytological aspects of reproduction and development and carried out a series of significant investigations. In the late 1880s, during his stays at the Zoological Station in Naples, he observed the unequal division of fertilized sea urchin eggs, a process also known as the formation of polar bodies. He called it reduction division because he recognized it as a mechanism to reduce chromatin quantity, which otherwise would double at each fertilization or mixing of paternal and maternal chromatin. More importantly were his theoretical conclusions, especially his conviction that the above-mentioned cytological accounts had something to do with heredity. This link, so self-evident from today's point of view, was far from being generally accepted in Weismann's time. On the contrary, they were among the most criticized aspects of his theory. The great majority of biologists rejected the idea of material carriers of inheritance.

As we have seen above, there were a few, like Darwin, who endorsed particulate theories. Another was Herbert Spencer (1820–1903). Similar to Darwin's gemmules, Spencer conceived of myriads of "molecules" ("physiological units") distributed all over the body. Contrary to Darwin's gemmules, for Spencer, the molecules of an organism were all qualitatively identical. In this way, he avoided Darwin's difficulties to elucidate how all qualitatively different gemmules from every part of the body should come together to form a complete set for the transmission to the next generation. Weismann, in fact, objected: "How could the gemmules of all the cells of an organism enter its germ-cells unless they are formed in the body-cells, migrate from there, circulate through the body, and come together in the germ-cells?" (Weismann 1893, p. 13). On the other hand, Spencer's theory was no solution, because it implied that the molecules were different for every species, race, and maybe even individual. Moreover, it was at odds with hybridization experiments, which showed that paternal and maternal characters mix.

As a response to these questions, from 1883 onward, Weismann did not abandon the idea of qualitatively different bearers of heredity but rejected the idea of their free-floating in the body, locating them instead inside the cell, or, more precisely, in the cell nucleus. Later he went even a step further and located them on the chromosomes. He, particularly, showed interest in Wilhelm Roux's (1850–1924) argumentation that the complicated facts of mitosis were only explainable assuming that chromosomes are not uniform and homogeneous but rather composed of qualitatively different "regions" that were divided and distributed to the daughter cells. This line of thought eventually led to Weismann's theory of the determinants. As we will see in the next section, ultimately, it was not the cell itself, which was responsible for the transmission, but the determinants contained in the cell nucleus. Although these conceptions further emphasized the importance of cells in vital phenomena, looking back one may also interpret them as already containing the seed for the twentieth-century crisis of cytology, when the focus increasingly shifted away from cells toward (invisible) sub-cellular entities, above all genes.

A third leitmotiv of Weismann's complex theory was Darwinism. His life-long devotion to Darwin is illustrated by two events. In 1868, his inaugural lecture as a professor in Freiburg was on Darwinism, and in 1893, he dedicated the English translation of his main work, *The Germ-Plasm*, "To the memory of Charles Darwin." The German version had been dedicated to the 70th birthday of his master Rudolf Leuckart. Yet, his relationship to Darwinism was not as easy as it may seem.

As we saw in the introduction, Weismann is often praised as the scientist, who provided the definite scientific proof against Lamarckism. In fact, already in his lecture of 1883, *On Heredity*, he harshly criticized the inheritance of characters modified by use and disuse, which was one of Lamarck's principal arguments. By conceiving of organisms as mosaics of distinct characters, Weismann could consider selection (of the characters fixed in the determinants) as sufficient to explain organic evolution. It sketched a mechanism through which natural selection could maintain certain traits stably while also allowing them to undergo modification (Gayon 1998, 151).

In the same years, Hugo de Vries (1848–1935) advanced a similar idea (Stamhuis 2003). The attention of Neo-Darwinians, eager to understand evolutionary selection and speciation, thus switched from factors external to the organism, such as biogeographical isolation, to factors inherent to the organisms, and from the characters of entire organisms, organs and cells, to the invisible but supposedly determining factors at the cellular and subcellular levels. According to Gayon (1998, p. 176), Weismann's wish to preserve Darwin's heritage, had the consequence that he radicalize the principle of selection to the point of subordinating it to the principle of heredity.

As already outlined, Darwin, like most of his contemporary biologists, allowed more than once Lamarckian mechanism as external causes of variation and adaptation, and insisted that "variability is not a principal co-ordinate with life or reproduction, but results from special causes, generally from changed conditions acting during successive generations" (Darwin 1868, II, p. 371). As we will see in the following chapter, Weismann's position was not as clear, either.

2.4   THE GERM PLASM THEORY

Besides the above-mentioned research, Weismann also carried out a series of experimental investigations on the seasonal dimorphism in butterflies (1896), on the biology of freshwater animals, and some other minor studies. He was always keen to emphasize the empirical basis of his general conceptions. When the microscopic investigations ruined his eyesight, Ernst Haeckel (1834–1914) advised him to continue his battle for Darwinism in the field of theoretical biology. From now on, Weismann concentrated on elaborating a comprehensive theory of heredity, cytology, development, and evolution, and produced a series of papers, lectures, and works like *Das Keimplasma* (1892) and *Vorträge über Descendenztheorie* (1902–1903; 1904). Though highly speculative, often contradictory, and frequently modified in its details, Weismann's model represented a new viewpoint, which helped in focusing the general debate on these questions. Moreover, linking embryology with heredity and with cell genealogy, provided an innovative conceptual basis for understanding development, and a plausible conception of the underlying material basis of processes, which hitherto had been only described on the visible macroscopic and microscopic level. Even if many aspects were heavily criticized and scarcely confirmed by future experimental researchers, they exerted significant influence on the following generation of embryologists, experimental geneticists, and evolutionary biologists.

Since Weismann's accounts changed over time which were further altered by successive interpretations, it is advisable to distinguish them chronologically.

In his *Die Continuität des Keimplasma's* (1885), Weismann sketched his idea of the forces that drive ontogeny:

"The simplest hypothesis would be to suppose that, at each division of the nucleus, its specific substance divides into two halves of unequal quality, so that the cell-bodies would also be transformed; for we have seen that the character of a cell is determined by that of its nucleus. Thus in any Metazoon the first two segmentation spheres would be transformed in such a manner that one only contained the hereditary tendencies of the endoderm and the other those of the ectoderm, and therefore, at a later stage, the cells of the endoderm would arise from the one and those of the ectoderm from the other. [...] In the course of further division the nucleoplasm of the first ectoderm cell would again divide unequally, e.g. into the nucleoplasm containing the hereditary tendencies of the nervous system, and into that containing the tendencies of the external skin." (Weismann [1885] 1889, p. 186)

Continuing the process of unequal cell divisions, all specific tissues gradually come into being in:

"a definitely ordered course [...] and the determining and directing factor is simply and solely the nuclear substance, the nucleoplasm, which possesses such a molecular structure in the germ-cell that all such succeeding stages of its molecular structure in future nuclei must necessarily arise from it." (ibid.)

In 1885, Weismann still did not have the idea of *Anlagen* or other distinct material particles that are segregated one by one. Rather, he thought of a diminishing complexity of the germ plasm:

"[T]he quantity of the nucleoplasm is not diminished, but only its complexity. [...] we must also guard against the supposition that unequal nuclear division simply means a separation of part of the molecular structure [...]. On the contrary, the molecular constitution of the mother-nucleus is certainly changed during division in such a way that one or both halves receive a new structure which did not exist before their formation." (ibid., p. 195)

In 1892, instead, in the wake of Wilhelm Roux's experiments on frog embryos and his mosaic theory of ontogeny (Coleman 1965, pp. 141–142, 152), Weismann's conceptions concerning the material basis of heredity and development considerably changed. In 1888, Roux had carried out his famous defect experiments with frog embryos. Taking embryos at the 2- and at the 4-cell stage and damaging one or two cells with a hot needle, he observed that the remaining blastomeres survived, but developed only into half embryos. These results seemed to confirm the view that the surviving cells contained only that part of the carriers of heredity which served for that special part of the body, and hence, the destiny of the cells was determined from the very first cleavage. Weismann thus stated:

> "[T]he hereditary substance of the egg-cell, which contains all the hereditary tendencies of the species, does not transmit them *in toto* to the segmentation cells, but separates them into various combinations, and transmits these groups to the cells." (Weismann [1892] 1893, p. 205)

He now even outlined a precise hierarchical structure of the germ plasm (Figure 2.1):

> According to my view, the germ-plasm of multicellular organisms is composed of ancestral germ-plasms or *ids*, -the vital units of the third order, -each nuclear rod or *idant* being formed of a number of these. Each id in the germ-plasm is built up of thousands or hundreds of thousands of *determinants*, -the vital units of the second order, -which, in their turn, are composed of the actual bearers of vitality ("Lebensträger"), or *biophors*, -the ultimate vital units. (Weismann [1892] 1893, p. 75; his emphasis)

Whereas the biophors—the most important material elements of the germ plasm, because they possessed the power of growth and multiplication, and determined a particular characteristic of the cell—were invisible, Weismann later identified the *idants* with the chromosomes. Why he assumed such a complicated hierarchy of heredity particles, finds an explanation in his primary interest in development. It was

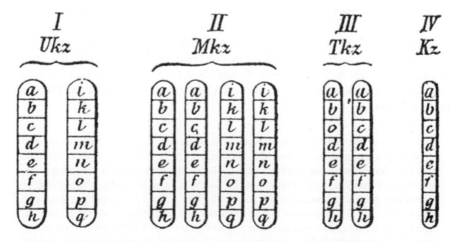

FIGURE 2.1 Weismann's illustration of ids (letters) located on the idants, and their behavior during the development of germ cells. After meiosis, the number of the ids of the germ cells (Kz) is half of those of the primordial germ cell (Ukz). (Courtesy of Weismann, A., *Amphimixis oder: Die Vermischung der Individuen*, Gustav Fischer, Jena, TH, p. 49, 1891.)

essential for his ideas about the structure of the hereditary substance, but also about the intracellular organization (i.e., the relationship between nucleus and protoplasm), and about the link between cells (and their characters) and the characters of the whole organism. Already Darwin had been aware, that certain characters were more linked with one another than others:

> There is another point on which it is useless to speculate, namely, whether all gemmules are free and separate, or whether some are from the first united into small aggregates. A feather, for instance, is a complex structure, and, as each separate part is liable to inherited variations, I conclude that each feather certainly generates a large number of gemmules; but it is possible that these may be aggregated into a compound gemmule. (Darwin 1868, II, p. 382)

Weismann resumed this line of thought and adapted it to his germ plasm theory:

> In the section on the control of the cell by the nuclear substance, I shall adopt what seems to me to be a remarkably happy idea on the part of de Vries, who supposes that material particles leave the nucleus, and take part in the construction of the body of the cell. These particles correspond to the "pangenes;" they are the 'bearers of the qualities' of the cell. [Yet they are] *primary* "bearers of qualities"; their mere presence in the hereditary substance gives no indication, or at most, only a very slight one, as to the character of the species. [...] In the course of his remarks, de Vries mentions the stripes of a zebra. How can these be hereditary if the different kinds of pangenes merely lie close together in the germ without being united into fixed groups, *hereditary as such*? There can be no "zebra pangenes," because the striping of a zebra is not a cell-character. There may perhaps be black and white colour of a cell; but the striping of a zebra does not depend on the development of these colours within a cell, but is due to the regular alternation of thousands of black and white cells arranged in stripes. (Weismann [1892] 1893, p. 16; his emphasis)

Weismann possessed a profound knowledge of embryology and of the complexity of development and was therefore well aware of a phenomenon later called the patterning of embryos, that is, the formation of complex structures, which require more than one cell and somehow superior coordination of more than one process. Weismann then continued:

> "For instance, the size, structure, veining, and shape of leaves, the characteristic and often absolutely constant patches of colour on the petals of flowers, such as orchids, may be referred to similar causes: these qualities can only arise by the regular co-operation of many cells. [...] they must be due to a *fixed grouping of pangenes*, or some other primary elements of the germ, which is *transferable from generation to generation*. [...] The idea which is here so clearly and decidedly expressed of the construction of innumerable species by various combinations of relatively few pangenes, shows that, even from de Vries's point of view, it is not the 'pangene *material*' as such, which is the main factor of determining the character of the species, but rather its *arrangement*, or, as I shall afterwards express it, *the architecture of the germ-plasm*." (Weismann [1892] 1893, pp. 17–18; his emphasis)

Only the germ cells possessed a complete set of determinants. In somatic cells, on the contrary, their distribution became qualitatively different. According to Weismann, the progressive subdivision of the *soma plasm* was responsible for cell differentiation. Fertilization was explained by the necessity to mix maternal and paternal determinants as the source of new combinations of variations. In his eponymous monograph of 1891, he called this process "Amphimixis" (Weismann 1891, p. 112). In order to avoid a continuous doubling of the number of determinants in every offspring, Weismann asserted that during maturation egg and sperm cells underwent a reduction division that exactly halved their number, selecting always a complete set. Variability was thus principally an internal phenomenon of recombination of maternal and paternal heredity factors.

In both principal variants of his theory, in 1885 and in 1892, Weismann conceived of development as a combination of three mutually influencing processes: the increasing differentiation of the cells, the decreasing complexity (1885) or composition (1892) of the nucleoplasm, and the increasing complexity of the nucleoplasm during phylogeny.

Weismann's localization of the "determining and directing factor" of development inside the nucleus, limited attention therefore to the events happening inside the cell. Notably, concentrating heredity on the germ plasm and differentiation on the progressive qualitatively different distribution of its determinants during the cell divisions, great difficulties arose when Weismann had to explain the post-embryological phenomena of regeneration and of tumor growths. Weismann's theory was meant to explain cell differentiation. It could also plausibly explain the capacity of certain organisms to regenerate lost parts. Yet a problem raised by the botanist De Vries was that many plant parts have the capacity to develop into a complete new individual. Weismann reacted to this objection by conceiving an auxiliary hypothesis that was later much criticized: certain cells also contain an amount of *passive* or latent germ plasm, which under certain circumstances could activate.

Another of Weismann's dilemmas was to reconcile his germ plasm theory with his concept of Darwinism as a gradual non-teleological process. Even in his *Germ-Plasm* monograph (1892), he still had not faced the possible consequences of his acceptance of natural selection on his theory. This demanded a further elaboration. The simple rearrangement of ancestral determinants could not explain the emergence of new adaptations and new species in the course of evolution. Indeed, Weismann's position regarding the production of variation underwent repeated changes. Around 1894, he introduced the additional hypothesis of germinal selection, that is, competition between the determinants of the same character. He also considered the potential transformation of determinants as a source of change.

Initially, like almost every biologist at that time, Weismann tended towards Lamarckian ideas of transformation induced by environmental changes. From about 1885 on, he was one of the very few to refute the possibility of transmission of acquired characters and to advocate natural selection as unique mechanism of evolution. According to him, neither the factors of somatic cells nor anything the body had learned could affect the constitution of the germ plasm. This conviction, outlined especially during his controversy with Herbert Spencer, made him one of the most prominent German Darwinists. However, at least since 1892, he did not completely exclude external influence. Differences, for example, in nutrition could affect the biophors and cause imperceptible fluctuations. These slight modifications were inherited and could then be amplified by recombination and thus, change the determinants. In this way, Weismann did not admit inheritance of acquired somatic characters but did admit acquired germ plasm variations.

Notwithstanding all earlier work of other scholars, Weismann's theory in its entirety represented a remarkable conceptual innovation. It therefore demanded also a new form of illustration. In 1892, he introduced his famous cell tree diagrams, which played a pivotal role in the publication. He admitted that they were a theoretical illustration, but they should be looked upon "as representing the phenomena as they occur in reality" (Weismann 1893, p. 103). His second tree diagram illustrated the germ-track through twelve cell generations of the threadworm *Rhabditis nigrovenosa* (Figure 2.2). The points of bifurcation corresponded to the moments of cell division. At each division, one of the two daughter cells entered the somatic line to become ectoderm (Ekt), endoderm (Ent) or mesoblast (Mes), whereas the other cell continued the germ track and finally led to the primitive germ cells (Kz). In order to highlight the germ track, the connecting lines were drawn as thick lines.

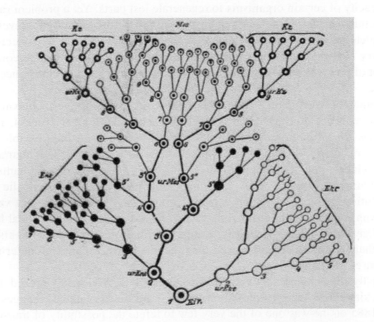

FIGURE 2.2 Weismann's illustration of the germ-track of *Rhabditis nigrovenosa*. (Courtesy of Weismann, A., *The Germ-Plasm: A Theory of Heredity*, Charles Scribner's Sons, New York, p. 196, 1893.)

2.5 THE RECEPTION AND MODIFICATION OF WEISMANN'S THEORIES

In the preface of his second last book, *Vorträge über Descendenztheorie* (1902–1903), Weismann complained that his germ plasm theory had not found the positive resonance it deserved:

> Notwithstanding much controversy, I still regard its fundamental features as correct, especially the assumption of "controlling" vital units, the determinants, and their aggregation into "ids"; but the determinant theory also implies germinal selection, which rejects the unfit and favours the more fit, is, to my mind, a mere torso, or a tree without roots. I only know of two prominent workers of our days who have given thorough-going adherence to my views: Emery in Bologna and J. Arthur Thompson in Aberdeen. But I still hope to be able to convince many others when the consistency and the far-reachingness of these ideas are better understood. (Weismann [1902–1903] 1904, I, p. VIII)

One reason for its lack of success may be, not the speculative nature of the theory by itself, but the scarcity of empirical evidence on the cytological level. Initially, the concept of the independence and the continuity of the germ plasm seemed promising. In 1888, Wilhelm Roux's experiments with frog embryos had brought strong support to Weismann's theory of differentiation. Yet Hans Driesch (1867–1941) soon challenged the outcome, because his experiments with sea urchin eggs brought diametrically opposed results. Single blastomeres separated from the early embryo succeeded in developing normal, though smaller, organisms. Other embryologists stepped in and applied different experimental strategies to verify the Roux-Weismann hypothesis.

One of the scholars, who set out to find these special germ cells in order to confirm or confute differential segregation, was Weismann's assistant Valentin Haecker (1864–1927). In 1892, he investigated the initial cleavage stages of the copepod *Cyclops* and observed an unequal cell division with one big cell migrating into the center of the embryo. This cell gave rise by division to both the germ cell and the somatic cells, which remained two distinct lines throughout the investigation stages (Haecker 1892). In the same year of 1892, Theodor Boveri (1862–1915) published a paper on the embryology of the parasitic worm *Ascaris megalocephala*. He noted that during the first cleavage stages, one of the two daughter cells underwent a visible chromatin reduction whereas the other remained normal. The normal cell again divided unequally, giving rise to one normal cell and one cell with reduced chromatin. In all, he observed a sequence of five unequal cleavage events. He, therefore, interpreted the one cell, which continued to have the complete chromatin set, as the primordial germ cell and the five cells with reduced chromatin as primordial soma cells, which continued multiplication to give rise to the various somatic tissues. It was hence possible to distinguish a soma line and a germ track. During the further differentiation of the primordial somatic cells, Boveri did not observe additional chromatin reductions. He nevertheless was convinced that he had empirical proof for Weismann's theory of determinant segregation, even if limited to the first five cell divisions (Boveri 1892, 117; see Dröscher 2014).

However, studies with other species failed to confirm these results. Concerning Boveri's results, it was later shown that the nematode *Ascaris* represents a very particular case, which provides no basis for generalizations. It is a classic example of very precocious segregation of the somatic from the germ line. While the lineage cells of the germ line retain their full chromosome complement, in the cells of the somatic line, pieces of chromosomes are lost. The loss amounts to about 27% of the total DNA of the cell. Moreover, nothing like a germ cell track was shown to exist in plants. Nor in mammals. On the contrary. After Waldeyer's pioneering works on the origin of germ cells in female mammals (see above) many others followed to demonstrate the somatic origin of germ cells. Studies with genetically marked cells showed in the 1970s that in mouse embryos each of the blastomeres of the 4- and 8-cell stage could become either a germ cell or a somatic cell. In fact, regarding this specific aspect of Weismann's theory, Edmund B. Wilson (1856–1939), the most influential cytologist in the English-speaking world, though a supporter of the germ plasm theory, was rather critical. In the third edition of his famous textbook, *The Cell in Development and Heredity* (1925), he underlined that normally no chromatin reduction takes place and that there is no neat distinction between the germ line and the somatic line:

> "The distinction between germ-cells and somatic cells, like that between 'germ-plasm' and 'somatoplasm' was however too sharply drawn by Weismann and his followers, and led to an opposition to his views in which the fundamental truth which they expressed often seemed to be lost sight of. A large body of evidence has accumulated in favor of the view that fundamentally any cell may be totipotent (i.e., contain the heritage of the species) and that the limitations of potency that it may display are due to secondary inhibitory conditions." (Wilson 1925, p. 310)

Germ plasm theory's main impact was probably on the Anglo-Saxon biologists. And the "fundamental truth" Wilson was talking about, concerned Weismann's conceptualizations of heredity and evolution, not that of development.

In the course of the early twentieth century, questions about the nature of the hereditary substance came to occupy the center stage. Weismann had sketched the first comprehensive concept of heredity and now, the supporters of the nascent discipline of genetics increasingly appreciated his theoretical outlines. Yet they extrapolated certain aspects, especially helpful for their own purposes. In the first decades of the century, although no definite empirical evidence could be brought in favor or against such a view, it was crucial to conceive of inheritance as transmitted by particular material carriers representing individual characters. Only in this form, the questions of heredity could be linked with nuclear cytology and with Mendel's laws. Weismann's theory contained both features. However, the more years went by, the more the growing group of geneticists concentrated on the mechanisms of the transmission of characters, separating it from the question of development, which had been central for Weismann and his generation.

Another distortion regards evolution. In Great Britain, and then in the United States, Weismann's conceptions found a more fertile ground than in the rest of Europe. In fact, all of his books and several of his essays were quickly translated into English.[1] One of his translators was Edward Bagnall Poulton, the colleague of Romanes. Late nineteenth-century Darwinians, like Romanes and Poulton, had set out to liberate Darwin's evolutionary theory from all traces of Lamarckian explanations, in particular, the inheritance of acquired characters. Weismann's theory of the continuity of the germ plasm provided an appropriate way to establish natural selection as the main feature of debate and the exclusive mechanism of evolution.

Putting all troublesome aspects on a shelf, Weismann's theory was simplified and transformed into a proof that external stimuli can in no case influence the genotype. James Griesemer and William Wimsatt recognized Edmund B. Wilson as one of the central figures, who have distorted Weismann's unified heredity-development theory (Griesemer and Wimsatt 1989; Griesemer 2005). Whilst for Weismann, the direct and exclusive succession through following generations referred to the germ *plasm*, because germ *cells* were somatic and thus open to modifications, Wilson (1896) transformed the germ *track* line into a germ *cell* line, further modifying the already graphically altered Weismann-diagram of Boveri (1892). In this way, Wilson stressed that characteristics acquired by the soma are not inherited because there is no developmental pathway by which they could influence the germ plasm (Figure 2.3) (Dröscher 2014). This extreme version later entered almost all textbooks as the "Weismann barrier." In the second half of the twentieth century, Weismannism became "molecular Weismannism," even though Weismann's thinking had nothing to do with the concept of (unidirectional) information flow from the DNA to RNA, and to proteins. In its molecular version, Weismannism became detached from the question of whether or not an independent germ track existed. Interest in the characteristics and the behavior of the germ cells properly shifted from main stage to sideshow.

As we have already seen above, Weismann's ideas about the exclusiveness of the role of natural selection were not that certain. In his 1892 book, *Das Keimplasma*, Weismann even stated: "The primary cause of *variation* is always the effect of external influences" (Weismann [1892] 1893, p. 463). These influences were mainly

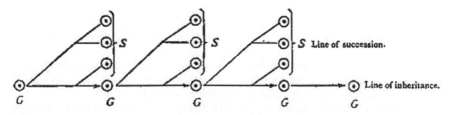

FIGURE 2.3 Wilson's diagram of the germ-line in animals. (Courtesy of Wilson, E.B., *The Cell in Development and Inheritance*, The Macmillan Company, New York, 1896, p. 11.)

---

[1] During Weismann's lifetime, besides his (Latin) doctoral thesis and an essay published in a French journal in 1881, only one work, the volume *Essais sur l'hérédité et la sélection naturelle* (1892), was published in another language than German or English. Actually, for instance, Italian philosopher Eugenio Rignano (1870–1930) invited Weismann to write for his journal *Rivista di Scienza*, but he declined the offer (Dröscher 2015, 395).

nutritional differences (ibid. p. 418). Winther (2001, pp. 526–528) limits Weismann's conviction of an uninfluenceable germ plasm to the period 1885–1892, but a letter from 1886 to myrmecologist Carlo Emery (1848–1925) shows that even in that period, Weismann was not very strict. Weismann responded to Emery's substantial arguments speaking in favor of the inheritance of acquired characters, and clarified:

"That is my opinion, too, yet I'm afraid to be slightly misunderstood on this point, probably because I expressed my opinion all too shortly. [...] Yet I think, that a direct modification of the structure of the germ plasm by external influences has indeed to be assumed; but only in those cases when these external influences remain the same for a long time. Just as the inferior unicellular organisms are directly modifiable, for me, also the germ cells of the polyplastids [a multicellular organism]. There are however good reasons to assume that these modifications do not appear easily and not as a result of every minor influence." (Weismann to Emery February 20, 1886; see Dröscher 2015, p. 399)

Romanes nicely portrayed the confusing situation of who around the turn of the century was a Darwinian, a Lamarckian, or a Weismannian:

Hence we arrive at this curious state of matters. Those biologists who of late years have been led by Weismann to adopt the opinions of Wallace, represent as anti-Darwinian the opinions of other biologists who still adhere to the unadulterated doctrines of Darwin. Weismann's *Essays on Heredity* (which argue that natural selection is the only possible cause of adaptive modification) and Wallace's work on Darwinism (which in all the respects where any charge of 'heresy' is concerned directly contradicts the doctrine of Darwin)—these are the writings which are now habitually represented by the Neo-Darwinians as setting forth the views of Darwin in their 'pure' form. The result is that, both in conversation and in the press, we habitually meet with complete inversions of the truth, which show the state of confusion into which a very simple matter has been wrought by the eagerness of certain naturalists to identify the views of Darwin with those of Wallace and Weismann. (Romanes 1895, p. 9)

REFERENCES

Balfour, F. M. 1878. On the structure and development of the vertebrate ovary. *Quart J Microsc Sci* 18: 1–81.

Boveri T. 1892. Über die Entstehung des Gegensatzes zwischen den Geschlechtszellen und die somatischen Zellen bei Ascaris megalocephela. *Sitzungsberichte der Gesellschaft für Morphologie und Physiologie in München* 8: 114–125.

Bulmer, M. 1999. The development of Francis Galton's ideas on the mechanism of heredity. *J Hist Biol* 32: 263–292.

Burnham, J. C. 1972. Instinct theory and the German reaction to Weismannism. *J Hist Biol* 5(2): 321–326.

Charpa, U. 2010. Darwin, Schleiden, Whewell, and the "London Doctors": Evolutionism and microscopical research in the nineteenth century. *J Gen Philos Sci* 41: 61–84.

Churchill, F. B. 1968. August Weismann and a break from tradition. *J Hist Biol* 1: 91–112.

Churchill, F. B. 1985. Weismann's continuity of the germ-plasm in historical perspective. *Freiburger Universitätsblätter* 24: 107–124.

Churchill, F. B. 1986. Weismann, hydromedusae and the biogenetic imperative: A "reconsideration." In *A history of embryology* (Eds.) T. J. Horder, J. A. Witkowski, and C.C. Wylie, pp. 7–33. Cambridge, UK: Cambridge University Press.

Churchill, F. B. 1987. From heredity theory to Vererbung: The transmission problem, 1850–1915. *Isis* 78: 337–364.

Churchill, F. B. 1999. August Weismann: A developmental evolutionist. In *August Weismann: Ausgewählte Briefe und Dokumente/Selected letters and documents* (Eds.) F. B. Churchill, and H. Risler. 2 Vols., pp. 749–798. Freiburg im Breisgau, Germany: Universitätsbibliothek Freiburg im Breisgau.

Churchill, F. B. 2015. *August Weismann: Development, Heredity, and Evolution*. Cambridge, MA: Harvard University Press.

Coleman, W. 1965. Cell, nucleus, and inheritance: An historical study. *Proc Amer Philos Soc* 109: 124–158.

Darwin, C. 1859. *On the Origin of Species by Means of Natural Selection or the Preservation of Favoured Races in the Struggle for Life*. London, UK: Charles Murray.

Darwin, C. 1868. *The Variation of Animals and Plants Under Domestication*. 2 Vols. London, UK: John Murray.

Darwin, C. 1875. *The Variation of Animals and Plants Under Domestication*. 2nd ed. revised, 2 Vols. London, UK: John Murray.

Deichmann, U. 2010. Gemmules and elements: On Darwin's and Mendel's concepts and methods in heredity. *J Gen Philos Sci* 41: 85–112.

Dröscher, A. 2008. Weismann, August Friedrich Leopold. *Encyclopedia of the Life Sciences*. Chichester, UK: John Wiley & Sons. doi:10.1002/9780470015902.a0002448.

Dröscher, A. 2014. Images of cell trees, cell lines, and cell fates: The legacy of Ernst Haeckel and August Weismann in stem cell research. *Hist Phil Life Sci* 36(2): 157–186.

Dröscher, A. 2015. Of germ-plasm and zymoplasm: August Weismann, Carlo Emery and the debate about the transmission of acquired characteristics. *Hist Phil Life Sci* 36(3): 394–403.

Dröscher, A. 2016. Lassen Sie mich die Pflanzenzelle als geschäftigen Spagiriker betrachten: Franz Ungers Beiträge zur Zellbiologie seiner Zeit. In *Einheit und Vielfalt: Franz Ungers (1800–1870) Konzepte der Naturforschung im internationalen Kontext* (Ed.) M. Klemun, pp. 169–194. Wien, Austria: Vienna University Press.

Endlicher, S. and Unger, F. 1843. *Grundzüge der Botanik*. Wien, Austria: Carl Gerold.

Galton, F. 1871. Experiments in pangenesis, by breeding from rabbits of a pure variety, into whose circulation blood taken from other varieties had previously been largely transfused. *Proc Roy Soc* 19: 393–410.

Gayon, J. 1998. *Darwinism's Struggle for Survival: Heredity and the Hypothesis of Natural Selection*. Cambridge, UK: Cambridge University Press.

Goette, A. 1874–1875. *Die Entwicklungsgeschichte der Unke als Grundlage einer vergleichenden Morphologie der Wirbeltiere*. Leipzig, Germany: Voß.

Griesemer, J. R. 2005. The informational gene and the substantial body: On the generalization of evolutionary theory by abstraction. In *Idealization XII. Correcting the Model. Idealization and Abstraction in the Sciences* (Eds.) M. R. Jones, and N. Cartwright, pp. 59–116. Amsterdam, the Netherlands: Rodopi.

Griesemer, J. R. and Wimsatt, W. C. 1989. Picturing Weismannism: A case study of conceptual evolution. In *What the Philosophy of Biology is. Essays Dedicated to David Hull* (Ed.) M. Ruse, pp. 75–137. Dordrecht, the Netherlands: Kluwer.

Haecker, V. 1892. Die Kerntheilungsvorgänge bei der Mesoderm- und Entodermbildung von Cyclops. *Arch mikr Anat* 39: 556–581.

Herbert, S. (Ed.). 1980. The red notebook of Charles Darwin. *Bull Brit Mus (Nat Hist) Hist series* 7: 1–164.

MacLaurin, J. 1998. Reinventing molecular Weismannism: Information in evolution. *Biol &
    Phil* 13: 37–59.
Mayr, E. 1985. Weismann and evolution. *J Hist Biol* 18: 295–329.
Meneghini, G. 1838. *Cenni sulla organografia e fisiologia delle alghe.* Padova, Italy: Tipi
    della Minerva.
Nussbaum, M. 1880. Zur Differenzierung des Geschlechtes im Thierreich. *Arch mikr Anat*
    18: 1–120.
Owen, R. 1849. *On Parthenogenesis, or the Successive Production of Procreating Individuals
    from a Single Ovum. A Discourse Introductory to the Hunterian Lectures on Generation
    and Development, for the Year 1849, Delivered at the Royal College of Surgeons of
    England.* London, UK: John van Voorst.
Romanes, E. D. (Ed.). 2011. *The Life and Letters of George John Romanes, Written and
    Edited by his Wife.* New edition. New York: Cambridge University Press.
Romanes, G. J. 1893. *An Examination of Weismannism.* Chicago, IL: The Open Court
    Publishing Company.
Romanes, G. J. 1895. The Darwinism of Darwin, and of the post-Darwinian schools. *Monist*
    6(1): 1–27.
Rupke, N. 2009. *Richard Owen: Biology without Darwin,* a revised edition. Chicago, IL: The
    University of Chicago Press.
Schleiden, M. J. 1848. *The Plant: A Biography in a Series of Popular Lectures,* translated by
    A. Henfrey. London, UK: Hyppolyte Bailliere.
Sloan, P. R. 1986. Darwin, vital matter, and the transformism of species. *J Hist Biol* 19(3):
    369–445.
Stamhuis, I. H. 2003. The reactions on Hugo de Vries's Intracellular Pangenesis: The discus-
    sion with August Weismann. *J Hist Biol* 36: 119–152.
Stanford, P. K. 2005. August Weismann's theory of the germ-plasm and the problem of
    unconceived alternatives. *Hist Phil Life Sci* 27: 163–199.
Waldeyer, W. 1870. *Eierstock und Ei: Ein Beitrag zur Anatomie und Entwicklungsgeschichte
    der Sexualorgane.* Leipzig, Germany: Wilhelm Engelmann.
Weismann, A. 1883. *Die Entstehung der Sexualzellen bei den Hydromedusen. Zugleich ein
    Beitrag zur Kenntniss des Baues und der Lebenserscheinungen dieser Gruppe.* Jena,
    TH: Gustav Fischer.
Weismann, A. 1885. *Die Continuität des Keimplasmas als Grundlage einer Theorie der
    Vererbung.* Jena, TH: Gustav Fischer.
Weismann, A. 1889. The continuity of the germ-plasm as the foundation of a theory of hered-
    ity. In *Essays Upon Heredity and Kindred Biological Problems,* pp. 161–249. Oxford,
    UK: Clarendon Press.
Weismann, A. 1891. *Amphimixis oder: Die Vermischung der Individuen.* Jena, TH: Gustav
    Fischer.
Weismann, A. 1892. *Das Keimplasma. Eine Theorie der Vererbung.* Jena, TH: Gustav
    Fischer.
Weismann, A. 1893. *The Germ-Plasm: A Theory of Heredity.* New York: Charles Scribner's Sons.
Weismann, A. 1896. New experiments on the seasonal dimorphism of Lepidoptera. *The
    Entomologist* 29: 29–39, 74–80, 103–113, 153–157, 173–208, 240–252.
Weismann, A. 1902–1903. *Vorträge über Descendenztheorie gehalten an der Universität zu
    Freiburg im Breisgau.* 2nd ed. Jena, TH: Gustav Fischer.
Weismann, A. 1904. *Evolution Theory.* Translated with the author's cooperation by J. Arthur
    Thomson and Margaret R. Thomson, 2 Vols. London, UK: Edward Arnold.
Weissman, C. 2010. The origins of species: The debate between August Weismann and
    Moritz Wagner. *J Hist Biol* 43: 727–766.

Wilson, E. B. 1896. *The Cell in Development and Inheritance*, New York: The Macmillan Company.

Wilson, E. B. 1925. *The Cell in Development and Heredity*, 3rd ed. New York: The Macmillan Company.

Wimsatt, W. C. and Griesemer, J. R. 2007. Reproducing entrenchments to scaffold culture: The central role of development in cultural evolution. In *Integrating Evolution and Development: From Theory to Practice* (Eds.) R. Sansom and R. N. Brandon, pp. 228–323. Cambridge, MA: MIT Press.

Winther, R. G. 2001. August Weismann on germ-plasm variation. *J Hist Biol* 34: 517–555.

3 Cell Lineages in Ontogeny and Phylogeny from 1900

Jane Maienschein

CONTENTS

3.1 INTRODUCTION

Today, the idea that it could be useful to trace cell lineages makes perfect sense, even if the work is difficult and not many researchers are willing to invest the tremendous dedicated energy required to carry out the early kinds of cell lineage studies. Following a cell lineage in its earliest sense means, tracking a cell in an embryo through each of its cell divisions as far as possible. The *Oxford English Dictionary* (*OED*) defines cell lineage as "The manner in which the parts of a multicellular organism develop from the blastomeres of the embryo; the line of descent of a cell from a blastomere or other embryonic precursor; a population of cells sharing such a line of descent." The *OED* also gives credit to the outstanding American cytologist Edmund Beecher Wilson for introducing the term in the late nineteenth century.

The idea of cell lineage study appeared before Wilson, but only barely. Before it made sense to ask about the lineage of cells through many generations of cell division, researchers needed a reason to care about the patterns and nature of the divisions. That, as Theodosius Dobzhansky later noted for all of biology, only made sense in light of evolution (Dobzhansky 1973). We don't think in terms of lineages if we aren't thinking in terms of descent. If cells are understood simply as dividing materially into more and more cells, but nothing about their historical background really matters, then why make the meticulous effort to trace the details of division over time?

Once scientists began thinking in terms of the evolution of species, it was a logical step to ask what kinds of observations can give evidence about past conditions and changes. How can we begin to "see" aspects of evolution when we were not there ourselves for the millions of years when evolution occurred? Perhaps cells, the patterns of their divisions, and their fates carry "ancestral reminiscences" that reflect the history and help us understand development in terms of evolution, as Wilson suggested. This paper looks at three different periods, which involve different ways researchers have explored cell lineages and their interpretations.

The first period occurred in the decades just before and after 1900. A number of researchers, especially at the Marine Biological Laboratory (MBL) in Woods Hole, Massachusetts, tracked cell lineages in the embryos of a variety of different animals. They started with questions about ancestral reminiscence and what they could learn about the evolutionary and phylogenetic relationships of animal groups. They also became interested in the cell divisions themselves, as a contributing factor in the development of individual embryos. That work was extremely time-consuming and challenging, and the researchers soon turned to other questions and other methods.

The second period of focus on cell lineage came with Sydney Brenner's idea that it should be possible to document every cell division in the nematode, *C. elegans*, and track its fate. Brenner, who later won a Nobel Prize along with John Sulston and Robert Horvitz for their work, sought an organism that was easy to observe, easy to cultivate, had a stable and predictable developmental pattern, and for which it was possible to correlate genetic mutations with structural effects.

More recently, the third wave of interest in cell lineage was discovered, which involves following the development of a lineage of particular cells from a starting point without consideration of the evolutionary past. We see this, for example, especially in efforts to interpret the causes and trajectories of cancers. Once a cancer cell is identified, questions that arise are: what does it do, where does it go, how does it develop, divide, or adapt? Are there cancer stem cells that develop into cancerous cells, and what does this even mean? These are lineage studies of identified cells as they develop in the body, but such studies typically do not look to the deeper evolutionary past to interpret the lineages. Another approach to following particular cells takes one or more cells out of the body altogether and traces the lineages of the cell lines in a culture dish. Again, the emphasis is on the particular cell and its development, division, and differentiation going forward rather than looking at the past. For this third type of cell lineage, researchers do not seek to identify or trace the ancestral factors of deep evolution that may be influencing present behavior. It's not that they would not love to be able to capture that evolutionary story, especially insofar as it would help inform understanding of the current situation. *But we don't yet have the tools to connect the individual cancer cells or cells that give rise to cell lines with evolutionary factors such as gene regulatory*

*networks. Not yet. It is nonetheless worth looking at this type of cell lineage work
and reflecting on what future research may bring.*

3.2 CELL LINEAGE AND ANCESTRAL REMINISCENCE AROUND 1900

A CHARLES OTIS WHITMAN

Though Wilson seems to have first called the study of the cleavage paths of animal
embryos "cell lineage," Charles Otis Whitman had already led the way in thinking
about the paths and patterns of early cell division. Whitman then took this research
emphasis to the new MBL, where he became the first director in 1888. There he
encouraged researchers and his own students to carry out cell lineage studies on
embryos of different animals, to allow comparisons across different species and animal groups or phyla.

Whitman's work began in the 1870s, a time of eagerness to understand cells and
early stages of development as far as gastrulation and germ layer formation. Ernst
Haeckel had suggested that germ layers provide the start of differentiation and organization in animal embryos, which led to considerable debate about what was happening, as well as what it meant. (See Richards 2008 for perspective on Haeckel
and his ideas.). In 1878, Whitman published his doctoral dissertation study of freshwater leeches (supervised by Rudolf Leuckart at the University of Leipzig) as "The
Embryology of Clepsine" (Whitman 1978). Whitman described the methods used,
which formed the starting point for his eventual volume on microscopical methods
of the day (Whitman 1885). He also gave a rich discussion of seemingly every detail
of egg preparation through early development. It is worth reviewing this first cell
lineage contribution more fully.

Whitman studied several different species of the freshwater leech *Clepsine*,
meticulously preparing his specimens and observing the changes over time. He
embedded his discussion in the context of previous studies and interpretations. In
the context of other often rather less comprehensive research of the time, Whitman's
attention to detail is noteworthy. He started with each step in the formation of eggs,
the role of the nucleus, the differences among his eggs, and those of other types of
organisms. He was clearly working to interpret what he saw in the very earliest developmental moments. For example, he referred to early "germ spots" of 0.0037 and
0.0025 mm, and it takes 3–4 days for a full sized "primary egg cell" to develop and
grow to about 0.55 mm. The mature egg takes about two weeks to develop. These
several pages of detail show that much is happening to prepare an egg for fertilization and eventually being deposited outside the adult.

Clepsines are hermaphroditic, Whitman explained, and therefore raise obvious
questions about how the egg undergoes "impregnation." He never observed a "sexual
union" (Whitman 1878, p. 8) but became convinced that the egg must be fertilized
while in the ovary, perhaps through self-fertilization. When the eggs are deposited,
they are mature and ready to begin development. Now comes the excitement of

cleavage, and "so far as yet known, these changes in the egg of Clepsine are unsurpassed in variety by those of any other egg" (Whitman 1878, p. 12). He carried on with details about each sequential step of the developmental process. Polar bodies, pellucid spot, polar globules, polar activity, polar rings, and pronuclei: these are all part of the complex organization that occurs prior to the first cleavage.

At this point, Whitman acknowledged that a diversity of interpretations existed for the phenomena he was observing. While some held that a cell consists of protoplasm with a single nucleus, he could see more than one nucleus in preparation for cell division. These "free nuclei" puzzled Whitman, who was not sure about their role or whether they had an ability to cause additional cells to coalesce and/or divide. Others thought they saw additional nuclei, and Whitman noted that these might be nucleoli instead, yet questions remained. Cell theory was not so tidy for Whitman as it later became, and we see his worries later in an essay on "The Inadequacy of the Cell Theory" (Whitman 1893).

Finally, about halfway through the 1878 paper, we get cleavage. "In the fecundated egg slumbers potentially the future embryo. While we cannot say that the embryo is predelineated, we can say that it is predetermined" (Whitman 1878, p. 49). Already, in his first publication on embryology, he pointed to comparisons with other species, apparently as a way to get at patterns in the ways cells related to each other. After offering comparisons, he asked whether the similarity of cell divisions and arrangements in birds and fish can be explained in the same way as in his invertebrates. "Since the process in both cases leads to similar results, it is natural to infer that it is controlled by the same general laws" (Whitman 1878, p. 94).

For Whitman, "The egg is, in a certain sense, a quarry out of which, without waste, a complicated structure is to be built up; but more than this, in so far as it is the architect of its own destiny. The raw material is first split into two, four, or more huge masses, and some or all of these into secondary masses, and some or all of these into tertiary masses, &c., and out of these more or less unlike fragments the embryonal building-stones are cut, and transported to their destined places" (Whitman 1878, p. 50). *This is the first declaration of cell lineages, despite the underlying uncertainty about precisely what cells are and precisely what mechanisms cause them to divide.*

Whitman then went on to describe what happens during division: the change of form, the plane of division and its movement, the timing of each division, the angles as cleavage progresses. We get ectoblasts, mesoblasts, neuroblasts, entoblasts, and discussion of movements of the cleavage products as they progress on their way to becoming the germ layers that many researchers of the time considered the starting point for an individual organism. Blastula gives way to gastrula, with its infolding and reshaping of the embryo. Whitman described the stages, gave the timing of each change, and used language that reflected already accepted definitions of each stage. He was identifying known parts and processes, not discovering them for the first time. That is, he was finding, in his leeches, the developmental processes seen elsewhere by others.

Differences did arise, however; Whitman noted that *Clepsines* do not pass through a morula (solid ball of cells) stage as many other species do. Some parts come from the upper pole in some forms and the lower pole in others. Yet in comparing the *Clepsine* neurula with that of vertebrates, he found a remarkable similarity in structure with some variations in detail and in the rate of change. *Fish, chicks, and leeches seemed to share the patterns of origin of the primitive streak, for example, and of other key developmental steps. Finding the parallels among leech, fish, chick, and other eggs reinforced the conviction that here was an "interesting remnant of the ancestral condition"* (Whitman 1878, p. 92).

Parts and organs emerge from the relevant germ layers. Ectoderm and mesoderm were clear, but he asked "Whence arises the entoderm?" that he would have expected (Whitman 1878, p. 66). He was asking about lineages and what cells gave rise to each germ layer and its subsequent parts. He needed more study to determine the origins of the entoderm, and he recognized that he might obtain clues by looking at what was known about other organisms. In his summary, Whitman explained that each cleavage had specific identifiable effects and regularities. In effect, the cells have lineages that lead from initial blastomeres to differentiated parts. "Thus it happens that, before a given ontogenetic stage is completed, the preliminary segregations and arrangements for the following stage are already more or less advanced. Thus the gastrula—and more rarely the blastula—is pre-stamped with the antimeric character of the ultimate bilateral form" (Whitman 1878, p. 79). Trace the cells backward, and it becomes clear that the later differentiated parts that makeup germ layers started in particular identifiable earlier cells predictably and reliably.

Some of the changes seem to come from physical pressures, as Wilhelm His had suggested in his emphasis on the efficacy of folding of parts in embryo formation (His 1874). Yet Whitman did not offer his own interpretative explanations but rather stuck with descriptions and comparison with the interpretations of others. As a result, his discussion comes across as entirely reliable, valuable, and significant in establishing details about the processes and progress of development. Roughly two decades later, around 1900, at the MBL, Whitman rallied a community of leading researchers to carry out more of this kind of work, or what came to be called cell lineage studies.

B WHY DID WHITMAN PURSUE THIS PROJECT?

Whitman had received his Bachelor's degree from Bowdoin College, then taught natural history in high school (See biography in Morse 1912). He joined other teachers in attending Louis Agassiz's Penikese Island School in natural history and there became inspired to pursue zoology professionally. That quest took him to the leading marine research station in the world, the Stazione Zoologica in Naples, and then to the University of Leipzig to study with Leuckart, who was renowned for his excellence in microscopic techniques and his meticulous attention to morphological detail. Best known later for his work in parasitology, Leuckart studied marine invertebrates and followed the life cycles of a number of organisms. His "Leuckart charts" graced the walls of classrooms around the world, including at the MBL, and can be accessed at http://legacy.mblwhoilibrary.org/leuckart/wall_charts.html.

Leuckart welcomed international scholars, including young Americans eager to gain research skills. Whitman grew from being a young student eager to learn into a professional biologist during his three years in Leipzig. He received his PhD in 1878 with the dissertation on *Clepsine*. During Whitman's time in Leipzig, Leuckart served as rector of the university for a year, and one can only speculate whether watching that activity in his advisor inspired Whitman as he took on his own administrative and leadership roles.

Whitman went from Leipzig to a position at the Imperial University of Japan, where he established zoology as a leading field in Japan and trained a generation of Japanese microscopists in the latest techniques. He returned to the United States and spent two years at Harvard as an assistant in zoology, then moved inland to direct the Allis Lake Laboratory 1886–1889, became the first head of zoology at the research-oriented Clark University 1889–1892, and finally went on to head biology at the new University of Chicago when it opened in 1892. In addition to those administrative roles, Whitman served as founding director of the MBL, accepting the position that offered no salary because it offered the opportunity to build a kind of Naples Stazione Zoologica in America. Both the Naples station and the MBL became leading places for bringing together the study of development, physiology, and evolution. Whitman wanted to educate and encourage research in an independent collaborative environment that served as an "assembling place" for modern biology. He also edited the *Journal of Morphology* and *Biological Bulletin* to provide an outlet for scholarship.

C   CELL LINEAGE AT THE MBL

As MBL director, Whitman started with a small group of seventeen instructors and students in the first year. Immediately, he began to recruit researchers to come to the MBL from the US Fish Commission, the government organization for the study of diverse fish species and their environments that was conveniently located just across the street from the MBL (Lillie 1944; Maienschein 1989). While a few other researchers in other countries carried out their own cell lineage studies, the group that Whitman assembled at the MBL carried out by far the largest and most concentrated study of cell lineages in early embryonic development. During the first decade of the MBL, Whitman recruited such outstanding biologists as Edwin Grant Conklin, Edmund Beecher Wilson, and Thomas Hunt Morgan. These three had all been graduate students at Johns Hopkins University and had visited marine stations with their advisor William Keith Brooks, including the Woods Hole Fish Commission. Each quickly found his way to the MBL, then continued to return there in summers for the rest of his life while also assuming leadership roles.

Wilson was the oldest of this Hopkins group at the MBL. He grew up in a small town of Geneva, Illinois, went to Antioch College for a year, worked while learning the basics required for Yale University, and received his Bachelor's degree from Yale (See Morgan 1940 for a biography). He then joined the U.S. Fish Commission in Gloucester, Massachusetts, in the summer of 1877, participating in dredging expeditions. From there he proceeded to the recently founded Johns Hopkins University while continuing his interest in marine studies. With a dissertation on the colonial polyp *Renilla*, for which he carried out the studies during three summers at the John Hopkins marine stations, Wilson traced the development of a single polyp into its colonial form.

Under Brooks's influence, Wilson was thinking in terms of evolution for that study; he was very much aware of the ideas of Charles Darwin, Ernst Haeckel, and others. He also learned to use serial histological sections to see inside complex organisms and cells. Wilson received his PhD in 1881 and remained at Hopkins as an assistant. In 1883, with support from a cousin, he visited England and then spent more time in Leipzig with Rudolf Leuckart, who continued to welcome American visitors as he had welcomed Whitman. From there he went to the Naples Stazione Zoologica, as Whitman had. This visit gave Wilson the chance to continue his biological research on development while also enjoying his serious interest in music. He played the cello as part of a quartet there and listened to the many concerts that the director Anton Dohrn brought to the Stazione. His visit reinforced Wilson's deep interest in the study of marine organisms, development, and evolution. Back in the US, he taught at Williams College, spent a year at MIT, co-authoring a textbook, took up a position at Bryn Mawr College in 1885 before moving to Columbia University in 1891. During the rest of his career at Columbia, Wilson traveled every summer to carry out research at the MBL.

At the MBL, Wilson learned from a colleague about the polychaete worm *Nereis*. Inspired by Whitman's earlier work on *Clepsine*, and in the hope of determining the origin of germ layers in annelid worms, Wilson reports that he had been looking for an appropriate organism. *Nereis* turned out to be useful, not just for the evolutionary question about origins, but also for following what Wilson first called cell lineage. At night, all it takes is a lantern to mimic the light of the moon and lure these colorful worms into the collecting net. From the dock on the Eel Pond, it was just a few steps to the well-equipped MBL laboratory, where Wilson could hurry into the laboratory and watch the *Nereis* eggs go through their developmental stages, including cleavages. *Nereis* eggs are transparent and relatively large, develop quickly, have visible structures to help identify which cell is which, and are easy to fix and stain (Wilson 1892, p. 363). There are good reasons that these eggs remain favorites for the MBL Embryology course today.

Wilson noted that the germ-layer theory had led to the considerable study of comparative embryology through the 1890s which, in turn, had yielded a surprising amount of disagreement in interpretations about both development and evolution. The theory held that at the point of gastrulation, an embryo develops germ layers that each lead to subsequent differentiation of different parts of the body (MacCord 2013). This raised new questions. What should count as a homology, for example: did the germ layers remain homologous with each other in different gastrulas, over time? A considerable debate centered on the significance of germ layers and their origins and Wilson concluded that the only way to resolve the various issues was *"by tracing out the cell-lineage of cytogeny of the individual blastomeres from the beginning of development"* (Wilson 1892, p. 367). Pointing to Whitman's "epoch-making" studies, Wilson saw early development as informing understanding of evolution through the impact of early variations rather than as simple mechanical proliferation of material. *"The very fact that the differentiation of the layers is effected in such a diversity of ways proves conclusively that these early stages of development are as susceptible to secondary modification as the later"* (Wilson 1892, p. 368).

Describing every step in detail, Wilson provided images and discussion intended to let any other researcher see the same thing he was seeing. He was looking for causes of the "organization of the egg." Those causes were at least in part hereditary, he determined, and he marveled at "the remarkable fact, and one which does not seem to be very clearly recognized" that the divisions of cells and the mechanical conditions that cause them to divide in the patterns they follow "has become hereditary" (Wilson 1892, 450; and see Guralnick 2002 for more discussion). This conclusion that mechanical conditions, rather than simply differences in form, drive the earliest cell divisions led him to further careful study of "germinal localization" in several species, and to a broader discussion of the idea of "ancestral reminiscence" in development. Those ideas received attention at the MBL when he presented a lecture on "Cell Lineage and Ancestral Reminiscence" as part of the *Biological Lectures* series in 1898, based on a paper presented earlier to the New York Academy of Sciences (Wilson 1898, 1899).

Wilson's essay resulting from that lecture captures the state of the field in the late 1890s. Each organism, Wilson noted, arises through the processes of a complicated mechanism and also of its past, including its evolutionary history. Sometimes the individual developmental ontogeny may seem to repeat the ancestral development or phylogeny, as Haeckel had argued. Wilson felt that the relationship was not one of repetition, however, but of reminiscences from past adaptations that are modified by environmental and other conditions. And, yes, cell lineage studies had shown that the reminiscences occur even at the earliest developmental stages. Cell cleavages follow an orderly and defined process, with "marvelous consistency" just as later developmental stages do. "The study of cell-lineage has thus given us what is practically a new method of embryological research" (Wilson 1899, p. 24). *In conjunction with close attention to the way the lineages give rise to the germ layers, comparative studies could show much about evolutionary relationships and about cell homologies that carry those ancestral reminiscences.*

Wilson's lecture took place in the context of work by other colleagues; studying the cell lineage of something was almost a requirement at the MBL in the 1890s. Even Thomas Hunt Morgan, known for work on regeneration, later on for research on chromosomes, and his Nobel Prize-winning work on genetics, looked closely at cell lineages and development. Morgan grew up in Kentucky, received his BS degree from the University of Kentucky, spent a year at Alpheus Hyatt's Annisquam Laboratory, then in 1886 went on to Johns Hopkins to study under Brooks. He spent 1888 in Woods Hole at the US Fish Commission, then moved across the street to the MBL in 1890. (See Sturtevant 1959 and Allen 1979 for biographies). Morgan did not carry out the detailed step-by-step cell lineage descriptions that Wilson did, but he was very much attuned to the importance of regularities and patterns as cleavage stages progressed through development.

Within the context of a look at relationships among sea spiders, Morgan focused his PhD dissertation on a particular sea spider with "A contribution to the embryology and phylogeny of the Pycnogonids" (Morgan 1891). Yet the year after he received his PhD, he also published on the larval Tornaria form of worm-like *Balanoglossus*, embryology of sea bass, and frogs. Over the next few years (1891–1904), Morgan resided as a faculty member at Bryn Mawr College, after Wilson had left the position vacated when he moved to Columbia University. While at Bryn Mawr, Morgan pursued research on teleosts, echinoderms, sea urchins, sea stars, fish, and others, looking at whatever organism seemed likely to produce some interesting phenomenon to explore or to make itself accessible to study (Maienschein 2015). He clearly knew about and learned from the cell lineage studies his colleagues were carrying out at the MBL.

As his biographer Garland Allen has emphasized, Morgan quickly adopted an experimental approach to embryology (Allen 1979). Rather than documenting in detail every step of cell lineage development, Morgan asked about how organisms function or what conditions cause changes. In this, he adopted the experimental approach that Wilhelm Roux had announced in the 1894 introduction to the new journal that he edited, *Archiv für Entwickelungsmechanik* (which later became *Roux's Archiv* and much later *Development Genes and Evolution*).

The MBL community engaged in lively discussion of what an experimental program meant for embryology. Morgan apparently listened hard; by 1895 he was publishing descriptions of what happens to cell development after various sorts of experiments. In 1898, he wrote a short note on developmental mechanics for *Science* and suggested that "Therefore, by means of an experiment, the student of the new embryology hopes to place the study of embryology on a more scientific basis" (Morgan 1898, p. 50). Morgan's visit to the experimentally-oriented Stazione Zoologica in 1894–1895, working at a table hosted by the Smithsonian Institution, surely reinforced his experimental emphasis.

This experimental turn for Morgan took him to studies of regeneration starting in 1898 that reached a peak with his book, *Regeneration* (Morgan 1901). Chopping off pieces of planarians, earthworms, and hydra, in particular, Morgan sought to determine what happened as a result. Study of regeneration would provide a window into how development normally works, as Mary Sunderland has discussed (Sunderland 2010). Thus, while Morgan did not carry out his own detailed studies to follow the lineage of cells throughout early development, such cell lineage work informed his own studies that followed sequences of cells in order to compare normal and experimental conditions.

Edwin Grant Conklin worked most closely and in parallel with Wilson so that the two referred to details of the other's work while comparing what was similar and different. Conklin grew up in a small town in Ohio, received his Bachelor's degree from Ohio Wesleyan University, taught at the historically black Rust University, then decided that he could become a professional biologist if he received a PhD. (See Harvey 1958 for a biography). And so, in 1888 he also went to Johns Hopkins University to study under William Keith Brooks, as a number of other leaders in the US in biology did as well. Brooks had connections with the Fish Commission in Woods Hole where Conklin went in 1889. Brooks suggested that Conklin study the siphonophores that he had studied at Hopkins but soon discovered that there were none in Woods Hole.

The need for a topic led him to look at many different species, while also working out a research question. Conklin settled on cell lineage of the slipper snail *Crepidula*. Like Wilson, he carried out extremely meticulous work and described in detail the changes with each cell cleavage. The snails behaved in some ways different, and in some ways the same as Wilson's worms and the comparisons enriched both of their studies. Conklin reported that Brooks was not at all convinced that this was a reasonable topic, nor a useful methodology. Staring at cells and the mechanics and structures of their changes during cleavage did not seem likely to have morphological significant, Brooks complained.

Yet the next summer working again at the Fish Commission, Conklin met Wilson working at the MBL and began close communication with him that led to a long friendship. Conklin completed his dissertation, and Brooks is said to have commented: "Well, Conklin, this university has sometimes given the doctor's degree for counting words; I think maybe it might give one degree for counting cells" (Harvey 1958, p. 63). Whitman approved and agreed to publish Conklin's dissertation in the *Journal of Morphology* that Whitman edited, even though the 226 pages, 9 plates, and 105 color figures almost bankrupted the journal. "What Is Money for?" Whitman asked in making it clear that of course, he would support publication of Conklin's work, as reported in an interview in the last days of Conklin's life. (Bonner and Bell 1984; the interview is deposited with the American Philosophical Society archives.)

In 1905 Conklin published one of the last major cell lineage studies with his "The Organization and Cell-Lineage of the Ascidian Egg" (Conklin 1905). There he noted that the system of nomenclature and descriptions of variations used for annelids and mollusks would not work for ascidians, and required adjustment, although it was not yet clear just how. The complexities help show why researchers largely set cell lineage studies aside in favor of other methods and questions.

While Wilson emphasized the dual influences of physical and mechanical factors and ancestral reminiscences, Conklin was a Darwinian first. Indeed, he had entered biology because of a fascination with Darwinian evolution, and his conviction that his Methodist beliefs were perfectly compatible with a proper understanding of evolution. The mechanics of development played a secondary, though necessary, role. Biology was at root evolutionary for Conklin.

Whitman's own graduate students carried out lineage studies as well. Frank Rattray Lillie, who became the second director of the MBL and the second chair of zoology at the University of Chicago after Whitman, reported having been recruited right away both to study with Whitman and specifically to carry out cell lineage work. Whitman assigned Lillie to work on the freshwater mussel *Unio*, which required him to lug his assemblage of buckets and waders to a pond nearby Falmouth. Aaron Treadwell studied the polychaete worm *Podarke Obscura* Verrill, and A. D. Mead looked at annelid worms, which provided valuable material for comparison.

D   Why Did Cell Lineage Work End—For a While?

Robert Guralnick has noted that although Conklin continued the longest with cell lineage work most others set the approach aside earlier. They had learned that there was much to learn, he suggests, and only Conklin was such a committed Darwinian that he saw cell lineage as supporting evolutionary biology and as worth pursuing to illuminate phylogenetic relationships (Guralnick 2002). Other factors have undoubtedly played a role in the move from cell lineage as well. Cell lineage work has always been hard, requiring many hours of tedious and careful observation, watching, describing, drawing, and sometimes preparing, preserving, fixing, staining, and so on. The researcher invests a tremendous amount of work before knowing what the results will be. Today's techniques for tracking cells and labeling have helped with part of the work, but a cell lineage researcher still has to invest time, energy, and attention to the work.

One student in the MBL Embryology course some years ago commented on how much harder she thought embryology is today, where "you have to know so many molecular techniques and work in the lab a lot." Yet when asked about what she thought Wilson had to do in order to carry out his cell lineage studies of *Nereis*, she imagined that he had to "watch an egg or maybe a few" that he had probably ordered from the supply department, and then "write down and draw the stages." The Wilson work, she imagined didn't sound so hard to her. More recently, the Embryology course director Alejandro Sánchez-Alvarado has had the students go down to the Eel Pond dock at night with collecting nets and lights, just as Wilson did. They take the worms back into the lab, identify the eggs that have been fertilized, and watch cell divisions. They watch all night, carefully observing to see the changes with each cell division. They try drawing and also learn that, in fact, much of the work is not just watching a few embryos develop and observing everything directly. Cells have to be collected, fixed, stained, sectioned, and so on. Each step requires more work and different skills. The students today acknowledge that cell lineage work is still demanding.

Wilson, Conklin, and the others did not have the molecular tools we have available today, but they drew on many other techniques. The MBL Archives still has boxes of slides that Conklin made for the 1939 Embryology course, and the many slides, each have many sections taken from hundreds of embryos. It becomes obvious that the work was highly skilled and difficult, and also that interpreting from all those many different individual images to discover what is "normal" involves careful interpretive work. Diversity across individual organisms and across types of organisms complicated the interpretations. Furthermore, Wilson's and Conklin's enthusiasm about their choice of organisms shows that actually following the lineages is much more successful in some species than in others, which were set aside as not so useful. The eggs have to be large and visible enough, develop fast enough, and have stable enough patterns: all factors that Wilson had pointed to in his first *Nereis* studies. The reasons just described are negative factors that may have pushed researchers away from further cell lineage work.

More positive factors also pulled researchers in new directions; these involved new methods that lured biologists both to ask different questions and to study them in different ways. Some have pointed to genetics as an alternative research program at the time, but cell lineage had already given way before genetics attracted many followers. Another consideration was that Wilson and Conklin, and others at the time, found other ways to study other details about cells, cell division, and the relative roles of nucleus and cytoplasm, for example. Wilson, in particular, published his *Atlas of the Fertilization and Karyokinesis* in 1895, which drew on photography to show what chromosomes were doing, step by step during early cell divisions (Wilson 1895).

A year later, the first edition of his magnificent textbook *The Cell in Development and Inheritance* appeared (Wilson 1896). With two follow up editions of *The Cell* and additional rich studies of cells and chromosomes, Wilson was recognized as the leading cell biologist for decades. This work drew on some of his cell lineage studies but even more importantly on other methods for fixing, staining, and observing the details inside cells. As a result of this body of work, the American Society of Cell Biology awards the E. B. Wilson Medal, and the MBL offers an E. B. Wilson History and Philosophy of Science Lecture. Conklin had a similar impact and adopted additional methods, and the Society for Developmental Biology recognizes Conklin for his contributions to embryology.

Most of those who wanted to study development took up experimental methods along with Morgan. Experimental methods allowed new phenomena to be discovered, and allowed comparisons by altering one or another factor and observing the different results. Much has been written about the history of experimental embryology, and we need not repeat it here. *The point to emphasize is that cell lineage studies were not so much seen as a dead end or worthless. Rather, the approach did not seem to address questions about developing embryos as well as experimental approaches at the time. Nor did it seem to illuminate understanding of evolution. In effect, the study of development and evolution had diverged by 1910, but not forever.* Cell lineage brought them back again in the 1960s with the nematode worm *C. elegans.*

3.3   CELL LINEAGE AND CAENORHABDITIS ELEGANS

The tiny nematode worm *C. elegans* is sometimes described in terms of negatives: it is not infectious or parasitic or hazardous or pathogenic. It also lives in soil and feeds on microbes. It is just the sort of organism that Wilson had sought with his cell lineage studies. It is very small (and 1 mm), has a short life cycle, it is prolific in producing eggs, it is easy and relatively inexpensive to cultivate in the laboratory. It is also transparent, so it is possible to watch the process of development as its cells divide. The worms typically reproduce through self-fertilization, but they can be crossed as well, which offers breeding advantages. We also know now that the cell lineage is essentially invariant under normal circumstances. An individual worm has a predictable 959 somatic cells in one sex and 1031 in the other, with 6 chromosomes, and over 100 million base pairs.

As Bruce Alberts wrote in his introduction to the 1222-page volume *C. elegans II,* studying this worm makes sense because "This simple creature is one of several 'model' organisms that together have provided tremendous insights into how all organisms are put together. It has become increasingly clear over the past two decades that knowledge from one organism, even one so simple as a worm, can provide tremendous power when connected with knowledge from other organisms. And because of the experimental accessibility of the nematodes, knowledge about worms can come more quickly and cheaply than knowledge about higher organisms" (Alberts in Riddle et al. 1997; also Brenner and Wood 1988). The worm network has developed a valuable resource for researchers to share results and methods through WormBase (Stein et al. 2001).

In 2002, the Nobel Prize Committee awarded the Prize in Medicine or Physiology to Sydney Brenner, H. Robert Horvitz, and John E. Sulston "for their discoveries concerning genetic regulation of organ development and programmed cell death" (Nobel Prize 2016). Their work brought a return to cell lineage studies. Why? Why was it thought to be productive to take up such studies?

Brenner had already established his reputation with his contributions to the discovery of messenger RNA, which led to a Lasker Award. Historian/philosopher of biology Rachel Ankeny explains very nicely how Brenner went on to establish *C. elegans* as a model organism for research in general, and the role of cell lineage studies in that work. Ankeny explains that the extremely creative Brenner wrote a letter saying that he felt that "nearly all the 'classical problems' of molecular biology have either been solved or will be solved in the next decade... the future of molecular biology lies in the extension of research to other areas of biology, notably development and the nervous system" (Ankeny 2001). Brenner saw *C. elegans* as promising for both development and neurobiology. He needed something tractable, with a small number of cells and the ability to track the cells. He sought a way to carry out cell lineage studies that would track each cell throughout its development (Brenner 1973).

Others argued that the nematode was not a good choice. Perhaps it is too simple. Perhaps the worm is just a tube of material without enough morphological structure to observe the differences and track cells. Perhaps its nervous system is not complex enough to track interesting features. Yet, Brenner documented in 1974 about 300 mutants and over 100 genes. Brenner started with the deceptively simple statement: "How genes might specify the complex structures found in higher organisms is a major unsolved problem in biology." With a universal genetic code, how does the sameness turn into difference? As yet, "we know very little about the molecular mechanisms used to switch genes on and off in eukaryotes. We know nothing about the logic with which sets of genes might be connected to control the development of assemblages of different cells that we find in multicellular organisms" (Brenner 1974, p. 71).

To study nervous system development requires both tracking structural effects of genetic differences to understand how genes specify the nervous system and also tracking how the nervous system produces behavior. Such research requires an organism with a simple enough system, much simpler than *Drosophila* or other favorites. Brenner concluded with understatement that "*C. elegans* is a favorable organism for genetic analysis" (Brenner 1974, p. 91). This was just the beginning, leading to the massive report in 1986 on "The structure of the nervous system of the nematode *Caenorhabditis elegans*," by Brenner and others (White et al. 1986). There we learned that 302 neurons work in a structure that does not vary across organisms, and they are coordinated with 5000 chemical synapses, 2000 neuromuscular junctions, and 600 gap junctions. The research required the kind of meticulous, time-consuming, and at times almost obsessive dedication to tracking every detail, just as the earlier cell lineage researchers had done.

Sometimes the researchers described the work in terms of mapping the architecture of the worm, mapping the genome, developing wiring diagrams. More recent studies have tied *C. elegans* systems to gene regulatory networks of the kind that Eric Davidson and Britten first introduced in 1969 (Britten and Davidson 1969).

Brenner's first emphasis was on the nervous system. Meanwhile, John Sulston worked on development more generally in a paper published in 1977 showed that the post-embryonic cell divisions carry out a very precise and predictable sequence, with very strictly specified cell fates (Sulston and Horvitz 1977). The early embryonic stages had already been laid out, and Sulston and Horvitz sought to complete the picture. They observed division after division and tracked the details, just as they acknowledged that Wilson and colleagues had done.

As Sulston and Horvitz put it, based on their observation of living nematodes: "As in embryogenesis, the pattern of these divisions is rigidly determined; essentially invariant postembryonic cell lineages generate fixed numbers of neurons, glial cell, muscles, and hypodermal cells of rigidly specified fates. These lineages reveal the ancestral relationships among specific cells of known structure and function; they thus complement the classical embryology, which defined the ancestral relationships among different organs" (Sulston and Horvitz 1977, p. 110).

Their work was very much in the tradition of cell lineage work around 1900. But where that earlier work had been largely set aside for reasons discussed, now these researchers saw the way forward in answering a wide range of developmental questions. As the authors noted, because they could observe every cell all the way through development, they could track cell migration in details. They could observe synapse formation and other functional steps with cell differentiation. Programmed cell death would be observable, and they could track the effects of particular mutations to get at genetic effects. All the exciting possibilities for cell lineage from around 1900 seemed accessible with this tiny worm in the 1970s. The full sequencing of the *C. elegans* genome in 1998 opened even more opportunity for study and secured this nematode as an NIH-approved model organism (*C. elegans* Sequencing Consortium 1998).

Now we come back to intersections of development with evolution. David Fitch and W. Kelley Thomas wrote the chapter "Evolution" for the massive 1997 volume *C. elegans II*. They reminded us why this is an excellent organism for studying evolution as well as development. They noted that "many evolutionary changes are similar to mutant phenotypes, suggesting that much of evolution may proceed by changes at the kinds of regulatory loci defined by genetic studies" (Fitch and Thomas in Riddle et al. 1997, p. 815). Following a review of phylogenetic relationships more generally, they then provide summaries of the evolution of various characters, looking at homologies, the different developmental stages, and other factors.

Fitch and Thomas note in their conclusion that study of worms is not likely to "divulge the precise developmental genetic changes that transformed our hominoid ancestors into humans only a few millions of years ago." Yet "it will provide models for how evolution works with development to make living forms. From models arise predictions. Only then can we evaluate and incorporate notions about general mechanisms into the body of explanatory principles being built by integrative approaches in biology" (Fitch and Thomas in Riddle 1997, p. 850). We have, as they note, just begun the search to understand that evolutionary past. Or as Wilson might have put it, we still see ancestral reminiscences but have not yet worked out causal and explanatory connections even though we have so many more tools and have made so much progress since Wilson's day; thanks to advancing tools available to address old questions.

3.4   CELL LINES

The third type of cell lineage work looks at lineages of particular cells, not at all cells from the earliest embryonic stages. Lineages of particular cells in the body give rise to one kind of research. Cancer cells are one of the favorite examples. Most of the time, obviously, we do not pay much attention to the particular individual cells in our body. We let them do their work and assume they are doing the job right. When cells become cancerous, however, we are concerned both about where they go next for clinical purposes, and where they came from for research purposes. Tracing the lineages forward can occur by watching metastatic growth throughout the body. For example, a cancer cell or cells can produce a tumor, then migrate to other places to produce other tumors, and so on. This is a kind of lineage, even though it is usually not possible to trace every cell division over time. Cancer stem cells also initiate new lines of cells—cancer cells—which has significant implications for what therapeutic approaches are likely to prove effective, as Lucie Laplane discusses in her 2016 book *Cancer Stem Cells. Philosophy and Therapies* (Laplane 2016).

In his 2007 book on the *Dynamics of Cancer: Incidence, Inheritance, and Evolution,* Stephen Frank offered insight into this first type of research approach to cancer cell lineages with his discussion of cell lineage history in Chapter 14. Frank sought not just to study cancer cells and their future cell divisions, but also to understand the accumulation of past mutations and heritable changes that have led to the cancer cells being studied. He acknowledged that the "present studies remain crude, but hint at what will come" (Frank 2007). Genetic sequences would provide just a start, Frank suggested and also were starting to be correlated with methylation patterns, microsatellite sequences, and other factors to understand what has shaped the cells to date and drives them forward in particular ways. This approach to somatic mapping and evolution of cell lineages has made some advances and continues to hold considerable promise. Yet to make still more progress, we will need more researchers to embrace the study of evolution for medicine, and to develop new techniques for identifying and connecting the various factors involved in the complex adaptive systems of cancers.

Much more widely adopted are techniques for culturing cells outside the body. Instead of trying to trace those cancer cells or their effects inside the body, put the cells in a culture dish where they are relatively easy to watch, track changes, and follow lineages. This approach had already begun in 1907, when Ross Granville Harrison took neuroblast cells out of a frog and placed them in a medium of frog lymph, then watched. He wanted to determine whether the cells would be able to grow and differentiate into nerve cells, and he assumed that if they did that they would be following the patterns of normal nerve cells. The story of Harrison's first ever tissue culture research, which was also the first ever stem cell research, has often been told, so we need not repeat it (Maienschein 1983; Witkowski 1985). Here the important message is that, over a century ago, researchers began culturing cells. Harrison himself did not pursue tissue culture research, but others immediately recognized its tremendous promise and pursued different ways to culture tissue, clusters of cells, and also individual cells.

Hannah Landecker has asked how we should think about those cultured cells. In *Culturing Life. How Cells Became Technologies* and more recent articles, Landecker argues convincingly that we have come to think of the biotechnological creation in the laboratory as life. Indeed the cell cultures are alive, they do consist of cells, and yet they are not and have never themselves been parts of organisms. The cells have, indeed, become technologies (Landecker 2007). Different kinds of such technologies include cell lines like HeLa cells that were taken initially from an individual's cancer cells, or cell lines starting from undifferentiated cells like human embryonic stem cells.

Thanks to the excellent bestselling book by Rebecca Skloot, a wide audience has heard about Henrietta Lacks and the HeLa cell line derived from her cervical cancer cells (Skloot 2010; Landecker 2007, Chapter 4). Back in the early twentieth century when Alexis Carrel was inspired by Harrison to culture cells from chicken hearts and had reached the conclusion that he could create what he came to believe were immortal cell lines, his results seemed both plausible and yet worrisome. Years later, when researchers realized that they could culture cancer cell lines "immortally," as from the cells of Henrietta Lacks, the possibilities seemed exciting and important. Laboratories in vastly different places could work with what were presumed to be the same cell lines and could then compare results. The HeLa cells became a "body of knowledge," as Landecker puts it. Standardized methods for culturing, freezing, sharing, and recording were set up to develop a network around the cells.

Yet already by the 1960s, some researchers had begun to worry about contamination in some of the cell lines being used extensively in biomedical research, and they raised questions. Are the HeLa cells around the world today really part of the same "immortal" line that started with Henrietta Lacks? Surely not. Mutations have occurred, environmental conditions have made a difference, people have made mistakes, things have happened. John Masters asked in a summary article in 2002 why researchers continue to accept ignorance about the quality of their cells, and perhaps even to engage in known fraud in knowingly putting forth cell lines such as HeLa as something they are not (Masters 2002). A more extensive historical reflection essay in 2009 carried the critique further and asked whether HeLa cells might not have evolved so far as to have become something else and need a new name (Lucey et al. 2009). Here we are looking at evolution that has occurred after the cell lines have been long established, not as ancestral reminiscences from the past in Wilson's or Conklin's terms. To some degree, the laboratory practices themselves are causing evolution to occur.

Stem cell lines raise similar questions, especially embryonic stem cell lines because they begin with undifferentiated or pluripotent stem cells (those defined as cells that have the capacity to become any kind of cells). Discovered in 1998 in humans, embryonic stem cell lines have achieved considerable notoriety for the ethical questions they have raised and the political posturing surrounding them. (For more discussion, see Maienschein 2014). Actually, culturing the stem cells, or using them for any research or medical purposes necessarily involves manipulating or even killing the embryo from which they came, which worries those who regard the embryo as something we should not be manipulating. For our purposes here, there are questions about the cell lineages that result when the stem cells are harvested from a blastocyst that plays the central role. What do these cultured cells tell us about life? What do they tell us about development or heredity or evolution?

Landecker argues rather forcefully that, "Biotechnology changes what it is to be biological" (Landecker 2007, p. 232). She made that comment initially with respect to the change in attitudes following cloning of Dolly the sheep and cryopreservation techniques involved. Cloning and freezing change the timescale. Not only can we culture cells for long periods of time; we can also freeze them and use them much later, which makes them asynchronous with respect to the organisms from which they came in the first place. In this light, cells began to seem highly plastic and changeable. Is life equally plastic under normal conditions, and what does it even mean to be normal? Landecker concludes her book by asking, "Once we have a more specific grasp on how altering biology changes what it is to be biological, we may be more prepared to answer the social questions that biotechnology is raising: What is the social and cultural task of being biological entities—being simultaneously biological things and human persons—when 'the biological' is fundamentally plastic?" (Landecker 2007, p. 235).

In another line of reasoning that is not immediately related to our story about cell lineages here but is nonetheless worth noting, both Landecker and also philosophers John Dupré and Maureen O'Malley point to cell-lineages as clusters of cells that indicate organisms. They focus on questions about what turns cells and cell lineages into organisms, and they all see metabolism as the force that makes those cells into an organism and a living individual. Dupré and O'Malley explain that "Our central argument is that life arises when lineage-forming collaborate in metabolism" (Dupré and O'Malley 2009). Landecker's study of metabolism follows similar lines in suggesting that we need a much richer understanding of metabolism and of what it means to be biological and alive. These reflections suggest one way in which the study of cell lineages informs and helps shape our understanding of life more generally, and are worth exploring further in other contexts.

3.5   CONCLUSION

We have come a long way from the cell lineage work of Whitman, Wilson, and Conklin, and yet perhaps not so far. In the 1890s, it looked like cell lineages might be quite predictable and follow reliable patterns so that each cell had a knowable fate. Experimental work by Morgan and others quickly showed that individual organisms, and the cells in them, respond to experimental conditions in ways that can change the normal patterns while still leading to functional living organisms. The patterns may be predictable under normal circumstances, but also subject to adaptation and revision. Despite their ancestral reminiscences, individual cells, and the organisms that result from the sequence of cell divisions are highly responsive to changing conditions. Indeed, perhaps the adaptability and plasticity also result from ancestral reminiscences of changing conditions and the advantages of the ability to adapt. The more-in-depth evolutionary background directs the normal patterns as well as the diversity of adaptive responses to changing conditions in ways that call for further investigation into how and why.

C. elegans studies show a very high degree of predictability in both normal and experimental adaptations. The expected patterns, as well as the plasticity, all reflect what it is to be biological, to start as an inherited egg cell that has some organization already in place, is then fertilized, and responds to particular conditions at hand through a lineage of cell divisions that are more or less predictable and fixed.

We have looked at three different approaches to cell lineage here: documentation of cell lineages in the early sense at the MBL, documentation of the details and patterns of every cell through every division revealed with work on *C. elegans*, and understanding of the forces and factors that allow cell lines to continue through many generations. These are three different approaches to understanding life. The researchers involved have brought different underlying assumptions to their work, and they have made different choices—or organisms for study as well as of questions to ask or methods to use. Whitman and his cell lineage crew had an evolutionary past in mind, providing what Wilson called those ancestral reminiscences. Brenner and the worm crew were focused on the phenomena in front of them, tracing how cells make up a functioning nervous system. Clinical studies of cancer cell lines emphasize the cells and how they behave after they become cancerous. As we see, there are different reasons to follow cell lineages. Taken together, they help give us a rich understanding of the complexities of life, as well as of the way in which research can make those complexities accessible through the hard work of paying close attention to one detail after another, as one cell divides into another.

REFERENCES

Allen, G. E. 1979. *Thomas Hunt Morgan. The Man and His Science*. Princeton, NJ: Princeton University Press.

Ankeny, R. 2001. The natural history of *Caenorhabditis elegans* research. *Nature Rev Genet* 2: 474–479.

Bonner, J. T. and Bell, W. J. 1984. What is money for? An interview with Edwin Grant Conklin, 1952. *Proc Am Philos Soc* 12: 79–84.

Brenner, S. and Wood, W. B. (Ed.). 1988. *The Nematode Caenorhabditis Elegans*. New York: Cold Spring Harbor Press.

Brenner, S. 1973. The genetics of behavior. *Br Med Bull* 29: 269–271.

Brenner, S. 1974. The genetics of *Caenorhabditis elegans*. *Genetics* 77: 71–94.

Britten, R. and Davidson, E. H. 1969. Gene regulation for higher cells: A theory. *Science* 165 (3891): 349–357.

C. elegans Sequencing Consortium. 1998. Genome sequence of the nematode *C. elegans*: A platform for investigating biology. *Science* 282: 2012–2018.

Conklin, E. G. 1905. The organization and cell-lineage of the ascidian egg. *J Nat Acad Sci Phila* 13: 1–119.

Dobzhansky, T. 1973. Nothing in biology makes sense except in the light of evolution. *Am Biol Teach* 35 (3): 125–129.

Dupré, J. and O'Malley, M. 2009. Life at the intersection of lineage and metabolism. *Philos Theor Biol*. doi:10.3998/ptb.6959004.0001.003.

Frank, S. A. 2007. *Dynamics of Cancer: Incidence, Inheritance, and Evolution*. Princeton, NJ: Princeton University Press. Available through Creative Commons licensing through the NIH at: http://www.ncbi.nlm.nih.gov/books/NBK1553/. Accessed August 10, 2016.

Guralnick, R. 2002. A recapitulation of the rise and fall of the cell lineage research program: The evolutionary-developmental relationship of cleavage to homology, body plans, and life history. *J Hist Biol* 35: 537–567.

Harvey, E. N. 1958. Edwin grant Conklin. *BMNAS* 1958: 54–91.

His, W. 1874. *Unsere Körperform und das physiologische Problem ihrer Entstehung Briefe an einen befreundeten Naturforscher.* Leipzig, Germany: F. C. W. Vogel.

Landecker, H. 2007. *Culturing Life. How Cells Became Technologies.* Cambridge, MA: Harvard University Press.

Laplane, L. 2016. *Cancer Stem Cells. Philosophy and Therapies.* Cambridge, MA: Harvard University Press.

Lillie, F. R. 1944. Founding and early history of the marine biological laboratory Chapter III in *The Woods Hole Marine Biological Laboratory.* Chicago, IL: University of Chicago Press.

Lucey, B. P., Nelson-Rees, W. A., and Hutchins, G. M. 2009. Henrietta Lacks, HeLa cells, and cell culture contamination. *Arch Pathol Lab Med* 133: 1463–1467.

MacCord, K. 2013. Germ layers. Embryo Project Encyclopedia. https://embryo.asu.edu/ pages/germ-layers 8/28/2016. Accessed August 12, 2016.

Maienschein, J. 1983. Experimental biology in transition: Harrison's embryology, 1895–1910. *Stud Hist Biol* 6: 107–127.

Maienschein, J. 1989. *100 Years Exploring Life, 1888–1988.* Boston, MA: Jones and Bartlett Publishers.

Maienschein, J. 2014. *Embryos Under the Microscope. The Diverging Meanings of Life.* Cambridge, MA: Harvard University Press.

Maienschein, J. 2015. Garland Allen, Thomas Hunt Morgan, and development. *J Hist Biol* 49: 587–601.

Masters, J. R. 2002. HeLa cells 50 years on: The good, the bad and the ugly. *Nat Rev Cancer* 2: 315–319.

Morgan, T. H. 1891. A contribution to the embryology and phylogeny of the Pycnogonids. *Johns Hopkins University Studies of the Biological Laboratory*, Baltimore, MD: Isaac Friedenwald, pp. 1–76.

Morgan, T. H. 1940. Edmund Beecher Wilson. *BMNAS* 1940: 315–342.

Morgan, T. H. 1898. Developmental mechanics. *Science* 7: 156–158.

Morgan, T. H. 1901. *Regeneration.* New York: Macmillan Publishers.

Morse, E. S. 1912. Charles Otis Whitman. 1842–1910. *BMNAS* 1912: 269–288.

The Nobel Prize in Physiology or Medicine 2002. Nobelprize.org. Nobel Media AB 2014: http://www.nobelprize.org/nobel_prizes/medicine/laureates/2002/. Accessed August 7, 2016.

Richards, R. 2008. *The Tragic Sense of Life: Ernst Haeckel and the Struggle over Evolutionary Thought.* Chicago, IL: University of Chicago Press.

Riddle, D. L., Blumenthal, T., Meyer, B. J., and Priess, J. R. (Eds.). 1997. *C. Elegans II.* Plainview, NY: Cold Spring Harbor Laboratory Press.

Skloot, R. 2010. *The Immortal Life of Henrietta Lacks.* New York: Crown Publishing.

Stein, L., Sternbert, P., Durbin, R., Mieg, J. T., and Spieth, J. 2001. WormBase: Network access to the genome and biology of *Caenorhabditis elegans. Nucleic Acids Res* 29: 82–86.

Sturtevant, A. H. 1959. Thomas Hunt Morgan. *BMNAS* 1959: 283–325.

Sulston, J. E. and Horvitz, H. R. 1977. Post-embryonic cell lineages of the nematode, Caenorhabditis elegans. *Dev Biol* 56(1): 110–156.

Sunderland, M. 2010. Regeneration: Thomas Hunt Morgan's window into development. *J Hist Biol* 43: 325–361.

White, J. G., Southgate, E., Thomson, J. N., and Brenner, S. 1986. *The structure of the nervous system of the nematode Caenorhabditis elegans.* Philos Trans R Soc Lon B 314: 1–340.

Whitman, C. O. 1878. The embryology of Clepsine. *Q J Microsc Sci* 18: 215–315.

Whitman, C. O. 1885. *Methods of Research in Microscopical Anatomy and Embryology.* Boston, MA: Cassino Publishers.

Whitman, C. O. 1893. The inadequacy of the cell-theory of development. *J Morphol* 8: 639–658.

Wilson, E. B. 1895. *The Atlas of the Fertilization and Karyokinesis of the Ovum.* New York: Macmillan and Company.

Wilson, E. B. 1898. Considerations on cell-lineage and ancestral reminiscence, based on a reexamination of some points in the early development of annelids and polyclades. *Ann NY Acad Sci* 11(1): 1–27.

Wilson, E. B. 1892. The cell-lineage of Nereis. A contribution to the cytogeny of the annelid body. *J Morphol* 6: 361–480.

Wilson, E. B. 1896. *The Cell in Development and Inheritance.* New York: Macmillan and Company.

Wilson, E. 1899. Cell lineage and ancestral reminiscence. *Biol Lect Mar Biol Lab* 1899: 21–42.

Witkowski, J. A. 1985. Ross Harrison and the experimental analysis of nerve growth: the origins of tissue culture. In *History of Embryology* (Eds.) T. Horder, J. A. Witkowski, and C. C Wylie. Cambridge, UK: Cambridge University Press.

4 Protists and Multiple Routes to the Evolution of Multicellularity

Vidyanand Nanjundiah, Iñaki Ruiz-Trillo, and David Kirk

CONTENTS

Complex multicellular eukaryotic forms independently evolved at different moments in life's history, most of them around approximately 800 Myr ago (Sebé-Pedrós et al. 2017).

4.1 INTRODUCTION

The aim of this chapter is to ask what aspects pertaining to the evolutionary origin of multicellular life might be inferred from a survey of otherwise dissimilar protists that display one or both of two features: a unicellular-to-multicellular transition as part of their normal life cycle, or membership of a closely related group that contains both unicellular and multicellular members. Accordingly, we highlight aspects of multicellular development in three different "supergroups" of eukaryote protists: Dictyostelid or cellular slime molds (CSMs) (supergroup *Amoebozoa*, Mycetozoa; Section 4.2); Choanoflagellates, Filastareans, and Icthyosporeans (*Opisthokonta*, unicellular Holozoa; Section 4.3); and Volvocine green algae (*Archaeplastida*, Chlorophyceae; Section 4.4); their last common ancestor is believed to lie at the very root of the evolutionary tree of eukaryotes (Burki 2014). In the first two, multicellularity is achieved by the aggregation of single cells, in the third, by the products of cell division staying together. A striking difference from metazoan multicellularity is that none of the life cycles contain an obligatory sexual phase. Because not many species have been studied in detail, the information to be presented comes from a very small number of (what one hopes are) representative cases. As it happens, the most interesting evolutionary implications pertain to discordances between studied features, which means that they are likely to be robust. A brief summary concludes each of the three sections that follow. A general discussion comes at the end of this chapter.

4.2 EVOLUTION OF MULTICELLULARITY VIA AGGREGATION: THE CELLULAR SLIME MOLDS

This section is restricted to the CSMs and some other amoeboid organisms that display facultative multicellularity with similar life cycles. It is organized around four themes:

1. Aggregative multicellularity followed by differentiation into a fruiting body is seen in six of the seven "supergroups" of eukaryotic life
2. In the best-studied forms, the CSMs, (which belong to the supergroup Amoebozoa), there is a disconnection between morphological similarity and phylogenetic relationship
3. In *Dictyostelium discoideum*, the CSM about which we know the most, much of the genetic repertoire required for multicellular development appears to have preexisted in unicellular ancestors
4. A small number of elementary cell behaviors and self-organization may be sufficient to account for the varied multicellular morphologies that are observed

Themes 1 through 3 are based on published findings whereas (4) has a strong speculative element. All four themes have been discussed recently from a developmental perspective (Nanjundiah 2016); the emphasis here is on evolution. This contributes the speculative element to what we say, especially in the context of theme (4). The focus is on how multicellularity may have originated. The evolution of traits *within* the CSMs has been discussed by several writers (Bonner 1982, 2003, 2013a; Loomis 2014, 2015; Schaap 2011; Schilde et al. 2014) and is not our main concern.

The evolutionary transition from unicellularity-to-multicellularity is believed to have taken place independently on at least 25 occasions, in both prokaryotes and eukaryotes (Grosberg and Strathman 2007; Sebé-Pedros et al. 2017). Phylogenetic reconstruction shows that in six of the seven supergroups of eukaryotes in which the transition occurred, it did so via the aggregation of free-living cells (Figure 4.1). The Archaeplastida, which includes the red algae, green algae, and land plants, is the only eukaryotic supergroup in which no evidence of aggregative multicellularity has been found so far (though multicellularity did evolve, as discussed toward the end).

Going by the extant species that have been studied, five of the six independent origins of aggregative multicellularity involve amoeboid forms and a sorocarpic life cycle. The sixth is found in the Alveolata, where *Sorogena*, a ciliate, forms multicellular groups by aggregation. In the other five cases, the descendants are free-living amoebae that aggregate when they run out of food and organize themselves into an integrated multicellular mass. The mass goes through changes in morphology and the cells comprising it differentiate, which results in a sorocarp or fruiting body, containing stress-resistant spores, supported by an upright stalk. As we shall see, there are variations within the broad contour just sketched. Still, as Brown (2010) said, "… in all major lineages of amoeboid protists there appears to be at least a single example of the cellular slime mold habit." A brief description of that "habit" follows.

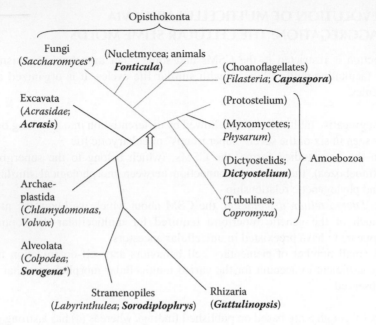

FIGURE 4.1 Independent origins of multicellularity and fruiting body formation via aggregation (indicated by bold lettering) in the major eukaryotic supergroups. Except for *Sorogena*, all begin as amoeboid single cells; *Sorogena* is a ciliate and *Saccharomyces*, a yeast, can form multicellular structures under synthetic conditions. The large arrow indicates the putative root of the tree, which has been dated to between 1866 and 1679 Myr ago. (From Parfrey et al. 2011; Brown, M.W. and Silberman, J.D., The non-dictyostelid sorocarpic amoebae, in *Dictyostelids. Evolution, Genomics and Cell Biology*, Romeralo, M. et al. (Eds.), Springer, Heidelberg, Germany, pp. 219–242, 2013; Romeralo, M. and Fiz-Palacios, O., Evolution of dictyostelid social amoebas inferred from the use of molecular tools, in *Dictyostelids. Evolution, Genomics and Cell Biology* Romeralo, M. et al. (Eds.), Springer, Heidelberg, Germany, pp. 167–182, 2013; Burki, F., *Cold Spring Harb. Perspect. Biol.*, 6, a016147, 2014; Modified from Nanjundiah, V., Cellular slime mold development as a paradigm for the transition from unicellular to multicellular life, in *Multicellularity: Origins and Evolution,* Niklas, K.J. and Newman, S.A., (Ed.), MIT Press, Cambridge, MA, pp. 105–130, 2016. With permission.)

A   The Cellular Slime Mold Life Cycle

CSMs are predatory amoebae found worldwide in a variety of soils, in bird droppings, and animal dung—namely wherever their prey, bacteria, and yeasts, are found (for details of CSM development, including references to the original literature, see Bonner 1967, Raper 1984 and Kessin 2001). As long as food is available, single amoebae keep going through vegetative cycles of feeding, growing, and doubling by mitosis, followed by cytokinesis. In a few cases that have been examined, postmitotic amoebae are haploid. When the food supply is exhausted, they adopt one of four strategies to mitigate the stress of starvation. (1) Some amoebae remain *solitary* in an apparent attempt to wait it out until they encounter food once more (Dubravcic et al. 2014). (2) An amoeba *encysts* itself and forms a dormant *microcyst*. (3) A large number of amoebae *aggregate* and develop into a fruiting body in which some differentiate into dormant *spores*. (4) If the aggregate contains cells of opposite mating (sexual) types, a pair of them can fuse to form a diploid cell. The diploid is cannibalistic and feeds on the remaining cells, eventually forming a dormant giant cell, a protozygote known as the *macrocyst*. The return of food supply, either in the same location or wherever the dormant form has (passively) dispersed, induces the release of an amoeba from the microcyst or spore. In the case of the macrocyst, there is a meiotic division, following the release of products as haploid amoebae. A released amoeba proceeds to feed, grow, and divide; and the life cycle is reinitiated. Once starvation has set in, the number of cells can remain unchanged (in 1, 2 and 3), or decrease, whether slightly, on account of cell death (in 1 and 3), or drastically, following cannibalism (in 4). A characteristic of the CSM life cycle, which the other sorocarpic amoebae appear to share, is that growth (which includes an increase in cell number) and development are clearly separated in time.

We have listed the consequences of starvation that have been observed in one or more CSM species. A form of responding to starvation may not be seen under laboratory conditions (e.g., microcysts have not been found in some species), but because we are largely ignorant of CSM ecology, we cannot say whether or not it is found in the wild. Similarly, even though conditions that favor microcyst formation, or favor the macrocyst pathway, are known, we do not know precisely how a cell fixes the relative probabilities of responses 1 to 4. By sampling spores from the same fruiting body, we can infer whether the preceding aggregate, and possibly the feeding group that gave rise to it, was clonal or polyclonal (one has to verify that a polyclonal aggregate has not split into two or more clonal fruiting bodies). Amoebae and spores of different species undoubtedly co-occur frequently, but either interspecies aggregation does not occur, or, if it does, cells sort out and form "almost species-pure" groups (Raper and Thom 1941, Bonner and Adams 1958, Jack et al. 2008, Sathe et al. 2014). For this reason, we will not consider multispecies groups any further. Therefore, we ignore the fascinating case of interspecies predation, to date known only in *D. caveatum*: its amoebae coaggregate with those of other species, form a chimeric multicellular mass, and then start to feed on the others. Instead, we will concentrate on developmental pathway (3). It is the one that involves aggregative multicellularity, a form of collective self-protection against the threat posed by starvation, and is found across supergroups.

B Similarities with Life Cycles of Other Sorocarpic Amoebae

Life cycles resembling those of the CSMs are found in the Amoebozoa (*Copromyxa* in Tubulinea, and Dictyostelia), the Opisthokonta (*Fonticula*), the Excavata (*Acrasis/Pocheina* in Heterolobosea), the Rhizaria (*Guttulinopsis*, Cercozoa), and the Stramenopila (the amoeboid labyrinthulid *Sorodiplophrys*) in

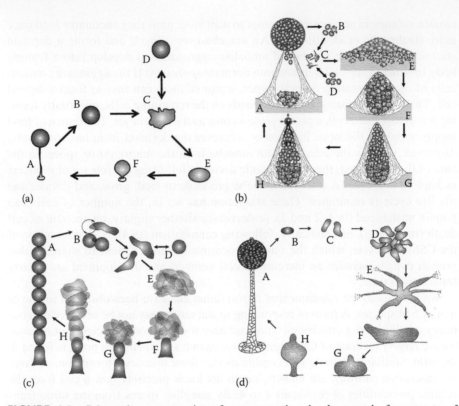

FIGURE 4.2 Schematic representation of post-starvation development in four groups of sorocarpic amoebae (name of supergroup within parentheses). (a) *Protostelium mycophaga* (Amoebozoa); (b) *Fonticula alba* (Opisthokonta); (c) *Acrasis helenhemmesae* (Excavata); (d) *Dictyostelium discoideum* (Amoebozoa). In (a) a single amoeba differentiates into a fruiting body; in the others, many amoebae do so following aggregation. See text for details. (From Brown, M.W., Placing the Forgotten Slime Molds (*Sappinia*, *Copromyxa*, *Fonticula*, *Acrasis*, and *Pocheina*), using molecular phylogenetics, 258p. PhD thesis dissertation, University of Arkansas, Fayetteville, AR, 2010. With permission.)

the Chromalveolata; as discussed in the next section, aggregative multicellularity per se is also seen in unicellular holozoans. Consider the developmental cycles illustrated in Figure 4.2 (reproduced from Brown 2010 with the kind permission of the author).

a. *Protostelium mycophaga* (Amoebozoa; Shadwick et al. 2009 and Figure 4.2, p. 41 in Brown 2010) provides a striking example of the same stereotypic morphological transitions, as in the CSMs but without any multicellularity at all. The free-living amoeba (C) can either transform reversibly into a dormant cyst (D), or it can become a spore (B) that is on top of a fruiting body (E→F→A) with an extracellular stalk.

b. In *Fonticula alba* (Opisthokonta; Brown et al. 2009 and Figure 4.1, p. 134), the sequence of events is similar except that free-living amoebae (C) can either form solitary cysts (D) or aggregate (E) into groups. Eventually, a slime sheath (F) surrounds the aggregate mound, and the cells inside begin secreting an extracellular stalk. The amoebae that are towards the top of the stalk become spores (G). Some are pressed upwards into a bulge (H) as the stalk matures; others remain outside the sorus (A), and yet other cells stay at the bottom of the fruiting body as undifferentiated amoebae.

c. Newly germinated limax amoebae of *Acrasis helenhemmesae* (Excavata; Brown et al. 2010 and Figure 4.1 on p. 168 of Brown 2010) emerge from spores (B); they encyst reversibly (D) or aggregate (E); one cell within the mound encysts itself to form a stalk cell (F), a second cell follows to become another stalk cell (G). The amoebae remaining in the mound stay on top, become aligned (H) and differentiate into spores, thereby, completing the fruiting body (A). Stalk cells and spores can both germinate and release amoebae.

d. Within the same broad sequence, the aggregative life cycle of *Dictyostelium discoideum* (Amoebozoa; Bonner 1967 and Figure 4.4, p. 43 of Brown 2010) displays more elaborations: spores (B) germinate, release amoebae (C) and begin to aggregate in pulsating waves formed by inwardly streaming cells (D→E). The finished aggregate forms a motile worm-like structure, the slug (F), which migrates to the top of the soil or dung substratum and erects itself (G→H) to form a fruiting body (A) that consists of a spore mass on top of a stalk of dead cells. (Note the similarity in appearance between the *D. discoideum* and *P. mycophaga* fruiting bodies.) Other Dictyostelid species display modifications on this theme: the stalk may be laid down concomitantly with the formation of the slug; the fruiting body may be branched and contain secondary stalks, each with its own spore mass; aggregate size may decide whether the stalk is made up of dead cells or is an extracellular exudate.

Three inferences can be drawn from the striking example of convergent morphogenesis illustrated in Figure 4.2 (the fraction of cells that die during development varies a lot, and we will comment on it at the end). First, *Protostelium* shows that a fruiting body can develop in the absence of aggregation; a single cell can construct one too, and it resembles that formed by an aggregate of cells. Tice et al. (2016) report that *Acanthamoeba pyriformis* (earlier known as *Protostelium pyriformis*), another unicellular relative of the CSMs, can also do so (and they propose that at least one "*Protostelium*," namely *P. pyriformis*, should be classified as an *Acanthamoeba*). Bonner (1967) points out that there is a complementary situation: amoebae of *Hartmannella astronyxis* form multicellular aggregates, but thereafter, the cells encyst without making a fruiting body (Ray and Hayes 1964; according to Brown 2010, *Copromyxa*, not *Hartmannella*, is the more appropriate genus name). Second, the developmental trajectories in the four cases show an extraordinary degree of convergence. This makes it likely that the unicellular ancestors of each of the supergroups in question, and perhaps a common unicellular ancestor of all of them, already possessed much of the genetic and behavioral repertoire needed to form a fruiting body. Third, the fact that the sequence of intermediate stages in the four cases is so similar suggests that they are consequences of a "generic" morphogenetic mechanism (Newman and Comper 1990) that can be implemented in the same way by cells whose ancestries are very different. A perspective on the evolution of multicellularity in myxobacteria (Arias del Angel et al. 2017) contains interesting speculations along similar lines. Molecular phylogenetic studies on the CSMs indirectly support the last two inferences, and we turn to them next.

C   Poor Correlation between Phylogeny and Morphology in the CSMs

Scattered observations in the literature indicate that the fruiting body of one species can sometimes resemble that of another (Bonner 1967, Olive 1975, Raper 1984). For example, Acytostelium forms an extracellular stalk whereas in *Dictyostelium lacteum* the stalk is made up of dead cells; but in very small fruiting bodies of *D. lacteum* the stalk can be partly acellular (Bonner and Dodd 1962). Shaffer found that sparse cultures of *Acytostelium* sometimes contained single-celled fruiting bodies similar to those of *Protostelium* (cited in Bonner 1967). Raper noted the mirror-image case in *Protostelium*: fruiting bodies are mostly unicellular but occasionally consist of two to four cells (Raper 1984). In working with *D. discoideum*, now and then one comes across a fruiting body that mimics a *Polysphondylium* fruiting body by having a lateral stalk in addition to the main stalk. Thus, CSM fruiting body phenotypes are plastic, but the fact becomes evident only rarely or when there is a significant alteration in size (Romeralo et al. 2013). Indeed, based on observations of developmental compatibility in intra- and interspecies mixes, Bonner and Adams (1958) went so far to conclude that "specific differences between strains of one species are as great, or even greater, than those between different species" (p. 352, referring to *D. mucoroides, D. discoideum, D. purpureum, and P. violaceum*).

An unexpected discovery from recent studies of molecular phylogeny in the CSMs makes us see the observations cited above in a new light (Schaap et al. 2006). The CSMs fall into four major clades ("groups"), with their most recent common ancestor dated to ~600 Myr ago (Figure 4.1 in Schilde et al. 2016; the most recent common ancestor with *Physarum polycephalum*, a myxogastrid in which amoebae fuse to form syncytia, is dated to ~650 Myr ago; and that with the unicellular *Acanthamoeba castellani* to ~850 Myr ago). An examination of the constructed phylogeny shows that developmental similarities, especially about fruiting body morphology, are poor indicators of recent common ancestry. There are many illustrations of this. The genus *Acytostelium* used to be characterized by the secretion of an extracellular stalk (Raper and Quinlan 1958), but in fact, the trait is polyphyletic and extends across species that were believed to belong to different genera. An extracellular stalk is found in *Acytostelium* species from two different clades, one of which also contains species with cellular stalks (Schaap et al. 2006; Romeralo et al. 2013). Species in the same clade (Group 4) can form fruiting bodies with an unbranched stalk and spores at the apex (e.g., *Dictyostelium mucoroides* and *D. discoideum*), or with spore masses distributed along the stalk (e.g., *Dictyostelium rosarium*), or with more than one transverse branch of the stalk at the same horizontal level and a spore mass at the tip of each (*Polysphondylium violaceum*, though it has been placed on a small side clade in between the clades formed by Group 3 and Group 4). Among members of the same clade, a cellular stalk can be released by the slug during migration and rise directly from the substrate (*D. giganteum*), or it can form after migration is completed and rest on a disc-shaped base (*D. discoideum*); in the clade comprising Group 2, the stalk can be cellular (*D. oculare*) or acellular (*Acytostelium leptosomum*). One member of the Group 3 clade preys on other CSM species (*D. caveatum*); two others, not known to be predatory on CSMs, form stalk that is cellular (*D. tenue*) or occasionally acellular (*D. lacteum*). Romeralo et al. (2013) reinforce the point in a study whose main finding is that across 99 CSM species, only a weak correlation exists between phenotype and molecular phylogeny. According to them, there is "a fairly scattered distribution of character states … with many states reappearing multiple times in different clades" (by character state they mean mainly the size and shape of multicellular structures; Discussion, p. 7 in Romeralo et al. 2013). They further observe that the morphology and distribution of fruiting body types (unbranched versus branched, solitary versus clustered) changes as a function of the density at which starved amoebae are plated prior to the onset of aggregation.

Schilde et al. (2016) found CSM nuclear genome sizes ranging between 31 and 35 Mb (*D. discoideum, D. fasciculatum and P. pallidum*); at 23 Mb, *D. lacteum* was the smallest of the four. The same study estimated 12,319 protein-coding genes in *D. discoideum* (10,232 in *D. lacteum*). There are significant overlaps with "metazoan" gene products and regulatory pathways, for example, the ones involving beta integrin (Cornillon et al. 2006), btg, and retinoblastoma (RB) (Conte et al. 2010), and Wnt and STAT (Sun and Kim 2011). Attempts have been made to use sequence comparisons and temporal gene expression profiles to identify novel protein-coding genes, or novel patterns of gene expression, that might be linked to the origin of multicellularity in the CSMs. Glöckner et al. (2016) compared 385 "developmentally essential genes" (DEGs) in *D. discoideum* with potentially similar genes in other CSMs. The genes were chosen on the basis that when their function was disrupted, multicellular development was affected, but feeding, growth, or cell division were not, or at least not grossly (the comparison rested on reports from studies carried out by different groups, and the original data did not refer to regulatory genes or noncoding DNA sequences, etc.; for instance, it was not based on saturation mutagenesis). They found that homologues of a significant proportion of the DEGs were present in all four major CSM subclades (i.e., besides *D. discoideum* itself, in *D. fasciculatum*, group 1; *P. pallidum*, group 2; *D. lacteum*, group 3). Significantly, 80% of protein-coding genes essential for (multicellular) CSM development were predicted to exist in their unicellular relatives. The overlaps were 76% with *Physarum polycephalum*, 46% with *Acanthamoeba castellani*, and 19% with *Entamoeba histolytica*. The authors point out that the low values of the latter two proportions are misleading and could be the consequence of secondary gene losses. Also, homologs of 72% of DEGs are found in non-Amoebozoan species—including in the Opisthokonta, other eukaryotes and prokaryotes. Conceivably the prokaryote links could originate from horizontal gene transfer, which is believed to have been common in prokaryotes (Thomas and Nielsen 2005) and rare but not unknown in eukaryotes (Danchin 2016).

The 305 DEGs that were also found in members of the Amoebozoa are predicted to encode protein kinases, nucleotide binding proteins, and a range of cytosolic and nuclear proteins. The remaining 80 DEGs did not yield Amoebozoan homologs (in this study) and are predicted to encode secreted proteins and proteins with an extracellular exposure, both being classes that would be expected to have roles in sensing the environment (including other cells) and mutual recognition. Interestingly, 37 DEGs are not shared by all CSMs, and among them, 26 appear to encode proteins that are secreted or likely to have an extracellular face, once again suggestive of a role in intercellular recognition and signaling, also possibly in mediating interspecies recognition. Evidently, "the cellular slime mould habit" does not depend on the presence of at least these 37, since one or the other is missing in some CSM; but the absence of any among them results in the failure of multicellular development in *D. discoideum*. The functions of the remaining DEGs, which seem to be exclusive to one or more CSMs but not all four, are known or conjectured to be related to the regulation of group size, cell-type proportioning, the timing of aggregation and, in one case, normal fruiting body formation.

The study by Schilde et al. (2016) compared (in the same 4 species) 186 genes of *D. discoideum* that were identified as vital for multicellular development based on similar but slightly different criteria. Their expression was upregulated by at least threefold during normal development. Additionally, they were known (from the published literature) to induce aberrant development when manipulated in some manner. It transpired that 33 of the 186 lacked an ortholog in representatives of at least one of the other three CSM groups and 20 lacked an ortholog in all three (the same three species mentioned above were used as representatives). Four hundred and sixty-six other genes, known to induce aberrant development when modified, were not upregulated significantly in development. In other words, they functioned quasi-constitutively, as would be expected of putative "housekeeping" genes. Tellingly, among the 2,352 to 2,395 genes that were developmentally upregulated in the four species taken together, roughly one half of the number in any one species had no ortholog in any of the remaining three.

D POSSIBLE ROUTES FOR THE UNICELLULAR-TO-MULTICELLULAR TRANSITION

Having prepared the ground with that sketch of comparative morphologies and genetic similarities, we proceed to speculate on how multicellularity could have evolved from an ancestor whose entire life was spent as a single cell to a descendant with the "CSM habit." From what we have seen, much of the required genetic repertoire would have been present in the unicellular ancestor, though the gene product in question may have played some other role. Further, *D. discoideum* shares not only genes, but regulatory mechanisms and morphogenetic pathways with the "higher" Metazoa (Kawata 2011; Loomis 2014, 2015; Santhanam et al. 2015), and very likely so do other sorocarpic amoebae. With regard to *D. discoideum*, and the genes homologous to those coding for beta integrin, btg, and RB, Wnt and STAT mentioned earlier, one can add the following gene products, genes, or regulatory pathways for which a homology with metazoans has been claimed: beta catenin (Coates et al. 2002), presenilin/γ-secretase (McMains et al. 2010), fused kinase (Tang et al. 2008), Src homology 2 (SH2) domain proteins (Sugden et al. 2011), tyrosine kinase phosphorylation (Sun and Kim 2011), and homeobox genes (Mishra and Saran 2015). Besides, noncoding RNAs have been found to regulate the development of *D. discoideum* (Avesson et al. 2011). No doubt more instances will crop up over time. Therefore, irrespective of how the transition may have taken place, it would have been facilitated by preadaptations.

Multicellularity through step-by-step adaptations via the acquisition of new genes. The conventional way of thinking about the evolutionary origin of multicellularity is to view it in terms of a series of cumulative steps, each being the consequence of a random genetic mutation that spread in the population because it happened to be adaptive. The mutation could be in the coding or regulatory portion of an existing gene, which, thereby, acquired an additional function. Alternatively, a functional role could have followed the duplication of a preexisting gene or the acquisition of a foreign gene by horizontal transfer.

In parallel with the evolution of multicellularity, sophisticated communication systems for the coordination of cellular activities must have come into play (Bonner 2001). Cell–cell adhesion would be of obvious advantage, as would contact-dependent signaling and communication via diffusible, long-range signals, and signal transduction mechanisms. Intercellular signaling may have preceded the evolution of a multicellular fruiting body. Starvation-induced encystment occurs near-simultaneously in clusters of *Hartmannella astronyxis* (Ray and Hayes 1954) and is said to begin with a single amoeba in *Acanthamoeba castellani*, from where it moves outward in the form of a wave-like propagated signal (Jahn and Bovee 1967, Pickup et al. 2007).

We do not know what modes of intercellular signaling operated during the evolution of multicellularity from a unicellular ancestor, but many have been identified in *D. discoideum* and some other CSM species. Released quorum-sensing factors enable a starved cell to decide whether there are enough other cells in its neighborhood for aggregation to be initiated. Cyclic AMP (or another chemoattractant) fosters aggregation. Calcium-independent cell–cell adhesion is present in vegetative amoebae, and other adhesion mechanisms come into play during and following aggregation. Later in development, cAMP and other signals induce cell type differentiation and the coordination of fruiting, also in species that do not use cAMP as the aggregation pheromone (reviews in Coates and Harwood 2001; Schaap 2011; Loomis 2014).

In addition to intercellular communication, one can speculate on the adaptive steps that may have accompanied the evolution of aggregative multicellularity as displayed by sorocarpic amoebae: (1) encystation, (2) coming together by clumping, (3) the development of adhesiveness, (4) aggregation via chemotaxis, (5) morphogenesis leading to an erect fruiting body, (6) differentiation within the fruiting body (with the constituent cells possibly differing in viability depending on their locations), (7) the secretion of an extracellular stalk as the fruiting body is being formed and the death of a subset of cells, leading to a fruiting body in which viable cells are supported by a dead cellular stalk (Nanjundiah 1985; Bonner 1998). These steps are conjectured based on what is known from studies on CSMs. However, as we have seen, molecular phylogenies show that steps 1 to 6 do not reflect evolutionary relationships—rather, there is no single "true" sequence in which morphology and genealogy go in parallel. All the same, it can be argued that taken individually, each of steps 1 to 8 is adaptive.

The adaptive value of step 1 seems obvious (survival in a dormant state the face of starvation, as frequently observed in unicellular organisms too), and so does that of Steps 5–7 (improved chances of dispersal because of elevation above the substrate). Steps 2–4 would be beneficial if being situated close to other cells and later, forming a cohesive group, enhanced the probability of long-term survival for each cell. For instance, coming together may improve the chance of surviving predatory attacks simply because of an increase in size. Boraas et al. found that when the unicellular green alga *Chlorella vulgaris* was cultured for many generations with the (predatory) flagellated protist *Ochromonas vallescia*, it evolved to a ~8-celled multicellular stably propagating form in which cells were protected against predation (Boraas et al. 1998). Similarly, Sathe and Durand (2015) found that in the presence of their natural predator *Peranema trichophorum*, single cells of *Chlamydomonas reinhardtii* protected themselves against predation by aggregating (as an interesting aside, the aggregates tended to be chimeric, not necessarily clonal). Kapsetaki (2015) has summarized the case of size increase in response to predation as a plausible adaptation in the evolution of multicellularity. However, group encystment may also be an automatic consequence of starvation setting in after cells have foraged collectively. (After coming together, cooperative foraging can improve the range of accessible prey and feeding efficiency, as seen in myxobacteria and myxomycetes; see Bonner 1998. But that is unlikely to be relevant in the situation we are considering.) If a single cell can form a fruiting body (as in *Protostelium*), the mechanical stability of the stalk increases, and the energetic cost per cell decreases, if more than one cell makes a fruiting body in the same place (Kaushik and Nanjundiah 2003).

Step 6, which has to do with spatiotemporal patterning and division of labor, and more so step 8, which involves cell death, have attracted so much attention in the CSM literature that they demand an extended discussion (also see Nanjundiah 2016). It has been argued that differential reproductive success, particularly, the death of some cells within the group, can be a stable evolutionary outcome if and only if the cells' genetic interests overlap substantially—for instance, via the formation of clonal groups. Thus the evolution of "altruistic" death in cells that form the stalk is accounted for on the hypothesis of kin selection, the reasoning being that the contribution by a stalk cell to spore fitness more than makes up for abandoning the possibility of its own reproduction (Kaushik and Nanjundiah 2003). The supporting evidence comes from three sorts of observations on *D. discoideum*. First, high relatedness is found to be a safeguard against "cheating," meaning the exploitation of the group by an amoeba that—in the extreme case—invariably differentiates into a spore and never contributes to the stalk (Gilbert et al. 2007). Second, polymorphism at the *lag* locus is correlated with the propensity for one clone to sort out from another, and this could be a form of implementing kin selection by excluding non-kin from a group (Benabentos et al. 2009). Third, *D. discoideum* can be found in large clonal patches in nature (Gilbert et al. 2009).

On the other hand, one can make a case for the hypothesis that reproductive division of labor is a concomitant of inter-individual competition for maximizing relative fitness within the multicellular group and has nothing to do with kinship as such (Atzmony et al. 1997). To begin with, as illustrated in Figure 4.1, cell death is indeed an obligatory part of the pathway leading to fruiting body formation in some soro-carpic amoebae but not in all. In *Fonticula alba*, which belongs to the Opisthokonta, a sister group to the Amoebozoa, all amoebae differentiate into spores and produce an extracellular stalk (Worley et al. 1979). The Acrasids belong to the Excavata, yet another supergroup. Their fruiting bodies have the appearance of amoebae piled on top of each other, possibly with secondary branches; again, all amoebae are via-ble (e.g., *Acrasis rosea*; Olive and Stoianovitch 1960). But cell death is present in *Guttulinopsis vulgaris* (Olive 1965), classified under a more distant sister group, the Rhizaria. Evidently, developmental cell death is not essential for spore disper-sal. When it does occur, its extent varies across species. Under standard laboratory conditions, the proportion of aggregated amoebae contributing to the stalk is ~50% in *D. giganteum* and ~20% in *D. discoideum*. In considering what these numbers might imply, one should bear two things in mind. First, the same CSM clade can contain some species that have a cellular stalk and some that do not: evolutionarily, the trait of stalk cell death can be a gain or a loss. Second, CSM species live in the same microenvironment, co-occur on the same speck of soil, can feed on the same bacteria, possibly have similar chances of being preyed upon and very probably their spores are dispersed over similar distances. Under these circumstances, it is difficult to think of differences in the fraction of cells that die while forming a stalk as dif-ferent specific adaptive outcomes at all, let alone as underpinned by kin selection—unless there are unknown factors behind fruiting body viability and spore dispersal, and for reasons that remain to be discovered, some CSMs are more likely to form clonal aggregations than others.

Besides, there is the finding, also involving *D. discoideum*, that unlike what one might expect, successful social exploitation and relatedness within the group do not show a straightforward correlation (Saxer et al. 2010). Other studies, involving *D. giganteum* and *D. purpureum*, show that naturally occurring CSM social groups can be genetically heterogeneous (Filosa 1962), with up to 9 clones in a single spore mass (Kaushik and Nanjundiah 2003, Sathe et al. 2010). It has been proposed that rather than kinship, nonlinear interactions (for want of a better term) and the social context decide who ends up "cheating" whom. In the long term, this could enable genetically heterogeneous "guilds" of a species to coexist (Kaushik et al. 2006, Nanjundiah and Sathe 2011, Sathe et al. 2014). Mathematical models have explored the consequences of a starved cell being allowed to choose whether to join a group or remain solitary. The results show that for cooperative group behavior to persist stably across generations, it is sufficient if some cells that participate in group liv-ing (i.e., in forming an aggregate and fruiting body) in one generation give rise to some cells that form a group in the next generation (De Monte and Rainey 2014). Also, fitness trade-offs that are built on the basis of intercellular differences and depend on the ecological context (Tarnita et al. 2015; Wolf et al. 2015), or varia-tions in the kinetic parameters that underlie components of intercellular signalling, can leave cell-type proportions unchanged (Uchinomiya and Iwasa 2013). They can give rise to the illusion that one genotype is "cheating" another, or that a "cheater" is predisposed to allocate cells to the stalk and spore pathways in proportions that are different from that of the wild type. To sum up, there are two ways to look at the evolution of differential reproductive success among the cells that make up an aggregate, one in terms of shared genes and the other in terms of cellular phenotypes and intercellular interactions. Future work should clarify the situation.

Multicellularity through self-organization. As we have seen, 72% of the 385 "developmentally essential" genes tested in *D. discoideum* are present in non-amoebozoan unicellular relatives (Glöckner et al. (2016), and ~50% of 2352-plus genes that are upregulated during development in at least one of four CSM species, each belonging to a different group, have no homolog in any of the others (Schilde et al. 2016). The proportions and numbers are no doubt provisional and subject to change. However, as more gene sequences and gene expression profiles are analyzed, two sets of figures seem likely to remain either as they are, or go up: concordances between a "developmental gene" in a CSM species and in at least one constitutively unicellular organism, and discordances between "developmental genes" among different CSM species. But even as they stand, some inferences are inescapable: (a) DNA sequences and gene expression profiles do not display substantial differences among CSMs, (b) intergroup differences within CSMs are blurred by the variations in developmental pattern that can be elicited from a single species, and (c) a substantial proportion of CSM "developmental genes" are likely to have homologs in unicellular organisms. The inferences are based on experiments carried out in the laboratory under controlled environmental conditions. It is on the cards that a similar investigation, conducted under diverse ecological circumstances, will show up an even weaker correlation between common ancestry and developmental signposts—not just in the CSMs, but also among other protists that show aggregative sorocarpic development.

That being the case, it may be worth thinking of features of the life cycles we have been discussing, in particular, fruiting body morphologies that look very different, as examples of what Bonner has termed neutral phenotypes (Bonner 2013b, 2015). If so, they could be viewed as different outcomes of self-organization. The same underlying system of physicochemical interactions among identical cells (apart from stochastic variations) could lead to alternative stable spatiotemporal patterns that arise spontaneously (Newman and Bhat 2009). There is a precedent in *D. discoideum* itself: clonal suspensions of starved amoebae spontaneously form two sub-groups of cells with high and low calcium content, respectively; the groups go on to exhibit presumptive spore (low calcium cells) and stalk (high calcium cells) tendencies (Saran et al. 1994). Which of the alternatives is chosen in any given instance can depend on variations in a small number of parameters that specify cell behavior. The parameters can be thought of as reflecting preadaptations that were already present in a unicellular ancestor. Mathematical models and simulations show how groups of cells can aggregate, achieve the correct spatial distribution of cell types and turn into a fruiting body (Savill and Hogeweg 1997; Marée and Hogeweg 2001; see especially the film by A. F. M. Mareé, P. Hogeweg and N. J. Savill [https://www.youtube.com/watch?v=GyAQepksJLU] along with the description in the Ph.D. thesis of A. F. M. Marée, 2000). Other models show how interactions between oscillating units can lead to mutual dependence and organization in the form of "phase clusters" (Kaneko 2016).

To the extent that they are good models of the biological situation, such approaches lead to a different conceptual picture of the evolution of multicellularity. In this picture aggregation and the various morphogenetic alternatives displayed by sorocarpic amoebae are canalized outcomes analogous to the canalized phenotypes that develop in metazoans (Waddington 1942), with the cells that take part possessing preadaptations that equip them for multicellular life. Subsequent genetic changes would stabilize one or the other multicellular morphology once it has come into being (Newman 2002). Studies on prokaryotes and eukaryotes have shown that once cells are forced to live in close proximity, multicellularity and division of labor follow more or less automatically under laboratory conditions (see for example Sťovíček et al. 2012; Włoch-Salamon 2013; Hammerschmidt et al. 2014). Also, complementary metabolic pathways could potentiate multicellularity (Wintermute and Silver 2010). The nature of the participating units would be less important than the capacities that they manifest. In ending, we point out that aggregative multicellularity may be a more plausible route to the origin of multicellularity of the metazoan type than is usually thought (Newman 2011, Dickinson et al. 2012). Finally, the view here expressed lends itself to the further extension that preexisting developmental control systems can be adapted to serve as functional control systems (Robertson and Cohen 1972).

E   Summing Up

The life cycles of sorocarpic amoebae share broad features in common: aggregation in response to the stress of starvation, differentiation along with the construction of a fruiting body, long-lasting dormancy and, in the CSMs and some others, reproductive division of labor (i.e., a germ line-soma separation). Each life cycle involves a shift from unicellularity-to-multicellularity, and it is tempting to view it as mirroring the evolutionary transition from constitutive unicellularity-to-facultative multicellularity. "Higher" metazoans and *D. discoideum* have many genes and regulatory pathways in common, and of the genes that appear to be essential for multicellularity in the CSMs, a substantial proportion were already present in unicellular ancestors. To the extent that they are reflected in molecular phylogenies, genetic changes are poorly correlated with conventional taxonomic assignments. The discordance may be since the same set of cell signals and responses has the potential to lead to a number of developmental outcomes, one or the other of which gets frozen by secondary genetic changes.

4.3   THE ROUTE TO ANIMAL MULTICELLULARITY: THE UNICELLULAR HOLOZOANS PAVED THE WAY

Among all acquisitions of multicellularity that had occurred within eukaryotes, the animal one is probably one of the most relevant to understand the evolution of complex life forms (Ruiz-Trillo et al. 2007; Rokas 2008; King 2004; Knoll 2011). Indeed, the emergence of multicellularity in the animal branch of life gave rise to the whole diversity of animals that we see today with their unique, complex, and coordinated embryonic development. This transition to such complex body plans is even more surprising when we consider that the ancestors were unicellular protists.

To understand such transition, we first need to understand how the unicellular ancestor of animals was. Given that we do not have that ancestor among us anymore, we can only infer how it was by comparing animals with their extant closest unicellular relatives. However, which of all extant protists are the closest unicellular relatives of animals is something that only became clear a few years ago with the advent of phylogenomic (multigene phylogenetic) analyses. Based on morphology, a group of flagellate protists, known as choanoflagellates, had already been proposed on the nineteenth century to be the closest unicellular relatives to animals (King 2005; Leadbeater 2015). The reason for uniting choanoflagellates with animals was a suggested homology of choanoflagellates with a specific cell type of sponges (the choanocyte) (although the homology has been recently disputed; see Mah et al. 2014). Given that sponges were thought to be the earliest animal lineage, the homology was easily explained if choanoflagellates were the sister-group to animals. Indeed, the first ribosomal phylogenies (Medina et al. 2003), and subsequent multigene or phylogenomic analyses (Lang et al. 2002; Steenkamp et al. 2006; Ruiz-Trillo et al. 2004, 2006, 2008; Shalchian-Tabrizi et al. 2008; Carr et al. 2008), supported this view, so that it is now clear that choanoflagellates are the sister-group to animals. These trees showed that animals and choanoflagellates, together with fungi, shared a closer ancestor than with plants or algae, forming a clade known as the opisthokonts.

Further molecular data from other potential opisthokont protists demonstrated that the tree of opisthokonts had additional lineages (Torruella et al. 2012, 2015; Ruiz-Trillo et al. 2004; Shalchian-Tabrizi et al. 2008; Steenkamp et al. 2006). Three of those lineages, the Filasterea, the Ichthyosporea, and the Corallochytrea, were subsequently shown to be close relatives to Metazoa. The clade composed of animals and their closest unicellular relatives is known as the Holozoa (Lang et al. 2002) (Figure 4.3). The most taxon-rich and gene-rich phylogenetic analysis of the Holozoa shows that Teretosporea (Ichthyosporea + Corallochytrea) represents the earliest branching lineage, followed by Filasterea, and then Metazoa and Choanoflagellata (Torruella et al. 2015; Figure 4.3).

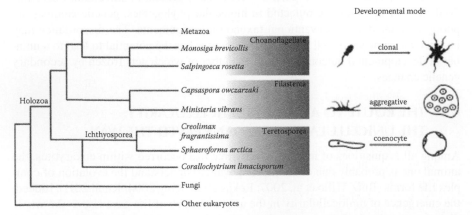

FIGURE 4.3 Schematic tree of the Holozoa with some known taxa and showing the three main lineages. On the right shows a scheme of their developmental modes. (Adapted from Figure 4.1 de Mendoza, A. et al., *Elife*, e08904, 2015. With permission.)

A THE THREE CLOSEST UNICELLULAR RELATIVES OF ANIMALS

Choanoflagellata, Filasterea, and Teretosporea are the key lineages to understand animal origins. Interestingly, these three lineages are very different morphologically and have distinct developmental modes (Figure 4.3). Choanoflagellates, with dozens of species described, are free-living unicellular flagellates that predate bacteria and leave in marine and fresh-water environments (King 2005; Leadbeater 2015). Some species can also form colonies by clonal division. Interestingly, bacteria seem to be responsible for the formation of the choanoflagellate colonies, at least in *Salpingoeca rosetta* (Alegado et al. 2012), which also have additional life stages, such as a sessile form and slow and fast swimmer stages (Dayel et al. 2011).

Filasterea, on the other hand, has only two described species: (1) *Capsaspora owczarzaki* and (2) *Ministeria vibrans* (see Paps and Ruiz-Trillo 2010 *for a review*). Both are amoeboid protist with filopodia. *C. owczarzaki* was isolated from the hemolymph of the freshwater snail *Biomphalaria glabrata*, while *M. vibrans* is marine and free-living. The life cycle of *C. owczarzaki* has been described, under laboratory conditions, and has a "multicellular" stage that forms by cell aggregation (Sebé-Pedrós et al. 2013b). In particular, three life stages have been described: (1) An adhaerent/filopodial stage in which cells crawl; thanks to their filopodia (actin-based cell protrusions) (see the video at https://www.youtube.com/watch?v=0Uyhor_nDts. The title of the video is Time-lapse video of the growth–maturation–dissemination stages of *Creolimax fragrantissima* (Ichthyosporea) by Hiroshi Suga and Iñaki Ruiz-Trillo); (2) A cystic stage in which there is no filopodia; and (3) An aggregative cell stage in which cells actively come together and form an aggregative structure that seems to have some kind of extracellular matrix (ECM) in between the cells (see time-lapse video of the agreggation of Capsaspora owczarzaki (Filasterea) by Arnau Sebé-Pedrós & Iñaki Ruiz-Trillo: https://www.youtube.com/watch?v=OvI6BvBucrc).

Finally, most of the Teretosporea go through a syncytial (multinucleate) development, some species with an amoeboid stage, some without (Mendoza et al. 2002; Glockling et al. 2013; Suga and Ruiz-Trillo 2013). A good example is the ichthyosporean *Creolimax fragrantissima*, originally isolated from the gut of different marine invertebrates (Marshall et al. 2008). Its life cycle starts with one cell with one nucleus and an external cell wall. The cell will go through several nuclear divisions, given rise to a mature, multinucleated coenocyte. The nuclei will cellularize and create amoeboid cells that will be released from the mature coenocyte. Those amoebas will crawl and encyst starting the life cycle again (Marshall et al. 2008; Suga and Ruiz-Trillo 2013) (see time-lapse video of the growth, maturation, and dissemination stages of Creolimax fragrantissima (Ichthyosporea) by Hiroshi Suga & Iñaki Ruiz-Trillo: https://www.youtube.com/watch?v=7Gvrg1I8jBA).

Thus, the three known lineages that are most closely related to animals have very different morphologies and developmental modes (Figure 4.3). Undoubtedly, if one aims to understand the emergence of metazoans from their unicellular ancestor, and given that ancestor is not present anymore, one should investigate those three extant lineages that are more closely related to animals. A few labs have investigated them from a genomic perspective providing important insights into the origin of animals.

B   Genome Data from Unicellular Holozoans Draws a Complex Unicellular Ancestor of Animals

The first genome sequence to be obtained from a unicellular holozoan was that of the choanoflagellate *Monosiga brevicollis* (King et al. 2008). A comparison of the genome sequence of *M. brevicollis* with the genomes of different animals showed that choano-flagellates had less genes (8,700) than, for example, sponges, cnidarians, or placozo-ans. The genome analysis also showed that choanoflagellates already have some genes involved in multicellularity functions, such as protein tyrosine kinases, cadherins, or the Myc transcription factor, all of them previously thought to be animal-specific. Interestingly, the analysis revealed that some other important "multicellular" genes were animal innovations. That was the case, for example, of integrins, one of the most impor-tant adhesion systems in animals, as well as several developmental transcription factors such as NF-kappaB, ETS, Smad, T-box, Runx and Grainyhead (King et al. 2008).

This initial comparative genomic analysis between animals and choanoflagellates was further updated and improved with the addition of new genomes from other taxa. Thus, the genomes of another choanoflagellate (*Salpingoeca rosetta*) and one filasterean (*C. owczarzaki*) provided a more complete view of the evolutionary history of the gene families with key functions in multicellularity and animal development (Fairclough et al. 2013; Suga et al. 2013). What those new genomes showed is that many other com-ponents were already present in *C. owczarzaki* and, therefore, in the unicellular ancestor of animals, but were secondarily lost in choanoflagellates (Suga et al. 2013; Sebé-Pedrós et al. 2010, 2011). This is the case of the integrin adhesome, as well as NFkappaB, T-box, Runx, and Grainyhead, all present in the genome of *C. owczarzaki* (Figure 4.4).

Additional work showed that not only the genes but also some transcription factor gene regulatory networks are conserved between animals and their unicellular rela-tives. A good example is *Brachyury*, a member of the T-box gene family that in bila-terian animals is involved in gastrulation. *C. owczarzaki* has a homolog of that gene that in *Xenopus* can rescue the endogenous function of Xenopus Brachyury (Sebe-Pedros et al. 2013a). Moreover, the downstream gene regulatory network seems to be conserved between *C. owczarzaki* and animals. An analysis of the Capsaspora-Bra downstream target network revealed genes involved in the establishment of cell polarity, phagocytosis, metabolism, transcription factors, and GPCR signaling genes (Sebé-Pedrós et al. 2016a). Moreover, the comparison of downstream orthologs of Capsaspora-Bra and mouse Brachyury targets shows that those orthologs common between the two taxa are significantly enriched in actin cytoskeleton and cell motility functions. That means the Brachyury downstream network was already present in the shared ancestor between animals and *C. owczarzaki* and likely regulating cell motility.

In general, what the genomes of unicellular holozoans (i.e., the closest unicel-lular relatives to animals) have told us is that the unicellular ancestor of animals already had a complex repertoire of genes involved in cell adhesion, cell signal-ing, and transcriptional regulation. Some of those genes were subsequently dupli-cated in the animal lineage, leading to extended gene families, but the complexity

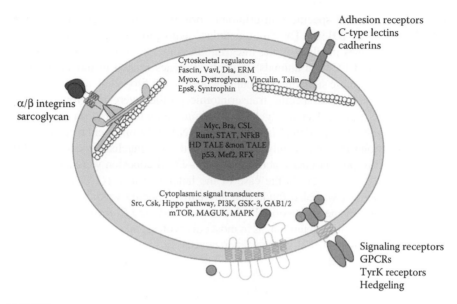

Adhesion receptors
C-type lectins
cadherins

Cytoskeletal regulators
Fascin, Vavl, Dia, ERM
Myox, Dystroglycan, Vinculin, Talin
Eps8, Syntrophin

α/β integrins
sarcoglycan

Myc, Bra, CSL
Runt, STAT, NFkB
HD TALE &non TALE
p53, Mef2, RFX

Cytoplasmic signal transducers
Src, Csk, Hippo pathway, PI3K, GSK-3, GAB1/2
mTOR, MAGUK, MAPK

Signaling receptors
GPCRs
TyrK receptors
Hedgeling

FIGURE 4.4 The potential cell of the unicellular ancestor that gave rise to animals, depicting some of the genes and pathways that were present. (Adapted from Suga, H. et al., *Nat. Commun.*, 4, 2325, 2013. With permission.)

at the gene level of the unicellular ancestor cannot be put into question (Figure 4.4). Moreover, the data has shown that the downstream networks of some developmental transcription factors evolved long before the advent of animals. This points to an important role of co-option at the origins of animals, in which ancestral genes that were working within a unicellular context were recycled to work within a multicellular organism.

C   REGULATION OF GENE EXPRESSION BY HOLOZOANS

The genome sequences from holozoans demonstrated that those taxa already had many genes involved in multicellularity. However, it remained unclear their capacity to regulate the expression of those genes. Recent transcriptomics analyses from one choanoflagellate (*S. rosetta*), one filasterean (*C. owczarzaki*), and one ichthyosporean (*C. fragrantissima*) have shown that those taxa tightly regulate gene expression to go from one life stage to another (Fairclough et al. 2013; Sebé-Pedrós et al. 2013b; de Mendoza et al. 2015).

In all cases, different cell types representing different life stages had different and specific transcriptomic profiles. In the case of *S. rosetta*, for example, septins and cadherins (which are involved in cell adhesion in animals) were found to be upregulated in the colonial stage (Fairclough et al. 2013). Similarly, the different life stages of *C. owczarzaki* had specific transcriptomic profiles involving specific functions (Sebé-Pedrós et al. 2013b). The "multicellular" aggregative stage, for example, is upregulated in genes involved in the integrin adhesome, as well as in proteins with domains typical of ECM in animals. In contrast, genes related to filopodia formation and tyrosine kinase signaling were upregulated in the filopodial, adhaerent stage. Completely different and specific transcriptomic profiles were also found for the amoeba and the multinucleated stage of *C. fragrantissima* (de Mendoza et al. 2015). For example, genes involved in DNA replication, RNA and amino acid metabolism, as well as translation were significantly upregulated in the multinucleate coenocyte, while genes involved in protein kinase activity and cell–ECM adhesion were significantly upregulated in the amoebas. In the case of the latter two taxa (*C. owczarzaki* and *C. fragrantissima*), it was also found that alternative splicing was also contributing to the regulation of gene expression. Both *C. owczarzaki* and *C. fragrantissima* were shown to have both intron retention, as in most eukaryotes, and exon skipping in some genes (Sebe-Pedros et al. 2013b; de Mendoza et al. 2015). Intron retention in those taxa were found to be differentially regulated between the different life cycle stages, probably contributing to the control of transcript levels. Interestingly, in *C. owczarzaki*, genes with differential exon skipping were found to be significantly enriched in protein kinase activity, suggesting the presence in *C. owczarzaki* of a regulated exon network linked to cell signaling, a feature that was thought to be animal-specific. In *C. fragrantissima* genes involved in exon skipping were significantly enriched in several biological functions, such as channel activity and histone modifications.

D   CELL TYPES IN HOLOZOANS

The data from transcriptomics point to the fact that premetazoan taxa have the capacity to differentiate into different cell types (each one with its own specific transcriptomic profile) in a temporal manner, within their life cycles. Two more recent analyses confirmed that the regulation of the different cell types (life stages) in *C. owczarzaki* is, in turn, regulated by a dynamic proteome and phosphoproteome remodeling, as animals do, as well as long noncoding RNAs and histone marks (Sebé-Pedrós et al. 2016a, b). The phospho-signaling regulation also affects some developmental transcription factors, such as Runx, P53, or CREB (Sebé-Pedrós et al. 2016b).

All these data suggest that the unicellular ancestor of animals not only had the gene repertoire needed for multicellular functions and animal development but also had most of the mechanisms to regulate cell-type differentiation in animals. In this case, those mechanisms were probably working in transitions from one life stage to another and were later co-opted to work spatially within a multicellular body plan.

E   CONCLUSION

The study of unicellular relatives of animals has clearly improved our understanding of how unicellular protists became multicellular animals. Thanks to those analyses.

We now know that the unicellular ancestor of animals was genetically much more complex than previously thought. Not only had that ancestor a rich repertoire of genes involved in multicellular functions, but it also had the capacity to strongly regulate the expression of those genes and had the mechanisms to perform cell differentiation. That means that many genes and gene regulatory networks present in extant animals and key for their multicellularity and development appeared in the premetazoans, being later recycled to work within a multicellular body. This, together with the acquisition of some novel genes and an important expansion of some gene families provided the basic metazoan genetic toolkit, allowing the evolution of spatial cell differentiation as well.

4.4 THE EVOLUTION OF MULTICELLULARITY AND CELLULAR DIFFERENTIATION IN GREEN ALGAE

"Few groups of organisms hold such a fascination for evolutionary biologists as the Volvocales. It is almost as if these algae were designed to exemplify the process of evolution...." (Bell 1985)

The chlorophytes (green algae in the class Chlorophyceae) are unrivaled champions at transitioning from unicellularity to multicellularity. Such a transition is said to have been made independently in more than two dozen different lineages (Bonner 1998; Grosberg and Strathman 2007); but nearly half of that many such transitions may have been made by the chlorophytes alone. Although most species of chlorophytes are unicellular flagellates, multicellular species are present in 9 of the 11 chlorophyte orders, and in each of those 9 orders multicellularity is believed to have arisen independently—and in some cases more than once (Melkonian 1990).

In most cases, such "multicellular" chlorophytes are multicellular in only the simplest sense; each individual is composed of more than one cell. Indeed, they usually consist of 2^n sister cells of identical type that have remained in association after being produced by n rounds of cell division (Figure 4.5). Such organisms are usually referred to by those who study them as "colonies" or "colonial organisms," and are characterized by the fact that each of their cells is capable of dividing n times to produce a new colony comprising 2^n sister cells (Starr 1980).

There can be little doubt that the propensity of the chlorophytes to form such colonies of sister cells is due in large measure to their shared capacity to produce and secrete two kinds of extracellular materials: (1) a glycoprotein-rich (cellulose-free) cell wall and (2) "mucilage," an amorphous sticky mixture of glycoproteins, acid mucopolysaccharides and other carbohydrates. In family after family of colonial chlorophytes, one finds that sister cells are held together by glycoprotein-rich walls that have either fused or have formed some sort of specialized junction between cells. And in case after case it is also found that any spaces between the cell walls are filled with mucilage (see Figure 4.5).

However, in addition to its use to refer to simple clusters of similar cells, the term "multicellular organism" has a second meaning. When the term "multicellular organism"

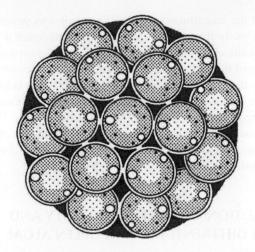

FIGURE 4.5 A drawing of a 32-cell colony of *Coelastrum microsporum*, a representative small, green Chlorophyte alga. Chlorophyte colonies may consist of 4, 8, 16, 32, 64, or 128 sister cells, depending on how many times the mother cell divided. Each cell is surrounded by a glycoprotein-rich cell wall that is shown here as a dark-light-dark tripartite ring surrounding each cell. The walls of neighboring cells are connected by specializations that were formed at points of contact. The cellular monolayer surrounds a central space filled with mucilaginous extracellular matrix (indicated in solid black). In such colonies, every cell is capable of dividing to produce a new colony of similar type. (Reprinted from Kirk, D.L., *Volvox: Molecular Genetic Origins of Multicellularity and Cellular Differentiation*, Cambridge University Press, Cambridge, UK, 1998. With permission.)

is used in the abstract, it most often conjures up images of more complicated beings—such as plants, animals, or fungi—consisting of multiple cell types that differ in both structure and function, and that must cooperate to survive and produce offspring. Elegant examples of this type of multicellular organism also have evolved in the Chlorophyceae.

A   THE GENUS *VOLVOX*: MULTICELLULAR ALGAE WITH A GERM-SOMA DIVISION OF LABOR

In a subset of chlorophytes known as the volvocine algae, the evolution of multicellularity in the simpler sense was followed by the evolution of cellular differentiation, resulting in algae such as *Volvox carteri* (Figure 4.6) that exhibit a germ-soma division of labor rather similar to that seen in most animals. The developmental biology of *V. carteri* has been reviewed frequently (Kirk 1998, Nishii and Miller 2010, Matt and Umen 2016). Here it will suffice to summarize key aspects of that biology.

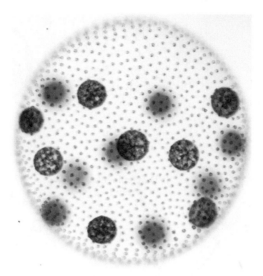

FIGURE 4.6 *Volvox carteri*, an alga with a striking germ-soma division of labor. Each individual (called a "spheroid") contains more than 2,000 small, biflagellate somatic cells embedded at the surface of a transparent sphere of glycoprotein- and mucopolysaccharide-rich extracellular matrix. 16 much larger cells, called gonidia, are located just internal to the somatic cells. The flagella of the somatic cells protrude from the surface and provide the spheroid with motility. The gonidia never have functional flagella; they serve as germ cells in the asexual reproductive cycle, by dividing to produce a new generation of spheroids with a similar cellular composition.

A *V. carteri* individual (called a "spheroid," because of its shape) contains two fully differentiated cell types: more than 2,000 tiny, biflagellate somatic cells, and about 16 large asexual reproductive cells called "gonidia." The somatic cells lie in a monolayer at the surface of a transparent sphere of ECM, with their flagella projecting from the surface and oriented so that their beating propels the spheroid through the water with a highly characteristic rolling motion. The gonidia lie just internal to the somatic cells in the ECM.

Gonidia lack flagella, are as much as 1,000 times the volume of somatic cells and are specialized for cell division. Once mature, each gonidium will divide rapidly 11 or 12 times, to produce all the cells that will be present in an adult of the next generation. In marked contrast, once the somatic cells have begun to differentiate, they have become postmitotic; they never divide again. The asexual life cycle of *V. carteri*, by which a new generation of spheroids is produced every two days, is diagrammed in Figure 4.7 and is described in the accompanying caption.

V. carteri also has a very interesting sexual reproductive cycle (Kirk 1998, Umen 2011), but space constraints prevent discussing it here.

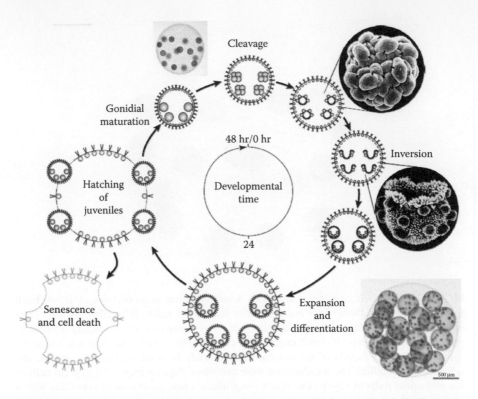

Cleavage

Gonidial
maturation

48 hr/0 hr

Inversion

Hatching
of
juveniles

Developmental
time

24

Senescence
and cell death

Expansion
and
differentiation

500 µm

FIGURE 4.7 A diagram of the asexual reproductive cycle of *Volvox carteri*. As indicated by the inner circle, one asexual life cycle takes precisely two days, if a synchronizing light-dark cycle (16L:8D) is used. Each life cycle comprises 5 phases of development that are labeled on the circumference, and in several cases illustrated with photomicrographs. These phases are as follows: (i) Cleavage. A mature gonidium executes a series of synchronous cell divisions. The first five divisions are symmetrical, producing 32 cells of similar size, arranged in a hollow sphere. In the sixth division cycle, the 16 cells in one hemisphere divide symmetrically again, but the 16 cells in the other hemisphere divide asymmetrically, producing 16 large-small sister-cell pairs (as shown in the micrograph, where arrowheads point from each large cell to its small sister cell). The large cell in each such pair becomes a gonidial initial, while its smaller sister becomes a somatic initial, as do all the cells of the other hemisphere that divided symmetrically. After being formed by asymmetric division in cycle six, the gonidial initials divide asymmetrically one or two more times, producing another small somatic initial in each division. Then the gonidial initials withdraw from the division cycles, while all the somatic initials go on dividing symmetrically, until they have completed a total of 11 or 12 divisions. (ii) Inversion. At the end of cleavage, the embryo contains all the cells that will be present in the adult. But they are arranged in a maladaptive orientation: all the somatic initials are oriented with their flagellar ends directed toward the interior space, and all the gonidial initials are on the outside of the sphere. This predicament, which would preclude spheroid motility, is resolved as the embryo turns itself completely inside-out, bringing the flagellar ends of the somatic initials to the exterior and moving the gonidia to the interior in the process known as inversion. Inversion occurs by a stereotyped series of changes in cell shape and cellular movement that have been analyzed extensively with regards to mechanical, cytological and genetic parameters (Viamontes et al. 1979; Nishii et al. 2003).   (*Continued*)

B THE VOLVOCINE LINEAGE: A SIMPLE, LINEAR PROGRESSION IN SIZE AND COMPLEXITY?

Volvox is the eponymous ("name-giving") member of the "volvocine algae," a group that encompasses the unicellular green flagellate, *Chlamydomonas reinhardtii*, a variety of multicellular green flagellates in the family Volvocaceae (including about 20 species of *Volvox*) and a few other small, green flagellates in two families, closely related to the Volvocaceae. A unifying feature is that they all have cells morphologically similar to *Chlamydomonas*. The notion that these algae constitute a monophyletic group of organisms that are related by what Darwin called "descent with modification" was accepted by many biologists for decades, in the absence of any supporting evidence beyond their common cytological features. A number of textbook authors went as far as to line up the various volvocine algae in order of increasing size and complexity, from *Chlamydomonas* to *Volvox*, and then imply that this probably resembled the pathway by which members of the group evolved: by a simple progressive increase in organismic size and complexity. The latter notion has been called "the volvocine lineage hypothesis" (Figure 4.8).

About a century ago, however, a markedly different hypothesis about aspects of volvocine evolution was proposed, based on the presence or absence of cytoplasmic connections in adults of various species of *Volvox* (Crow 1918). During their embryonic development, all members of the family Volvocaceae have cytoplasmic bridges that form between sister cells as a result of incomplete cytokinesis, thereby, linking all cells of the embryo into a syncytium. In many species of *Volvox* (including *V. carteri*, shown in Figure 4.6), and in all species of the other volvocacean genera, these cytoplasmic bridges break down at the

FIGURE 4.7 (Continued) At the end of inversion, the embryo has taken on the adult configuration and is called a juvenile spheroid. But at that stage, its presumptive somatic and gonidial cells differ in little but size. (iii) Cytodifferentiation and expansion. Under synchronizing growth conditions, the last stages of cleavage and inversion occur in the dark, after which nothing changes visibly until the lights come back on. As soon as the lights come on, both cell types begin actively translating mRNAs that had accumulated in the dark (Kirk and Kirk 1985), flagella grow outward from the somatic cells, the gonidia begin to enlarge, and the two cell types diverge progressively in their patterns of gene expression and morphology. Now the principal activity of the gonidia is growth, in preparation for the next round of embryogenesis, whereas the major activity of the somatic cells is synthesis and secretion of glycoproteins, mucopolysaccharides and other ECM components that will self-assemble and cause the spheroid to increase greatly in diameter. (iv) Hatching. A day and a half after their formation began with cleavage of the maternal gonidia, each juvenile spheroid digests an opening in the overlying parental ECM, creating a birth canal through which it then swims to the exterior, to become a free-living young adult. By this time, the somatic cells of the parent have become senescent, and they and the parental ECM soon undergo dissolution and disappear. (v) Gonidial maturation. The gonidia of the free-swimming young adults now do whatever is required to initiate a new round of embryogenesis. As those gonidia begin to divide, one asexual life cycle has been completed, and another begins. (Reprinted from Prochnik, S.E. et al., *Science*, 329, 223–226, 2010. With permission.)

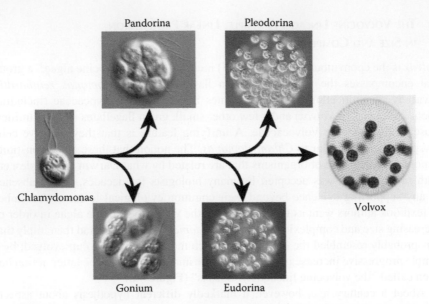

Pandorina   Pleodorina

Chlamydomonas

Volvox

Gonium   Eudorina

FIGURE 4.8   The volvocine lineage hypothesis. It is possible to line up *Chlamydomonas* and selected volvocacean genera in a conceptual series (as in this diagram) in which there is a progressive increase in cell number, the ratio of ECM to cellular volume, and the number of cells that are set aside as nondividing, sterile somatic cells. A typical colony of Gonium, one of the smallest volvocine algae, consists of a convex disc of 8 or 16 *Chlamydomonas*-like cells, each of which will eventually divide to produce a new colony of similar form. A colony of *Pandorina* consists of 16 or 32 cells that are initially packed together closely to form an elliptical ball: later however, the cells separate from each other somewhat as ECM accumulates. As in *Gonium*, every *Pandorina* cell is capable of dividing to produce a new colony. *Eudorina* colonies are spheres of 16, 32, or 64 cells (depending on the species and environmental conditions) in which cells are more widely separated than in the previous two genera. In a 64-cell *Eudorina* colony 4 cells at the anterior pole of the sphere may function as nonreproducing somatic cells and continue beating their flagella to keep the colony afloat while all the other cells lose their flagella, enlarge and divide to produce progeny. In *Pleodorina* there are usually 64 or 128 cells per individual. They are all initially biflagellate cells contributing to motility; but then cells in the posterior hemisphere resorb their flagella, grow and function as gonidia to produce progeny, while the remainder of the cells functions as terminally differentiated somatic cells. The more total cells there are in *Pleodorina*, the higher will be the ratio of somatic cells to gonidia. That trend continues in the genus *Volvox*. In species of *Volvox* having as few as 1,000 cells per spheroid, there may be as many as 75 gonidia. But in species with as many as 10,000 cells, there may be fewer than ten very large gonidia, with all the rest being terminally differentiated somatic cells (Kirk 1998). So, the following relationships can be generalized for *Eudorina*, *Pleodorina* and *Volvox*: as the total number of cells per colony or spheroid increases, the gonidia decrease in number but increase in size, while the somatic cells increase in number but decrease in size (Bell 1985, Koufopanou 1994). The Volvocine Lineage Hypothesis postulates that by lining up these algae in order of increasing size and complexity we get a good approximation of the way in which the group evolved: by a simple progressive increase in organismic size and complexity, from *Chlamydomonas* to *Gonium*, to *Pandorina*, to *Eudorina*, and so forth.

end of embryogenesis. As a result, adult cells in all those species lack intercellular connections and have a smooth, round profile when viewed in optical cross section. However, in another large section containing about half of all the recognized species of *Volvox*, the cytoplasmic bridges are retained and thickened in the adult, so that each adult cell is connected to all its neighbors by stout cytoplasmic bridges, which makes the cells appear stellate in optical cross section. Crow postulated that *Volvox* species, lacking cytoplasmic connections in the adult might have evolved from a *Chlamydomonas*-like ancestor—as others had previously suggested—but that those species of *Volvox* in which the adult cells are connected by bridges, had evolved independently from a very different unicellular green flagellate genus, *Haematococcus*—cells of which appear somewhat stellate in optical cross section.

C COMPARATIVE DNA SEQUENCING FALSIFIES THE VOLVOCINE LINEAGE HYPOTHESES

The first published comparison of the sequences of nuclear-encoded rRNAs of *Volvox* and *Chlamydomonas* (Rausch et al. 1989) reinforced the inference drawn from morphology that these two genera were closely related. A more extensive comparison of volvocine nucleic acid sequences (Kirk et al. 1990, later amplified in Larson et al. 1992) produced several interesting conclusions that have been corroborated by subsequent work: First, it falsified Crow's hypothesis, by showing that both the *Volvox* species that had adult cytoplasmic bridges and the species that lacked such bridges had rDNA sequences that were much more closely related to the rDNA sequences of *Chlamydomonas reinhardtii* than they were to those of *Haematococcus*. Second, it justified the choice of *C. reinhardtii* as a proxy for the unicellular ancestor of the Volvocaceae by showing that *C. reinhardtii* rDNA is much more closely related to that of all volvocaceans than it is to the rDNA sequence of another species of *Chlamydomonas*. Third, it produced results consistent with the hypothesis that the family Volvocaceae is a monophyletic assemblage of closely related organisms; they all share a common unicellular ancestor. Fourth, it falsified the volvocine lineage hypothesis by indicating clearly that the phylogenetic relationships among the genera represented in Figure 4.8 are much more complicated than the volvocine lineage hypothesis suggests. Fifth, it indicated that, although the volvocine algae exist as a group are monophyletic, the genus *Volvox* is not monophyletic, since the two species of *Volvox* that were studied were placed on two separate branches of the preliminary family tree that was produced in the study. Several of these conclusions were quickly supported by the results of a similar study by Buchheim and Chapman (1991).

Currently the best estimate is that *Chlamydomonas reinhardtii* and *Volvox carteri* last shared a common ancestor about 200 million years ago (Herron et al. 2009), which is several hundred million years more recent than animals, plants, or fungi, are thought to have shared a common ancestor with their unicellular forebears (Parfrey et al. 2011).

D THE FAMILY VOLVOCACEAE IS MONOPHYLETIC, BUT SEVERAL OF ITS GENERA ARE NOT

Since 1992 there have been a number of additional studies using comparative DNA sequencing of various nuclear and chloroplast genes from a growing set of volvocine algae, in an effort to establish a robust phylogeny for the group (e.g., Coleman 1999; Nozaki et al. 2000, 2014; Nozaki 2003; Nakada et al. 2008; Herron and Michod 2008; Herron et al. 2009, 2010). The number of taxa included in such studies has increased over the years, in part because new species—and even new genera—of volvocaceans are regularly being isolated from natural fresh water sources by Nozaki and his colleagues (examples: Nozaki and Coleman 2011; Isaka et al. 2012; Nozaki et al. 2014, 2015a,b, 2016.) Although various recent studies have differed in magnitude and methodology, they have generally supported three conclusions drawn from the earliest molecular–phylogenetic studies of the group: (1) The family Volvocaceae, taken as a whole, is monophyletic (i.e., its members share a common unicellular ancestor). (2) The phylogenetic relationships among the various genera and species of volvocaceans are more complex than the volvocine lineage hypothesis suggested. (3) The (so-called) genus *Volvox* is polyphyletic.[1] However, as such studies have included ever more taxonomic units and analyzed them with greater resolution, it has become increasingly apparent that *Volvox* is not the only polyphyletic genus in the family.

The latter point is clearly illustrated by a recent study, the results of which are summarized in a simplified form in Figure 4.9. There, in addition to finding species of *Volvox* on four different branches of the family tree, we find *Pleodorina* and *Eudorina* species on three branches each, and even different isolates of the so-called species *Eudorina elegans* are located on somewhat widely separated branches.

What conclusions should we draw from such results?

The first conclusion to be drawn is that several of the genus and species names that are currently used to classify volvocaceans identify *grades* of organizational complexity, *not clades* of closest relatives. This conclusion—that a single scientific name may sometimes group volvocine algae that are only somewhat distantly related at the genetic level is consistent with a number of earlier studies of natural populations. The clearest example is Annette Coleman's (1980) stunning analysis of *Pandorina morum* strains that had been isolated from ponds around the world. She found that various isolates of this morphologically monotypic "species" fell into at least 24 reproductively isolated mating groups, or syngens. Different syngens sometimes

---

[1] Spatial constraints here preclude full discussion of the range of developmental variations that occur among species currently assigned to the genus *Volvox* and the complex evolutionary relationships among those species. But for an extensive, authoritative, and enlightening discussion of such matters, see Herron et al. (2010).

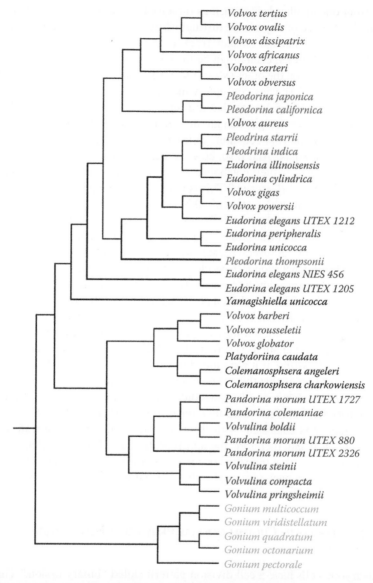

FIGURE 4.9 A dendrogram illustrating relationships among various volvocine genera and species. Adapted from a Bayesian inference tree that was based on the sequences of five chloroplast genes in 58 volvocine algae (Based on Nozaki, H. et al., BMC Evol. Biol., 14, 37–46, 2014). In this adaptation, the tree has been greatly simplified, using lines of arbitrary length to emphasize the branching patterns between various taxa. For quantitative information, such as genetic distances along various branches, posterior probabilities and bootstrap values, see the original.

differed from one another by as much as threefold in chromosome number (from 4 to 12), in the absence of any visible distinguishing characteristics. She also showed that although two *P. morum* isolates from a single pond might be reproductively isolated from one another, each might be interfertile with a *P. morum* coming from some particular pond on the other side of the world. Clearly, *Pandorina morum* does not fit the "biological species concept" which asserts that: "species are groups of interbreeding natural populations that are reproductively isolated from other such groups." By this criterion, algae that are morphologically indistinguishable from the type specimen of *P. morum* constitute at least 24 species. Should they all be given different Latin binomials? And if so, how are field biologists going to decide which Latin binomial to write in their notebooks when they find an alga in a pond that looks like good old-fashioned *P. morum*? This is a taxonomic quandary that will probably never be resolved to the satisfaction of all biologists.

A second conclusion to be drawn from recent molecular–phylogenetic studies of the volvocine algae is that transitions have apparently been made repeatedly—and in both directions—between several organizational grades (aka genera) of volvocaceans in the past. A third conclusion follows close behind the second, which is, the genetic changes required to transition from one volvocacean organizational level to the next must be quite modest. That conclusion is consistent with earlier observations that (for example) a single mutation in *Volvox powersii* changes its morphology enough that it would be called *Pleodorina*, if it were found in the wild (Vande Berg and Starr 1971), or that a pair of point mutations is sufficient to change *Volvox carteri* into a form that would be called *Eudorina*, if it were found in nature (Tam and Kirk 1991).

It is important to note that there is no evidence indicating that the important transition from *Chlamydomonas*-like unicellularity to simple, colonial (*Gonium*-like) multicellularity has occurred more than once in this group of algae. So far, all the available evidence is consistent with the notion that this important transition has occurred only once and, therefore, the family Volvocaeae is monophyletic.

To understand how that transition occurred, and why it never occurred more than once, we need to step back to consider the unusual pattern of cell division seen in *Chlamydomonas*.

E   MULTIPLE FISSION IN *CHLAMYDOMONAS* PROVIDED A FOUNDATION FOR MULTICELLULARITY

Most eukaryotic cells have a cell division pattern called "binary fission." That is to say, after a period of growth in which they approximately double in size, they divide in two, and then they repeat the cycle of growth and division. But *Chlamydomonas* and certain other protists have a very different pattern, called "multiple fission," or "palintomy." In palintomic organisms, growth and cell division are uncoupled; cells grow 2^n-fold without dividing and then divide n times rapidly (in the absence of further growth) to produce 2^n progeny cells. In vegetatively reproducing *Chlamydomonas* cultures, the value of n is usually between 2 and 4 (depending on culture conditions), resulting in the production of 4, 8, or 16 progeny cells per cycle.

All colonial volvocine algae, including *Gonium*, *Pandorina*, *Eudorina*, and *Pleodorina* (plus several other genera not specifically named and discussed here) exhibit the same multiple fission pattern as *Chlamydomonas* does, accounting for the fact that they typically contain 2^n cells per colony, with the value of *n* varying by genus. (In various *Volvox* species, however, the multiple-fission program becomes altered in various ways—or abandoned altogether—to arrange for production of gonidia of species-specific size and number; Herron et al. 2010.)

F TWO CHANGES WERE REQUIRED IN CONCERT TO CONVERT MULTIPLE FISSION TO MULTICELLULARITY

It has been proposed that the evolution of a *Volvox carteri*-like organism from a *Chlamydomonas*-like ancestor involve twelve substantial genetic and/or morphological changes (Kirk 2005). But two of those twelve changes were the *minimal* changes required to evolve a simple, multicellular colony from a *Chlamydomonas*-like unicell. Specifically, the minimal changes required for this critical transition were: (1) Incomplete cytokinesis, to form transient cytoplasmic bridges between sister cells that hold the cells in a fixed relationship during the division period. (2) Modification of the cell wall-forming process that follows cell division, to form some sort of attachment between adjacent cell walls, in order to hold the cells in the same fixed relationship after the cytoplasmic bridges have broken down.[1] Both of these important steps have been observed during *Gonium pectorale* development (Stein 1958, Iida et al. 2013); both of them occur, with some minor modifications, in all other colonial volvocaceans (Kirk 1998), and they also occur with certain additional modifications in all species of *Volvox* (Herron et al. 2010).

The importance of establishing firm cell wall, or ECM connections between neighboring cells before the cytoplasmic bridges break down, has been demonstrated most clearly with *V. carteri*. In normal development of *V. carteri* embryos, a glycoprotein called ISG self-assembles to form a layer over the surface of the newly inverted embryo. This ISG layer not only provides the first tenuous extracellular linkages between neighboring cells, but it also acts as a scaffold for assembling the other ECM components, including the fused cell walls that normally form solid connections between neighboring cells. So, when ISG assembly was prevented, the rest of the ECM never assembled properly, and as a result, as soon as the cytoplasmic bridges broke down, the embryo fell apart into a single-cell suspension (Hallmann and Kirk 2000). From such studies, we conclude that these two new features (incomplete cytokinesis and cell wall fusion) would need to have been achieved *in concert* to make the transition from a unicellular to a colonial body plan. And the improbability of two such morphological features evolving almost simultaneously may well account for the fact that the transition to multicellularity was made only once in this lineage.

In contrast, other volvocacean transitions (such as between the *Pandorina* and the *Eudorina* levels, or between the *Eudorina* and *Pleodorina* levels of size and

---

[1] *Gonium* exhibits other morphological differences from *Chlamydomonas* that are shared by all the other volvocaceans (Kirk 2005), but the two changes discussed here are the minimal changes that would be required to hold the post-division cells in a predictable spatial relationship.

complexity) appear to have involved only a single substantial change (Kirk 2005), which would have given them a much higher probability of occurring more than once over evolutionary time.

Next question: Can we discern what genetic changes underlie all the steps that were involved in the evolution of *Volvox carteri* from a *Chlamydomonas reinhardtii*-like ancestor?

G THE *CHLAMYDOMONAS* AND *VOLVOX* GENOMES ARE STRIKINGLY SIMILAR

Three years after the sequence of the *C. reinhardtii* genome had been determined (Merchant et al. 2007), the *V. carteri* genome was also sequenced (Prochnik et al. 2010). A primary motivation for sequencing the *V. carteri* genome was, of course, the hope that a detailed comparison of the two genomes would reveal clearly which important genetic changes had accumulated in going from *Chlamydomonas* to *Volvox*.

In retrospect, two earlier observations probably should have alerted us to the realization that this might very well be a false hope, because it had been found that two genes that are essential for normal *V. carteri* development had undergone so little change in the ~200 million years since Chlamydomonas and *Volvox* lineages diverged that the *C. reinhardtii* genes could substitute for their *V. carteri* counterparts.

This finding came from a genetic study of two very interesting processes in *V. carteri* embryonic development that have no known parallel in *C. reinhardtii*, namely: asymmetric division, by which germ cell and somatic cell precursors are set apart during cleavage, and inversion, by which the fully cleaved *Volvox* embryo turns itself inside out (Figure 4.7). Transposon mutagenesis was used to establish that when a *Volvox* gene called *glsA* is inactivated, no asymmetric divisions occur (Miller and Kirk 1999). Strikingly, the *Chlamydomonas* ortholog of *glsA* is fully capable of rescuing the *glsA* mutant and restoring normal asymmetric division and germ cell specification (Cheng et al. 2003). Similarly, after a transposon insertion was used to mutagenize *invA* (a *Volvox* kinesin-encoding gene, Nishi et al. 2003), and show that its product is required for normal inversion of the *Volvox* embryo, the *Chlamydomonas* ortholog of *invA was* shown to be fully capable of rescuing the *invA* mutant and restoring perfectly normal inversion (Nishii and Miller 2010).

In the light of such observations, it is not surprising, perhaps, that the *C. reinhardtii* and *V. carteri* genomes turn out to be strikingly similar (Table 4.1) (Prochnik et al. 2010, Umen and Olson 2012). Although the *V. carteri* genome is ~17% larger than the *C. reinhardtii* genome, most of the difference is accounted for by the greater abundance of repetitive sequences in *Volvox*. The two genomes contain very similar numbers of protein-coding genes and encode similar numbers of largely overlapping protein families. More than 9,000 (~64%) of the protein-coding sequences in the two algae encode proteins in families that are shared with many other eukaryotes, but 1,835 of the coding sequences (~12%) are volvocine-specific, in the sense that they are found in both algal genomes, but not in other sequenced genomes.

The volvocine-specific genes are of potential long-term interest because many of them exhibit asymmetric expansion/contraction patterns between these two algae. In most cases, however, the functions of the encoded proteins are unknown,

TABLE 4.1

A Comparison of the *Chlamydomonas reinhardtii* and *Volvox carteri* Genomes

Species	*Chlamydomonas*	*Volvox*
Genome size (Mbp)	118	138
Number of chromosomes	17	14
Interspersed repeats (millions)	14.8	28.2
Protein-coding loci	14,516	14,520
PFAM (protein family domains)	2,354	2,431
% coding	16.3	18.0
Introns per gene	7.4	7.05
Median intron length (bp)	174	35
Volvocine-specific genes	1,835	1,835
Pherophorins	27	45
Matrix metalloproteinases	8	42
D1 cyclins	1	4
Histone gene clusters	35	14
Ankyrin repeat proteins	146	80

precisely because homologs have not been found elsewhere. But there are two volvocine-specific families where the significance is fairly obvious; genes encoding pherophorins and matrix metalloproteinases, which are significantly more abundant in the *V. carteri* than in the *C. reinhardtii* genome (Table 4.1, below the dotted line). The pherophorins are hydroxyproline-rich glycoproteins that are related to certain *Chlamydomonas* cell wall proteins, and are major building blocks of the *Volvox* ECM. Moreover, the matrix metalloproteinases are thought to be intimately involved in fashioning, refashioning, and dissolving the ECM at various stages of the life cycle (Hallmann 2006). In view of the fact that the ECM volume is nearly 100X the cellular volume in an adult *Volvox*, but less than 1% of the cellular volume in *Chlamydomonas*, it is hardly surprising that these two gene families that encode major ECM constituents are expanded in the *Volvox* genome relative to the *Chlamydomonas* genome.

It is slightly less obvious why the D1 cyclin gene family should be expanded 4-fold in *Volvox*. In animals and land plants, D-type cyclins play an important role in the regulation of the cell cycle, by activating cyclin-dependent kinases that then phosphorylate RB proteins, and both proteins have been shown to play a key role in regulating cell cycle progression in *Chlamydomonas* (Umen and Goodenough 2001). So, it has been hypothesized that the expansion of the D1 cyclin family in *Volvox* might be related to the fact that *Volvox* exhibits many more stage-specific and mating-type specific cell cycle variants than *Chlamydomonas* does (Umen and Olson 2012). We will return to that hypothesis later.

In contrast to the expansion of genes encoding major ECM components in *Volvox*, which are rather easy to rationalize, it is difficult to rationalize the greater abundance of genes encoding histones and ankyrin-repeat proteins in *Chlamydomonas* (Table 4.1). The number of histone-encoding genes in *Chlamydomonas* (which is unusually high with respect to land plants and many other algae, as well as *Volvox*) might be rationalized if *Chlamydomonas* had much more rapid division cycles than *Volvox*; but it does not (Umen and Olson 2012).

The greater abundance of genes in *Chlamydomonas* is equally enigmatic. Ankyrin repeats are involved in protein–protein interactions in a wide variety of interesting proteins, such as transcriptional initiators, cell cycle regulators, cytoskeletal proteins, ion transporters, and signal transducers. But there is no obvious reason why any or all of those protein categories should be much more abundant in *Chlamydomonas* than in *Volvox*.

One important conclusion can safely be drawn from a comparison of these two genomic sequences: Major revision of the genome was not required to evolve from the *C. reinhardtii*-level to the *V. carteri*-level of size and developmental complexity.

H VOLVOX REINHARDTII AND V. CARTERI HAVE VERY DIFFERENT SMALL-RNA SYSTEMS

The finding that *C. reinhardtii* and *V. carteri* do not have as many differences in protein-coding genes as some might have expected is not without precedent. When it was first realized in the 1970s that humans and apes were extremely similar at the DNA level, many considered this finding paradoxical, and the challenge became " to explain how species that have such substantially similar genes can differ so substantially...." (King and Wilson 1975). The situation had not changed significantly by the time the complete genome sequences of both apes became available, as the title of one review article made very clear ("...Searching for needles in a haystack." Varki and Altheide 2005). The conclusion drawn from many such studies by a substantial number of developmental biologists has been that "...changes in morphology generally result from changes in the spatiotemporal regulation of gene expression during development." (Carroll 2008).

There are of course as many ways to control gene expression as there are steps in the conversion of a DNA coding sequence to a visible phenotype. But one of the most recently discovered categories of gene regulatory mechanisms, and the one that has been most studied by those interested in the regulation of volvocine gene expression is post-transcriptional regulation by two types of small, noncoding RNA molecules, namely: micro RNAs (mi-RNAs) and small interfering RNAs (siRNAs) (Carrington and Ambros 2003, Bartel 2004). Both types of small RNAs are usually 20–24 nucleotides in length, and both function to regulate gene expression at the post-transcriptional level by either interfering with the translation of mRNAs containing the complementary sequence or by triggering the destruction of such mRNAs by the Argonaute nuclease. But they differ in origin: miRNAs are derived from stem-loop regions of mRNAs, whereas siRNAs are derived from long double-stranded RNAs, but in both cases, they are released from their source molecules by a Dicer nuclease (Vaucheret 2006). It has been postulated that by fine-tuning gene expression, small RNAs have played a major role in macroevolution and the origin of morphological novelties (Peterson 2009). Therefore, they have attracted the interest of several groups interested in the evolution of the volvocine algae.

The first unicellular organism that was found to possess miRNAs, as well as Dicer and Argonaute nucleases was *C. reinhardtii* (Molnar et al. 2007). Prior to that time, it had been thought that the miRNA system was present in multicellular organisms only and that the system had evolved together with multicellularity, independently and convergently in the multicellular plant and animal lineages (Allen et al. 2004). Molnar et al. isolated more than 2,000 nonredundant sRNAs and identified many thousand candidate genes of origin. They then established that at least some of the miRNAs they had characterized were capable of directing site-specific cutting of target mRNAs encoding known proteins, with the cleavage being consistent with the action of an Argonaute nuclease.

Zhao et al. (2013) isolated several thousand small RNAs of unique sequences in *Chlamydomonas*, of which about twenty were judged to be miRNAs or candidate-miRNAs and were studied further. They were found to be capable of directing cleavage of their target sequences, and while some were more abundant, others were less abundant in gametes than in vegetative cells. None of these RNAs had sequence homologs in *Ostreococcus* (another unicellular green alga), in *V. carteri*, or in land plants or animals.

Subsequently, this same research group performed parallel studies with *V. carteri* RNAs (Li et al. 2014) and characterized 174 miRNAs in 160 different families. They then used methods similar to those used with *C. reinhardtii* to identify many potential target mRNAs that encode proteins involved in a variety of metabolic pathways and found evidence of miRNA-directed mRNA cleavage in 60% of the 243 potential target mRNAs that were studied. Only one *Volvox* miRNA exhibited a significant degree of sequence similarity to a *Chlamydomonas* miRNA.

Studies of the relative abundance of various miRNAs in somatic cells and gonidia yielded interesting results: Of the 99 miRNAs studied, 50 were more abundant in somatic cells, and 49 were more abundant in gonidia. In most cases, these distribution asymmetries were relatively modest, but nearly a dozen miRNAs were found to be 10–20 times more abundant in one cell type than in the other. Further study of such asymmetrically distributed miRNAs is likely to be rewarding.

A more recent study of *V. carteri* took a slightly different approach by cloning and sequencing small RNAs that coimmunoprecipitated with the *V. carteri* Argonaute-3 protein, which led to the identification of 490 members of 324 miRNA families (Dueck e al. 2016). The genomic sources of these RNAs were highly varied, including a number from known transposons (*Jordan* and *Kangaroo*), others from protein-coding genes (both sense and antisense strands), intergenic regions, repetitive elements, and so on. As in the preceding study, some of these miRNAs were found to be more abundant in somatic cells, while others were more abundant in gonidia or in eggs. A global comparison of *V. carteri and C. reinhardtii* miRNA sequences with multiple sequence alignments revealed essentially no conservation of sequences between the two species. Rather similar observations were made with respect to the other classes of functional small RNAs. In summary, the authors state, "Taken together, our data identify an extended small RNA system in *V. carteri*, which appears to be as complex as in higher plants."

What role these small RNAs play in *V. carteri* development remains to be determined. But the fact that the small RNAs of *Volvox* are almost entirely different from those of *Chlamydomonas* raises the intriguing possibility that diversification of this category of gene-expression regulators may have played a crucial role in the evolution of volvocine multicellularity. It is hoped that such a hypothesis will be tested soon.

I A THIRD VOLVOCINE GENOME SEQUENCE PROVIDES AS MANY NEW QUESTIONS AS ANSWERS

The genome of a third volvocine alga, *Gonium pectorale* (one of the smallest colonial volvocine algae, Figure 4.8) has recently been sequenced (Hanschen et al. 2016). It is quite similar to the *Chlamydomonas* and *Volvox* genomes in terms of size, number of coding loci, number of introns per gene, and so forth (although all such numbers fluctuate to some extent as additional analyses of the genomes are performed with more sensitive or more stringent methods: Goodstein et al. 2011, Umen and Olson 2012, Hanschen et al. 2016).

The difference in the number of pherophorin-coding genes in these three genomes (31 in *Chlamydomonas*, 35 in *Gonium*, versus 78 in *Volvox*) is consistent with the fact that *Gonium* colonies produce a bit more ECM than *Chlamydomonas* does, but a great deal less than *Volvox*. A rather surprising finding was that *Gonium* and *Volvox* (the two multicellular forms) share far fewer volvocine-specific genes with one another (9) than either of them shares with *Chlamydomonas* (32 and 44, respectively; see Figure 4.10). This reinforces the notion that the evolution of multicellularity in this group appears not to have required a substantial number of new genes.

According to the title of the *Gonium* genome paper (Hanschen et al. 2016), as well as several statements throughout the text, what the evolution of multicellularity did require was "co-option of the RB cell cycle regulatory pathway." The RB cell-cycle pathway was analyzed in great detail in *C. reinhardtii* and shown to be in control of the unusual cell division pattern known as multiple fission, or palintomy (Olson et al. 2010). As noted above, multiple fission is also a hallmark of all the

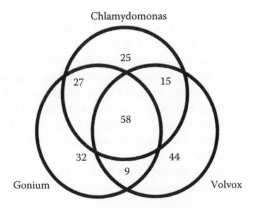

FIGURE 4.10 A Venn diagram indicating the numbers of volvocine-specific (i.e., "new") genes that are found in one, two or three of the sequenced volvocine genomes.

colonial volvocaceans, and all components of the RB pathway have been found to be present and functional in both *Gonium* and *Volvox* (Hanschen et al. 2016). So it appears that the volvocaceans have used the RB pathway that they inherited from their *C. reinhardtii*-like ancestor and continue to use it pretty much "as is," for its established function in controlling multiple fission. Whether that is an example of co-option, or simply of inheritance, is a moot point.

In any case, it is rather curious that in a paper that claims cell-cycle regulation plays such a centrally important evolutionary role, no mention is made of any attempt to evaluate cell-cycle parameters. Are the cell-cycle parameters of *Gonium* significantly different from those of *Chlamydomonas*? And when *Chlamydomonas* is transformed with the *Gonium* Rb gene (as will be discussed in the next section) do any of its cell-cycle parameters change? Such seemingly important questions are not addressed.

J ATTEMPTS TO EXPERIMENTALLY INDUCE VOLVOCINE MULTICELLULARITY

Three attempts to experimentally induce volvocine multicellularity have been reported in recent years, and are worthy of discussion here. However, we first need to discuss a phenomenon that is all too familiar to most *Chlamydomonas* investigators: the so-called palmelloid state. When an actively swimming population of *Chlamydomonas* cells in liquid culture is stressed in any one of a number of ways, the cells tend to resorb their flagella and secrete mucilage, which binds the cells together into amorphous clumps containing anywhere from few to several hundred immotile cells per clump (Schlösser 1976, Kirk 1998). This is called the palmelloid state because it resembles the normal growth form of a rather distantly related green alga, named *Palmella*. Some species of *Chlamydomonas* other than *C. reinhardtii*, alternate between the active-swimming phase and the palmelloid phase in every cell cycle (Schlösser 1976). But a few would assert that switching from the actively mobile to the palmelloid state is equivalent to having evolved multicellularity.

The assertion by Hanschen et al. (2016) that co-option of the RB pathway played a critical role in the evolution of volvocine multicellularity rests heavily on their observation that cell clusters (called "colonies") can be found in cultures of *C. reinhardtii* that had been transformed with the *Gonium* RB-encoding gene. The clusters varied in size from 2 to 16 cells, the four examples that were illustrated with very small, low-resolution photographs look different from one another and do not resemble any known colonial volvocacean. The authors in one sentence use the term "non-palmelloid colonies" for these cell clumps, but never indicate what criteria, if any were used to justify the use of the term "non-palmelloid." They do not indicate whether the clusters of transformed cells exhibited either of the two important differences that regularly distinguish dividing *Gonium* cells from dividing *Chlamydomonas* cells, namely: cytoplasmic bridges between sister cells that are the result of incomplete cytokinesis, and (somewhat later) attachments between neighboring cell walls (Stein 1958, Iida et al. 2013). Clumps of cells, which, as noted above, often develop in stressed *Chlamydomonas* cultures, do not necessarily constitute colonies.

Earlier, Ratcliff et al. (2013) had reported that multicellular variants of *C. reinhardtii* were generated by selecting for rapidly settling individuals in each transfer generation. To be more specific, for each three-day period of cultivation of *C. reinhardtii* in static liquid medium, the investigators selected for transfer to fresh medium cells located at the bottom of a tube that had been centrifuged briefly. Twenty parallel cultures were subjected to this serial-transfer protocol for 219 days. By the end of that period, one of the twenty cultures had established a population of cell clusters that were so large that they would settle rapidly to the bottom of the tube under earth's gravitational field. These amorphous clusters contained hundreds of immotile cells trapped in a transparent matrix. They bear no resemblance to any of the recognized genera of colonial volvocaceans. But what they do resemble is the palmelloid phase described just above. Time will tell how significant the results of this centrifugal-selection protocol are.

Meanwhile, at a recent international *Volvox* conference, Herron (one of the coauthors of the preceeding study) reported successful production of multicell versions of *C. reinhardtii* with a different selection scheme, namely: cocultivation with the predatory ciliate, *Paramecium* (Herron 2016). Expanding on the published note, he told me in a personal note that:

> "...the evolved isolates from the predation experiment look quite different from the ones from the centrifugation experiment. Instead of large, amorphous clusters of up to a couple of hundred cells (as in the centrifugation experiment), we see smaller, more structured clusters of 4, 8, 16, or 32 cells. Probably they result from a simple failure of daughter cells to escape from the mother-cell wall, sometimes for two generations (i.e., we sometimes see 'superclusters' made up of four 4-celled clusters). Most look a lot like Pandorina or the like, but they can't swim and almost certainly don't invert."
> (M. D. Herron, pers. commun., quoted verbatim, with permission.)

Photographs that Herron provided with that note appear to confirm his interpretation that these individuals "result from a simple failure of daughter cells to escape from the mother-cell wall." Failure to escape from the mother cell wall (as *C. reinhardtii* daughter cells normally do right after they have completed their last division), presumably makes the clusters too large to be consumed by *Paramecium*. But it hardly qualifies them as newly evolved multicellular organisms.

One of the major selective advantages of the transition from unicellularity-to-multicellularity that occurred in the volvocine lineage some 200 million years ago may very well have resulted from an increase in organismic size, which greatly reduced predation pressures. Nevertheless, it will very likely take a bit more than cocultivation with a predator such as a *Paramecium* to duplicate that historic transition in a modern laboratory.

4.5 DISCUSSION

We have considered how multicellularity may have evolved in three major groups of life, the Amoebozoa (CSMs), Opisthokonta (unicellular holozoans and metazoans) and Archaeplastida (volvocine green algae), where it must be stressed that the studied examples come from very few species. A number of tentative inferences can be drawn from features that are common to the three cases. (1) Most important, perhaps, is this: naïve ideas of what is simple (="primitive") and what is complex (="evolved"), primarily based on morphology, bear no relation to what are categorized on the basis of DNA-based phylogeny as ancestral and derived states. To repeat, *grades* of organizational complexity need not necessarily reflect *clades* of closest relatives. The inference hinges entirely on the assumption that deduced phylogenetic relationships reflect the true phylogeny, which is supposed to be based on descent with (possibly) modification. It would be invalid if lateral gene exchange is common among the taxa in question, something which does not appear to be the case as far as we know. (2) A related inference is of phenotypic plasticity. Cells with the same genome or similar genomes can become multicellular in more than one way, or go through multicellular phases differently, or display a variety of multicellular forms. In the CSMs, the same species occasionally mimics what was believed to be a different genus. The evidence from the unicellular holozoans is not as direct (though further studies may change the picture): choanoflagellates form clonal colonies, filastereans aggregate and teretosporeans form a coenocyte (therefore, strictly speaking, are unicellular but multinucleate).

In contrast, although cells of the prototypical unicellular volvocine alga, *Chlamydomonas*, can be caused to form loose aggregates under various experimental conditions, there is no evidence that such aggregates have ever played any role in the origins of true volvocine multicellularity. In every case that has been studied, multicellular volvocine algae arise by a failure of mitotic sister cells to separate fully at the end of the cell-division cycle, rather than by aggregation of free-living cells. Also, volvocine algae provide a dramatic example of temporal differentiation giving way to spatial differentiation beyond a critical size (=number of cells). The CSMs too show size-dependent morphologies and developmental patterns, though not as strikingly.

Given that single-celled ancestors seem to have possessed many of the protein-coding genes that were believed to be specific to metazoans, the evolutionary transitions to multicellularity may have been potentiated by minor changes in patterns of gene regulation. All that may have been required for a unicellular form to 'go multicellular' may have been an environmental trigger (e.g., an increase in atmospheric oxygen content) that permitted size increase that, among other things, was a defense against predation (Bonner 1998, 2001; Knoll 2011). Alternatively, environmental changes may have fostered multicellular forms arising on the basis of preexisting cellular interaction systems; genetic changes may have arisen secondarily by way of ensuring developmental reliability. In a subset of those cases in which embryonic development arose as well (i.e., embryophytes and metazoans), there could be a combination of all those causes, as well as the evolution of new major genomic regulatory capabilities, such as distal regulation (Sebé-Pedrós et al. 2016a). Clearly, additional data from more taxa, a better appreciation of the range of developmental forms consistent with a single genotype and an increased knowledge of the molecular basis of phenotypic plasticity will advance our understanding of how different unicellular organisms became multicellular.

ACKNOWLEDGMENTS

We wish to acknowledge the assistance of Sunil Laxman, Gavriel Matt, Pauline Schaap, Rüdiger Schmitt, and James Umen who commented on previous drafts of parts or all of this chapter. A special word of thanks to Matthew Brown for permitting the use of illustrations from his thesis, shown here as Figure 4.2. IRT acknowledges financial support from an European Research Council Consolidator grant (ERC-2012-Co -616960), and a grant (BFU2014-57779-P) from Ministerio de Economía y Competitividad (MINECO), the latter with FEDER funds.

REFERENCES

Alegado, R. A., Brown, L. W., Cao, S., Dermenjian, R. K., Zuzow, R., Fairclough, S. R., Clardy, J., and King, N. 2012. A bacterial sulfonolipid triggers multicellular development in the closest living relatives of animals. *Elife* 1: e00013. doi:10.7554/eLife.00013.

Allen, E., Xie, Z., Gustafson, A. M., and Carrington, J. C. 2004. microRNA-directed phasing during trans-acting siRNA biogenesis in plants. *Cell* 121: 207–221.

Arias Del Angel, J. A., Escalante, A. E., Martínez-Castilla, L. P., and Beniítez, M. 2017. An evo-devo perspective on multicellular development of myxobacteria. *J Exp Zool (Mol Dev Evol)* 328B: 165–178.

Atzmony, D., Zahavi, A., and Nanjundiah, V. 1997. Altruistic behaviour in *Dictyostelium discoideum* explained on the basis of individual selection. *Curr Sci* 72(2): 142–145.

Avesson, L., Schumacher, H. T., Fechter, P., Romby, P., Hellman, U., and Söderbom, F. 2011. Abundant class of non-coding RNA regulates development in the social amoeba *Dictyostelium discoideum*. *RNA Biol.* 8: 1094–1104. doi:10.4161/rna.8.6.17214.

Bartel, D. P. 2004. MicroRNAs: Genomics, biogenesis, mechanism, and function. *Cell* 116: 281–289.

Bell, G. 1985. The origin and early evolution of germ cells, as illustrated by the Volvocales. In *The Origin and Evolution of Sex* (Eds.) H. O. Halverson and A. Monroy, pp. 221–256. New York: Alan R. Liss.

Benabentos, R., Shigenori, H., Sucgang, R., Curk, T., Katoh, M., Ostrowski, E. A., Strassmann, J. E. et al. 2009. Polymorphic members of the lag gene family mediate kin discrimination in *Dictyostelium*. *Curr Biol* 19: 567–572.

Bonner, J. T. 1967. *The Cellular Slime Molds*, 2nd ed. Princeton, NJ: Princeton University Press.

Bonner, J. T. 1982. Evolutionary strategies and developmental constraints in the cellular slime molds. *Am Nat* 119: 530–552.

Bonner, J. T. 1998. The origins of multicellularity. *Integr Biol* 1: 27–36.

Bonner, J. T. 2001. *First Signals: The Evolution of Multicellular Development*. Princeton, NJ: Princeton University Press.

Bonner, J. T. 2003. Evolution of development in the cellular slime molds. *Evol Dev* 5: 305–313.

Bonner, J. T. 2013a. The evolution of the cellular slime molds. In *Dictyostelids: Evolution, Genomics, and Cell Biology* (Eds.) M. Romeralo, S. Baldauf, and R. Escalante, pp. 183–191. Berlin, Germany: Springer.

Bonner, J. T. 2013b. *Randomness in Evolution*. Princeton, NJ: Princeton University Press.

Bonner, J. T. 2015. The evolution of evolution: Seen through the eyes of a slime mold. *BioScience* 65: 1184–1187.

Bonner, J. T. and Adams, M. S. 1958. Cell mixtures of different species and strains of cellular slime moulds. *J Embryol Exp Morphol* 6: 346–356.

Bonner, J. T. and Dodd, M. R. 1962. Aggregation territories in the cellular slime molds. *Biol Bull* 122: 13–24.

Boraas, M. E., Seale, D. B., and Boxhorn, J. E. 1998. Phagotrophy by a flagellate selects for colonial prey: A possible origin of multicellularity. *Evol Ecol* 12: 153–164. doi:10.1023/A:1006527528063.

Brown, M. W. 2010. Placing the forgotten slime molds (*Sappinia, Copromyxa, Fonticula, Acrasis*, and *Pocheina*), using molecular phylogenetics. 258p. PhD thesis dissertation, University of Arkansas, Fayetteville, AR.

Brown, M. W. and Silberman, J. D. 2013. The non-dictyostelid sorocarpic amoebae. In *Dictyostelids. Evolution, Genomics and Cell Biology* (Eds.) M. Romeralo, S. Baldauf, and R. Escalante, pp. 219–242. Heidelberg, Germany: Springer.

Brown, M. W., Silberman, J. D., and Spiegel, F. W. 2010. A morphologically simple species of *Acrasis* (Heterolobosea, Excavata), *Acrasis helenhemmesae* n. sp. *J Eukaryot Microbiol* 57: 346–353.

Brown, M. W., Spiegel, F. W., and Silberman, J. D. 2009. Phylogeny of the "forgotten" cellular slime mold, *Fonticula alba*, reveals a key evolutionary branch within Opisthokonta. *Mol Biol Evol* 26: 2699–2709.

Buchheim, M. A. and Chapman, R. L. 1991. Phylogeny of the colonial green flagellates: A study of 18S and 26S rRNA sequence data. *BioSystems* 25: 85–100.

Burki, F. 2014. The eukaryotic tree of life from a global phylogenomic perspective. *Cold Spring Harb Perspect Biol* 6: a016147.

Carr, M., Leadbeater, B. S. C., Hassan, R., Nelson, M., and Baldauf, S. L. 2008. Molecular phylogeny of choanoflagellates, the sister group to metazoa. *Proc Natl Acad Sci USA* 105: 16641–16646. doi:10.1073/pnas.0801667105.

Carrington, J. C. and Ambros, V. 2003. Role of microRNAs in plant and animal development. *Science* 301: 336–338.

Carroll, S. B. 2008. Evo-Devo and an expanding evolutionary synthesis: A genetic theory of morphological evolution. *Cell* 134: 25–36.

Cheng, Q., Fowler, R., Tam, L.-W., Edwards, L., and Miller, S. M. 2003. The role of GlsA in the evolution of asymmetric division in the green alga *Volvox carteri*. *Devel Genes & Evol* 213: 328–335.

Coates, J. C. and Harwood, A. H. 2001. Cell-cell adhesion and signal transduction during *Dictyostelium* development. *J Cell Sci* 114: 4349–4358.

Coates, J. C., Grimson, M. J., Williams, R. S., Bergman, W., Blanton, R. L., and Harwood, A. J. 2002. Loss of the beta-catenin homologue aardvark causes ectopic stalk formation in *Dictyostelium Mech Dev* 116: 117–127.

Coleman, A. W. 1980. The biological species concept: Its applicability to the taxonomy of freshwater algae. In *Proceedings of the 2nd International Symposium on Taxonomy of Algae* (Ed.) T. V. Desikachary, pp. 22–36. Chennai, India: University of Madras.

Coleman, A. W. 1999. Phylogenetic analysis of "Volvocaceae" for comparative genetic studies. *Proc Natl Acad Sci USA* 96: 13892–13897.

Conte, D., MacWilliams, H. K., and Ceccarelli, A. 2010. BTG Interacts with retinoblastoma to control cell fate in *Dictyostelium*. *PLoS One* 5(3): e9676. doi:10.1371/journal.pone.0009676.

Cornillon, C., Gebbie. L., Benghezal, M., Nair, P., Keller, S., Wehrle-Haller, B., Charette, S. J., Brückert, F., Letourneur, F., and Cosson, P. 2006. An adhesion molecule in free-living *Dictyostelium* amoebae with integrin β features. *EMBO Reps* 7: 617–621. doi:10.1038/sj.embor.7400701.

Crow, W. B. 1918. The classification of some colonial Chlamydomonads. *New Phytol* 17: 151–159.

Danchin, E. G. J. 2016. Lateral gene transfer in eukaryotes: Tip of the iceberg or of the ice cube? *BMC Biol* 14: 101. doi:10.1186/s12915-016-0315-9.

Dayel, M. J., Alegado, R. A., Fairclough, S. R., Levin, T. C., Nichols, S. A., McDonald, K., and King, N. 2011. Cell differentiation and morphogenesis in the colony forming *Choanoflagellate Salpingoeca* Rosetta. *Dev Biol* 35: 73–82. doi:10.1016/j.ydbio.2011.06.003.

de Mendoza, A., Suga, H., Permanyer, J., Irimia, M., and Ruiz-Trillo, I. 2015. Complex transcriptional regulation and independent evolution of fungal-like traits in a relative of animals. *Elife* e08904. doi:10.7554/eLife.08904.

De Monte, S. and Rainey, P. B. 2014. Nascent multicellular life and the emergence of individuality. *J Biosci* 39: 237–248.

Dickinson, D. J., Nelson, W. J., and Weis, W. I. 2012. An epithelial tissue in *Dictyostelium* and the origin of metazoan multicellularity. *Bioessays* 34: 833–840. doi:10.1002/bies.201100187.

Dubravcic, D., van Baalen, M., and Nizak, C. 2014. An evolutionarily significant unicellular strategy in response to starvation stress in *Dictyostelium* social amoebae [version 1; referees: 2 approved]. *F1000Research* 3: 133. doi:10.12688/f1000research.4218.1.

Dueck, A., Evers, M., Henz, S. R., Unger, K., Eichner, N., Merkl, R., Berezikov, E. et al. 2016. Gene silencing pathways found in the green alga *Volvox carteri* reveal insights into evolution and origins of small RNA systems in plants. *BMC Genomics* 17: 853–867.

Fairclough, S. R., Chen, Z., Kramer, E., Zeng, Q., Young, S., Robertson, H. M., Begovic, E. et al. 2013. Premetazoan genome evolution and the regulation of cell differentiation in the cho-anoflagellate *Salpingoeca rosetta*. *Genome Biol* 14: R15. doi:10.1186/gb-2013-14-2-r15.

Filosa, M. F. 1962. Hetercytosis in cellular slime molds. *Am Nat* 96: 79–91.

Gilbert, O. M., Foster, K. R., Mehdiabadi, N. J., Strassmann, J. E., and Queller, D. C. 2007. High relatedness maintains multicellular cooperation in a social amoeba by controlling cheater mutants. *Proc Natl Acad Sci USA* 104: 8913–8917.

Gilbert, O. M., Queller, D. C., and Strassmann, J. E. 2009. Discovery of a large clonal patch of a social amoeba: Implications for social evolution. *Mol Ecol* 18: 1273–1281.

Glockling, S. L., Marshall, W. L., and Gleason, F. H. 2013. Phylogenetic interpretations and ecological potentials of the Mesomycetozoea (Ichthyosporea). *Fungal Ecol* 6: 237–247. doi:10.1016/j.funeco.2013.03.005.

Glöckner, G., Lawal, H. M., Felder, M., Singh, R., Singer, G., Weijer, C. J., and Schaap, P. 2016. The multicellularity genes of dictyostelid social amoebas. *Nature Communications* 7: 12085, 1–11. doi:10.1038/ncomms12085.

Goodstein, D. M., Shu, S., Howson, R., Neupane, R., Hayes, R. D., Fazo, J., Mitros T. et al. 2011. Phytozome: A comparative platform for green plant genomics. *Nucl Acid Res* 40: D1178–D1186.

Grosberg, R. K. and Strathman, R. R. 2007. The evolution of multicellularity: A minor major transition? *Annu Rev Ecol Evol Syst* 38: 621–654.

Hallmann, A. 2006. The pherophoriins: Common building blocks in the evolution of extra-cellular matrix architecture in volvocales. *Plant J* 45: 292–307.

Hallmann, A. and Kirk, D. L. 2000. The developmentally regulated ECM glycoprotein ISG plays an essential role in organizing the ECM and orienting the cells of *Volvox*. *J Cell Sci* 113: 4605–4617.

Hammerschmidt, K., Rose, C. J., Kerr, B., and Rainey, P. B. 2014. Life cycles, fitness decoupling and the evolution of multicellularity. *Nature* 515: 75–79. doi:10.1038/nature13884.

Hanschen, E. R., Marriage, T. N., Ferris, P. J., Hamaji, T., Toyoda, A., Fujiyama, A., Neme, R. et al. 2016. The *Gonium pectorale* genome demonstrates co-option of cell cycle regulation during the evolution of multicellularity. *Nature Commun* 7: 11370. doi:10:1038/ncomms11370.

Herron, M. D. 2016. Origins of multicellular complexity: *Volvox* and the volvocine algae. *Mol Ecol* 25: 1213–1223.

Herron, M. D. and Michod, R. E. 2008. Evolution of complexity in the volvocine algae: Transitions in individuality through Darwin's eye. *Evolution* 62: 436–451.

Herron, M. D., Desnitsky, A. G., and Michod, R. E. 2010. Evolution of developmental programs in *Volvox* (Chlorophyta). *J Phycol* 46: 316–324.

Herron, M. D., Hackett, J. D., Aylward, F. O., and Michod, R. E. 2009. Triassic origin and early radiation of multicellular volvocine algae. *Proc Natl Acad Sci USA* 106: 3254–3258.

Iida, H., Ota, S., and Inouye, I. 2013. Cleavage, incomplete inversion, and cytoplasmic bridges in *Gonium pectorale* (Volvocales, Chlorophyta). *J Plant Res* 126: 699–707.

Isaka, N., Kawai-Toyooka, H., Matuzaki, R., Nakada T., and Nozaki, H. 2012. Description of two new monoecious species of *Volvox* sect. *Volvox* (Volvocaceae, Chlorophyceae), based on comparative morphology and molecular phylogeny of cultured material. *J Phycol* 48: 759–767.

Jack, C. N., Ridgeway, J. G., Mehdiabadi, N. J., Jones, E. I., Edwards, T. A., Queller, D. C., and Strassmann, J. E. 2008. Segregate or cooperate—A study of the interaction between two species of *Dictyostelium*. *BMC Evol Biol* 8: 293. doi:10.1186/1471-2148-8-293.

Jahn, T. L. and Bovee, E. C. 1967. Motile behaviour of protozoa. In *Research in Protozoology* (Ed.) T.-T Chen. Oxford, UK: Pergamon Press.

Kawata, T. 2011. STAT signaling in *Dictyostelium* development. *Develop Growth Differ* 53: 548–557. doi:10.1111/j.1440-169X.2010.01243.

Kaneko, K. 2016. A scenario for the origin of multicellular organisms: Perspective from multilevel consistency dynamic. In *Multicellularity: Origins and Evolution* (Eds.) K. J. Niklas and S. A. Newman, pp. 201–223. Cambridge, MA: MIT Press.

Kapsetaki, S. E. 2015. Predation and the evolution of multicellularity. MSc thesis, University of Oxford. http://zoo-web02.zoo.ox.ac.uk/group/west/pdf/SteffiMastersThesis.pdf. Accessed March 3, 2018.

Kaushik, S. and Nanjundiah, V. 2003. Evolutionary questions raised by cellular slime mould development. *Proc Ind Natl Sci Acad* B69: 825–852.

Kaushik, S., Katoch, B., and Nanjundiah, V. 2006. Social behaviour in genetically heterogeneous groups of *Dictyostelium giganteum*. *Behav Ecol Sociobiol* 59: 521–530.

Kessin, R. H. 2001. *Dictyostelium—Evolution, Cell Biology, and the Development of Multicellularity*. Cambridge, UK: Cambridge University Press.

King, M. and Wilson, A. 1975. Evolution at to levels in humans and chimpanzees. *Science* 188: 107116.

King, N., Westbrook, M. J., Young, S. L., Kuo, A., Abedin, M., Chapman, J., Fairclough, S. et al. 2008. The genome of the Choanoflagellate *Monosiga brevicollis* and the origin of metazoans. *Nature* 451: 783–788. doi:10.1038/nature06617.

King, N. 2004. The unicellular ancestry of animal development. *Dev Cell* 7: 313–325.

King, N. 2005. Choanoflagellates. *Curr Biol* 15: R113–114.

Kirk, D. L. 1998. *Volvox: Molecular Genetic Origins of Multicellularity and Cellular Differentiation*. Cambridge, UK: Cambridge University Press.

Kirk, D. L. 2005. A twelve-step program for evolving multicellularity and a division of labor. *BioEssays* 27: 299–310.

Kirk, D. L., Kirk M. M., Stamer, K., and Larson, A. 1990. The genetic basis for the evolution of multicellularity and cellular differentiation in the volvocine green algae. In *The Unity of Evolutionary Biology* (Ed.) E. C. Dudley, pp. 568–581. Portland, OR: Discorides Press.

Kirk, M. M. and Kirk, D. L. 1985. Translational regulation of protein synthesis, in response to light, at a critical stage of development. *Cell* 41: 419–428.

Knoll, A. H. 2011. The multiple origins of complex multicellularity. *Ann Rev Earth Planet Sci* 39: 217–239.

Koufopanou, V. 1994. The evolution of soma in the volvocales. *Amer Nat* 143: 907–931.

Lang, B. F., O'Kelly, C., Nerad, T., Gray, M. W., and Burger, G. 2002. The closest unicellular relatives of animals. *Curr Biol* 12: 1773–1778.

Larson, A., Kirk, M. M., and Kirk, D. L. 1992. Molecular phylogeny of the volvocine flagellates. *Mol Biol Evol* 9: 85–105.

Leadbeater, B. S. C. 2015. *The Choanoflagellates*. Cambridge, UK: Cambridge University Press. doi:10.1017/CBO9781139051125.

Li, J. R., Wu, Y., and Qi, Y. 2014. MicroRNAs in a multicellular green alga *Volvox carteri*. *Chin Sci Life Sci* 57: 36–45.

Loomis, W. F. 2014. Cell signaling during development of *Dictyostelium*. *Dev Biol* 391: 1–16. doi:10.1016/j.ydbio.2014.04.001.

Loomis, W. F. 2015. Genetic control of morphogenesis in *Dictyostelium*. *Dev Biol* 402: 146–161. doi:10.1016/j.ydbio.2015.03.016.

Mah, J. L., Christensen-Dalsgaard, K. K., and Leys, S. P. 2014. Choanoflagellate and Choanocyte collar-flagellar systems and the assumption of homology. *Evol Dev* 16: 25–37. doi:10.1111/ede.12060.

Marée, A. F. M. 2000. From pattern formation to morphogenesis. Multicellular coordination in Dictyostelium discoideum PhD thesis, Department of Theoretical Biology and Bioinformatics, Utrecht University, the Netherlands. http://theory.bio.uu.nl/stan/Thesis/.

Marée, A. F. M. and Hogeweg, P. 2001. How amoeboids self-organize into a fruiting body: Multicellular coordination in *Dictyostelium discoideum*. *Proc Natl Acad Sci USA* 98: 3879–3883. doi:10.1073/pnas.061535198.

Marshall, W. L., Celio, G., McLaughlin, D. J., and Berbee, M. L. 2008. Multiple isolations of a culturable, motile Ichthyosporean (Mesomycetozoa, Opisthokonta), *Creolimax fragrantissima* N. Gen., N. Sp., from marine invertebrate digestive tracts. *Protist* 159: 415–433.

Matt, G. and Umen, J. 2016. *Volvox*: A simple algal model for embryogenesis, morphogenesis and cellular differentiation. *Dev Biol* 419: 99–113.

McMains, V. C., Myre, M., Kreppel, L., and Kimmel, A. R. 2010. *Dictyostelium* possesses highly diverged presenilin/γ-secretase that regulates growth and cell-fate specification and can accurately process human APP: A system for functional studies of the presenilin/γ-secretase complex. *Dis Models Mech* 3: 581–594. doi:10.1242/dmm.004457.

Medina, M., Collins, A. G., Taylor, J. W., Valentine, J. W., Lipps, J. H., Amaral-Zettler, L., and Sogin, M. L. 2003. Phylogeny of opisthokonta and the evolution of multicellularity and complexity in fungi and metazoa. *Int J Astrobiol* 2: 203–211.

Melkonian, M. 1990. Phylum chlorophyta. In *Handbook of Protoctista* (Eds.) L. Margulis, J. O. Corliss, M. Melkonian, and D. J. Chapman, pp. 597–660. Boston, MA: Jones & Bartlett.

Mendoza, L., Taylor, J. W., and Ajello, L. 2002. The class mesomycetozoea: A Heterogeneous group of microorganisms at the animal-fungal boundary. *Annu Rev Microbiol* 56: 315–344.

Merchant, S. S., Prochnik, S. E., Vallon, O., Harris, E. H., Karpowicz, S. J., Witman, G. B., Terry, A. et al. 2007. The *Chlamydomonas* genome reveals the evolution of key animal and plant functions. *Science* 318: 245–250.

Miller, S. M. and Kirk, D. L. 1999. *glsA*, a *Volvox* gene required for asymmetric division and germ cell specification, encodes a chaperone-like protein. *Development* 126: 649–658.

Mishra, H. and Saran, S. 2015. Classification and expression analyses of homeobox genes from *Dictyostelium discoideum*. *J Biosci* 40: 241–255. doi:10.1007/s12038-015-9519-3.

Molnar, A., Schwach, F., Studholme, D. J., Thunemann, E. C., and Baulcombe, D. C. 2007. miRNAs control gene expression in the single-cell alga *Chlamydomonas reinhardtii*. *Nature* 447: 1126–1129.

Nakada, T., Misawa, K., and Nozaki H. 2008. Molecular systematics of Volvocales (Chlorophyceae, Chlorophyta) based on exhaustive 18S rRNA phylogenetic analysis. *Mol Phylogenet Evol* 48: 281–291.

Nanjundiah, V. 1985. The evolution of communication and social behaviour in Dictyostelium discoideum. *Proc Indian Acad Sci* 94: 639–653.

Nanjundiah, V. 2016. Cellular slime mold development as a paradigm for the transition from unicellular to multicellular life. In *Multicellularity: Origins and Evolution* (Eds.) K. J. Niklas and S. A. Newman, pp. 105–130. Cambridge, MA: MIT Press.

Nanjundiah, V. and Sathe, S. 2011. Social selection and the evolution of cooperative groups: The example of the cellular slime molds. *Integr Biol* 3: 329–342.

Newman, S. A. 2002. Putting genes in their place. *J Biosciences* 27: 97–104.

Newman, S. A. 2011. Animal egg as evolutionary innovation: A solution to the "embryonic hourglass" puzzle. *J Exp Zool B Mol Dev Evol* 316: 467–483. doi:10.1002/jez.b.21417.

Newman, S. A. and Bhat, R. 2009. Dynamical patterning modules: A "pattern language" for development and evolution of multicellular form. *Int J Dev Biol* 53: 693–705.

Newman, S. A. and Comper, W. D. 1990. "Generic" physical mechanisms of morphogenesis and pattern formation. *Development* 110: 1–18.

Nishii, I. and Miller, S. M. 2010. *Volvox*: Simple steps to developmental complexity? *Curr Opin Plant Biol* 13: 646–653.

Nishii, I., Ogihara, S., and Kirk, D. L. 2003. A kinesin, invA, plays an essential role in *Volvox* morphogenesis. *Cell* 113: 743–753.

Nozaki, H. 2003. Origin and evolution of the genera *Pleodorina* and *Volvox* (Volvocales). *Biologia (Bratislava)* 58: 425–431.

Nozaki H. and Coleman, A. W. 2011. A new species of *Volvox* sect Merrillosphaera (Volvocales, Chlorophyceae) from Texas. *J Phycol* 47: 673–679.

Nozaki, H., Matsuzaki, R., Yamamoto, K., Kawachi, M., and Takahashi, F. 2015a. Delineating a new heterothallic species of Volvox (Volvocaceae, Chlorophyceae) using new strains of "*Volvox africanus*". *PLoS One* 10(11): e0142632. doi:10.1371/journal.pone.0142632.

Nozaki, H., Misawa, K., Kajita, T., Kato, M., Nohara, S., and Watanabe, M. M. 2000. Origin and evolution of the colonial Volvocales (Chlorophyceae) as inferred from multiple chloroplast gene sequences. *Mol Phylogenet Evol* 17: 256–268.

Nozaki, H., Ueki, N., Isaka, N., Saigo, T., Yamamoto, K., Matsuzaki, R., Takahashi, F., Wakabayashi, K. I. and Kawachi, M. 2015b. A new morphological type of *Volvox* from Japanese large lakes and recent divergence of this type and *V. ferrisii* in two different freshwater habitats. *PLoS One* 11(11): e0167148. doi:10.1371/journal.pone.0167148.

Nozaki, H., Ueki, N., Isaka, N., Saigo, T., Yamamoto, K., Matsuzaki, R. et al. 2016. A new morphological type of Volvox from Japanese large lakes and recent divergence of this type and V. ferrisii in two different freshwater habitats. *PLoS One* 11: e0167148. doi:10.1371/journal.pone.0167148.

Nozaki, H., Yamada, T. K., Takahashi, F., Matsuzaki, R., and Nakada, T. 2014. New "missing link" genus of the colonial volvocine green algae gives insights into the evolution of oogamy. *BMC Evol Biol* 14: 37–46.

Olive, L. S. 1965. A developmental study of *Guttulinopsis vulgaris* (Acrasiales). *Am J Bot* 52: 513–519. doi:10.2307/2440268.

Olive, L. S. and Stoianovitch, C. 1960. Two new members of the Acrasiales. *Bull Torrey Bota Club* 87: 1–20.

Olive, L. S. 1975. *The Mycetozoans*. New York: Academic Press.

Olson, B. J. S. C. and Nedelcu, A. 2016. Co-option during the evolution of multicellular and developmental complexity in the volvocine green algae *Curr Opin Genet Develop* 39: 107–115.

Olson, B. J. S. C., Oberholzer, M., Li, Y., Zones, J. M., Kohli, H. S., Bisova, K., Fang, S. C., Meisenhelder, J., Hunter, T., and Umen, J. G. 2010. Regulation of the *Chlamydomonas* cell cycle by a stable, chromatin-associated retinoblastoma tumor suppressor complex. *Plant Cell* 22: 3331–3347.

Paps, J. and Ruiz-Trillo, I. 2010. Animals and their unicellular ancestors. In *Encyclopedia of Life Sciences* (*ELS*). doi:10.1002/9780470015902.A0022853.

Parfrey, I. W., Lehr, D. J., Knoll, A. H., and Katz, L. A. 2011. Estimating the timing of early eukaryotic diversification with multigene molecular clocks. *Proc Natl Acad Sci USA* 108: 13624–13649.

Peterson, K. J., Dietrich, M. R., and McPeek, M. A. 2009. MicroRNAs and metazoan macroevolution: Insights into canalization, complexity, and the Cambrian explosion. *BioEssays* 31: 736–747.

Pickup, Z. L., Pickup, R., and Parry, J. D. 2007. A comparison of the growth and starvation responses of *Acanthamoeba castellanii* and *Hartmannella vermiformis* in the presence of suspended and attached *Escherichia coli K12*. *FEMS Microbiol Ecol* 59: 556–563.

Prochnik, S. E., Umen, J., Nedelcu, A. M., Hallmann, A., Miller, S. M., Nishii, I., Ferris, P. et al. 2010. Genomic analysis of organismal complexity in the multicellular green alga *Volvox carteri*. *Science* 329: 223–226.

Raper, K. B. 1984. *The Dictyostelids*. Princeton, NJ: Princeton University Press.

Raper, K. B. and Quinlan, M. S. 1958. *Acytostelium leptosomum*: A unique cellular slime mould with an acellular stalk. *J Gen Microbiol* 18: 16–32.

Raper, K. B. and Thom, C. 1941. Interspecific mixtures in *the Dictyosteliaceae*. *Am J Bot* 28: 69–78.

Ratcliff, W., Herron, M. D., Howell, K., Pentz, J. T., Rosenzweig, F., and Travisano, M. 2013. Experimental evolution of an alternating uni- and multicellular life cycle in *Chlamydomonas Reinhardtii*. *Nat Commun* 4: 2742. doi:10.1038/ncomms3742.

Rausch, H., Larsen, N., and Schmitt, R. 1989. Phylogenetic relationships of the green alga *Volvox carteri* deduced from small-subunit ribosomal RNA comparisons. *J Mol Evol* 29: 255–265.

Ray, D. L. and Hayes, R. E. 1954. *Hartmannella astronyxis*: A new species of free living amoeba. *J Morphol* 95: 159–188.

Robertson, A. and Cohen, M. H. 1972. Control of developing fields. *Ann Rev Biophys Bioengg* 1: 409–464.

Rokas, A. 2008. The origins of multicellularity and the early history of the genetic toolkit for animal development. *Annu Rev Genet* 42: 235–251.

Romeralo, M. and Fiz-Palacios, O. 2013. Evolution of dictyostelid social amoebas inferred from the use of molecular tools. In *Dictyostelids. Evolution, Genomics and Cell Biology* (Eds.) M. Romeralo, S. Baldauf, and R. Escalante, pp. 167–182. Heidelberg, Germany: Springer.

Romeralo, M., Skiba, A., Gonzalez-Voyer, A., Schilde, C., Lawal, H., Kedziora, S., Cavender, J. C., Glöckner, G., Urushihara, H., and Schaap, P. 2013. Analysis of phenotypic evolution in Dictyostelia highlights developmental plasticity as a likely consequence of colonial multicellularity. *Proc R Soc Lond B* 280(1764): 20130976.

Ruiz-Trillo, I., Roger, A. J., Burger, G., Gray, M. W., and Lang, B. F. 2008. A phylogenomic investigation into the origin of metazoa. *Mol Biol Evol* 25: 664–672.

Ruiz-Trillo, I., Lane, C. E., Archibald, J. M., and Roger, A. J. 2006. Insights into the evolutionary origin and genome architecture of the unicellular opisthokonts *Capsaspora owczarzaki* and Sphaeroforma arctica. *J Eukaryot Microbiol* 53: 1–6.

Ruiz-Trillo, I., Burger, G., Holland, P. W., King, N., Lang, B. F., Roger, A. J., and Gray, M. W. 2007. The origins of multicellularity: A multi-taxon genome initiative. *Trends Genet* 23: 113–118.

Ruiz-Trillo, I., Inagaki, Y., Davis, L. A., Sperstad, S., Landfald, B., and Roger, A. J. 2004. *Capsaspora owczarzaki* is an independent opisthokont lineage. *Curr Biol* 14: R946–R947. doi:10.1016/j.cub.2004.10.037.

Santhanam, B., Cai, H., Devreotes, P. N., Shaulsky, G., and Katoh-Kurasawa, M. 2015. The GATA transcription factor GtaC regulates early developmental gene expression dynamics in *Dictyostelium Nat Comm* 6: 7551. doi:10.1038/ncomms8551.

Saran, S., Azhar, M., Manogaran, P. S., Pande, G., and Nanjundiah, V. 1994. The level of sequestered calcium in vegetative amoebae of *Dictyostelium discoideum* can predict post-aggregative cell fate. *Differentiation* 57: 163–169.

Sathe, S. and Durand, P. 2015. Cellular aggregation in *Chlamydomonas* (Chlorophyceae) is chimaeric and depends on traits like cell size and motility. *Eur J Phycol* 51: 129–138.

Sathe, S., Kaushik, S., Lalremruata, A., Aggarwal, R. K., Cavender, J. C., and Nanjundiah, V. 2010. Genetic heterogeneity in wild isolates of cellular slime mold social groups. *Microb Ecol* 60: 137–148. doi:10.1007/s00248-010-9635-4.

Sathe, S., Khetan, N., and Nanjundiah, V. 2014. Interspecies and intraspecies interactions in social amoebae. *J Evol Biol* 27: 349–362. doi:10.1111/jeb.12298.

Savill, N. J. and Hogeweg, P. 1997. Modeling morphogenesis: From single cells to crawling slugs. *J Theor Biol* 184: 229–235.

Saxer, G., Brock, D. A., Queller, D. C., and Strassmann, J. E. 2010. Cheating does not explain selective differences at high and low relatedness in a social amoeba. *BMC Evol Biol* 10: 76.

Schaap, P. 2011. Evolution of developmental cyclic AMP signalling in the Dictyostelia from an amoebozoan stress response. *Dev Growth Differ.* 53: 452–462. doi:10.1111/j.1440-169X.2011.01263.x.

Schaap, P., Winckler, T., Nelson, M., Alvarez-Curto, E., Elgie, B., Hagiwara, H., Cavender, J. et al. 2006. Molecular phylogeny and evolution of morphology in the social amoebas. *Science* 314(5799): 661–663. doi:10.1126/science.1130670.

Schilde, C., Lawal, H. M., Noegel, A. A., Eichinger, L., Schaap, P., and Glöckner, G. 2016. A set of genes conserved in sequence and expression traces back the establishment of multicellularity in social amoebae. *BMC Genomics* 17: 871. doi:10.1186/s12864-016-3223-z.

Schilde, C., Skiba, A., and Schaap, P. 2014. Evolutionary reconstruction of pattern formation in 98 *Dictyostelium* species reveals that cell-type specialization by lateral inhibition is a derived trait. *EvoDevo* 5: 34.

Schlösser, U. W. 1976. Entwicklungsstadien-und sippenspezifische Zellwand-Autolysine bei der Freisetzung von Fortplanzungszellen in der Gattung *Chlamydomonas*. *Ber Deutsch Bot Ges* 89: 1–56.

Sebé-Pedrós, A., Ariza-Cosano, A., Weirauch, M. T., Leininger, S., Yang, A., Torruella, G., Adamski, M. et al. 2013a. Early evolution of the T-Box transcription factor family. *Proc Natl Acad Sci USA* 110: 16050–16055. doi:10.1073/pnas.1309748110.

Sebé-Pedrós, A., Roger, A. J., Lang, F. B., King, N., and Ruiz-Trillo, I. 2010. Ancient origin of the Integrin-mediated adhesion and signaling machinery. *Proc Natl Acad Sci USA* 107(22): 10142–10147. doi:10.1073/pnas.1002257107.

Sebé-Pedrós, A., Irimia, M., del Campo, J., Parra-Acero, H., Russ, C., Nusbaum, C., Blencowe, B. J., and Ruiz-Trillo, I. 2013b. Regulated aggregative multicellularity in a close unicellular relative of metazoa. *Elife* 2: e01287. doi:10.7554/eLife.01287.

Sebé-Pedrós, A., de Mendoza, A., Franz Lang, B., Degnan, B. M., and Ruiz-Trillo, I. 2011. Unexpected repertoire of metazoan transcription factors in the unicellular Holozoan *Capsaspora Owczarzaki*. *Mol Biol Evol* 28: 1241–1254. doi:10.1093/molbev/msq309.

Sebé-Pedrós, A., Ballaré, C., Parra-Acero, H., Chiva, C., Tena, J. J., Sabidó, E., Gómez-Skarmeta, J. L., Di Croce, L., and Ruiz-Trillo, I. 2016a. The dynamic regulatory genome of *Capsaspora* and the origin of animal multicellularity. *Cell* 165: 1224–1237. doi:10.1016/j.cell.2016.03.034.

Sebé-Pedrós, A., Peña, M. I., Capella-Gutiérrez, S., Antó, M., Gabaldón, T., Ruiz-Trillo, I., and Sabidó, E. 2016b. High-throughput proteomics reveals the unicellular roots of animal phosphosignaling and cell differentiation. *Dev Cell* 39: 186–197. doi:10.1016/j.devcel.2016.09.019.

Sebé-Pedrós, A., Degnan, B. M., and Ruiz-Trillo, I. 2017. The origin of Metazoa: A unicellular perspective. *Nature Rev Genet*. doi:10.1038/nrg.2017.21.

Shadwick, L. L., Spiegel, F. W., Shadwick, J. D. L., Brown, M. W., and Silberman, J. D. 2009. Eumycetozoa = Amoebozoa?: SSUrDNA phylogeny of protosteloid slime molds and its significance for the supergroup Amoebozoa. *PLoS One* 4(8): e6754.

Shalchian-Tabrizi, K., Minge, M. A., Espelund, M., Orr, R., Ruden, T., Jakobsen, K. S., and Cavalier-Smith, T. 2008. Multigene phylogeny of Choanozoa and the origin of animals. *PLoS One* 3: e2098.

Starr, R. C. 1980. Colonial chlorophytes. In *Phytoflagellates* (Ed.) E. R. Cox, pp. 147–163. Amsterdam, the Netherlands: Elsevier.

Steenkamp, E. T., Wright, J., and Baldauf, S. L. 2006. The protistan origins of animals and fungi. *Mol Biol Evol* 23: 93–106.

Stein, J. 1958. A morphological and genetic study of *Gonium pectorale*. *Am J Bot* 45: 388–397.

Sťovíček, V., Váchová, L., and Palková, Z. 2012. Yeast biofilm colony as an orchestrated multicellular organism. *Commun Integr Biol* 5: 203–205. doi:10.4161/cib.18912.

Suga, H. and Ruiz-Trillo, I. 2013. Development of Ichthyosporeans sheds light on the origin of metazoan multicellularity. *Dev Biol* 377: 284–292. doi:10.1016/j.ydbio.2013.01.009.

Suga, H., Chen, Z., de Mendoza, A., Sebe-Pedros, A., Brown, M. W., Kramer, E., Carr, M. et al. 2013. The Capsaspora genome reveals a complex unicellular prehistory of animals. *Nat Commun* 4: 2325. doi:10.1038/ncomms3325.

Sugden, C., Ross, S., Annesley, S. J., Cole, C., Bloomfield, G., Ivens, A., Skelton, J., Fisher, P. R., Barton, G., and Williams, J. G. 2011. A *Dictyostelium* SH2 adaptor protein required for correct DIF-1 signaling and pattern formation. *Dev Biol* 353(2): 290–301. doi:10.1016/j.ydbio.2011.03.003.

Sun, T. and Kim, L. 2011. Tyrosine phosphorylation-mediated signaling pathways in *Dictyostelium*. *J Signal Transduct*. Article ID 894351. doi:10.1155/2011/894351.

Tam, L.-W. and Kirk, D. L. 1991. The program for cellular differentiation in *Volvox carteri* as revealed by molecular analysis of development in a gonidialess/somatic regenerator mutant. *Development* 112: 571–580.

Tang, L., Franca-Koh, J., Xiong, Y., Chen, M.-Y., Long, Y., Bickford, R. M., Knecht, D. A., Iglesias, P. A., and Devreotes, P. N. 2008. Tsunami, the *Dictyostelium* homolog of the fused kinase, is required for polarization and chemotaxis. *Genes Dev* 22: 2278–2290. doi:10.1101/gad.1694508.2278.

Tarnita, C. E., Washburne, A., Martinez-Garcia, R., Sgro, A. E., and Levin, S. A. 2015. Fitness tradeoffs between spores and nonaggregating cells can explain the coexistence of diverse genotypes in cellular slime molds. *Proc Natl Acad Sci USA* 112: 2776–2781.

Thomas, C. M. and Nielsen, K. M. 2005. Mechanisms of, and barriers to, horizontal gene transfer between bacteria. *Nat Rev Microbiol* 3: 711–721.

Tice, A. K., Shadwick, L. L., Fiore-Donno, A. M., Geisen, S., Kang, S., Schuler, G. A., and Spiegel, F. W. 2016. Expansion of the molecular and morphological diversity of Acanthamoebidae (Centramoebida, Amoebozoa) and identification of a novel life cycle type within the group. *Biol Direct* 11: 69. doi:10.1186/s13062-016-0171-0.

Torruella, G., Derelle, R., Paps, J., Lang, B. F., Roger, A. J., Shalchian-Tabrizi, K., and Ruiz-Trillo, I. 2012. Phylogenetic relationships within the Opisthokonta based on phylogenomic analyses of conserved single-copy protein domains. *Mol Biol Evol* 29: 531–544. doi:10.1093/molbev/msr185.

Torruella, G., de Mendoza, A., Grau-Bové, X., Antó, M., Chaplin, M. A., del Campo, J., Eme, L. et al. 2015. Phylogenomics reveals convergent evolution of lifestyles in close relatives of animals and fungi. *Curr Biol* 25: 2404–2410. doi:10.1016/j.cub.2015.07.053.

Uchinomiya, K. and Iwasa, Y. 2013. Evolution of stalk/spore ratio in a social amoeba: Cell-to-cell interaction via a signaling chemical shaped by cheating risk. *J Theor Biol* 336: 110–118. doi:10.1016/j.jtbi.2013.07.024.

Umen, J. G. 2011. Evolution of sex and mating loci: An expanded view from Volvocine algae. *Curr Opin Microbiol* 14: 634–641.

Umen, J. G. and Goodenough, U. W. 2001. Control of cell division by a retinoblastoma protein homolog in *Chlamydomonas*. *Genes Dev* 15: 1652–1661.

Umen, J. G. and Olson, B. J. S. C. 2012. Genomics of volvocine algae. *Adv Bot Res* 6: 186–245.

Vande Berg, W. J. and Starr, R. C. 1971. Structure, reproduction and differentiation in *Volvox gigas* and *Volvox poswersii*. *Arch Protitenkd* 113: 195–219.

Varki, A. and Altheide, T. K. 2015. Comparing the human and chimpanzee genomes: Searching for needles in a haystack. *Genome Res* 15: 1746–1758.

Vaucheret, H. 2006. Post-transcriptional small RNA pathways in plants: Mechanisms and regulations. *Genes Dev* 20: 759–771.

Viamontes, G. I., Fochtmann, L. J., and Kirk, D. L. 1979. Morphogenesis in *Volvox*: Analysis of critical variables. *Cell* 75: 719–730.

Waddington, C. H. 1942. Canalization of development and the inheritance of acquired characters. *Nature* 150(3811): 563–565. doi:10.1038/150563a0.

Wintermute, E. H. and Silver, P. A. 2010. Emergent cooperation in microbial metabolism. *Mol Syst Biol* 6: 407. doi:10.1038/msb.2010.66.

Włoch-Salamon, D. M. 2013. Sociobiology of the budding yeast. *J Biosci* 38: 1–12. doi:10.1007/s12038-013-9344-5.

Wolf, J. B., Howie, J. A., Parkinson, K., Gruenheit, N., Melo, D., Rozen, D., and Thompson, C. R. 2015. Fitness trade-offs result in the illusion of social success. *Curr Biol* 25: 1086–1090.

Worley, A. C., Raper, K. B., and Hohl, M. 1979. *Fonticula alba:* A new cellular slime mold (Acrasiomycetes). *Mycologia* 71: 746–760.

Zhao, T., Li, G., Mi, S., Hannon, G. J., Wang, X. J., and Qi, Y. 2013. A complex system of small RNAs in the unicellular green alga *Chlamydomonas reinhardtii*. *Genes Dev* 21: 1190–1203.

5 Symbiosis in Eukaryotic Cell Evolution

Genomic Consequences and Changing Classification

Shinichiro Maruyama and Eunsoo Kim

CONTENTS

"If the theory is correct all eukaryotic cells must be seen as multi-genomed systems. This implies that a goal of cellular chemistry is understanding the way in which all biochemical reactions are coded off the nucleic acid of the nucleus and the subcellular organelles."

–Lynn Sagan (1967)

5.1 INTRODUCTION

Symbioses abound in nature—we humans associate with tens of trillions of microorganisms in our gut (DeSalle et al. 2015), while ecologically important reef-building corals sequester tiny photosynthetic algae into their cells for the generation of organic nutrients (Gilbert et al. 2010). Symbioses are also a major force of cellular and genetic innovations, several of which have persisted since the distant origin of eukaryotic cells. This chapter addresses these older symbioses, in particular, those concerning the symbiotic origin of eukaryotes and those aspects of their diversification driven by plastid-generating symbioses. We start off by discussing what constitutes eukaryotic cells and then describe the roles that endosymbioses have played in the evolution of organelles—the mitochondrion and plastid—and their genomes. While these sections provide a general review of our current understanding of the topics, we also investigate the philosophical context that led to some significant consecutively held and discarded models of the evolution of eukaryotic photosynthesis. Therefore, the last section addresses the theoretical background to our changing views on plastid evolution.

5.2 WHAT IS A EUKARYOTIC CELL?

Eukaryotes are usually contrasted against prokaryotes, the latter comprising bacteria and archaea. However, the distinction between eukaryotic versus prokaryotic cells is becoming less clear than previously thought, as many of what were once considered eukaryote-specific traits, such as the cytoskeleton, turn out to have their origins in prokaryotes. This is in agreement with our growing recognition that eukaryotes, at least in their modern forms, are chimeras between an archaeon and at least one bacterium. Nevertheless, there are eukaryote-specific traits, like the nucleus, that suggest a single origin of the eukaryotic cell body plan.

A THE ENDOMEMBRANE SYSTEM

Eukaryotic cells are distinguished from prokaryotic cells by having the vast majority of their genetic material compartmentalized within a membrane-bound organelle known as the nucleus. This is in fact reflected in their name: the word *eukaryote* is derived from the Greek *eu*, meaning "good" or "true" and *karyon*, meaning "nut" or "kernel (=nucleus)." By comparison, prokaryotic cells—as their name implies (Greek *pro*, meaning "before")—do not have a nucleus and their genetic material lies in the cytoplasm (Woese et al. 1990).

The nucleus (specifically the nuclear envelope) is part of the larger endomembrane system, which is unique to eukaryotes (Gould et al. 2016), and includes the endoplasmic reticulum (ER), Golgi apparatus, lysosome, and peroxisome. Of these, the ER can be considered the centerpiece of the endomembrane system, from which other endomembrane systems may have originated. Together with the nucleus, the ER is present in all extant eukaryotes. In contrast, other organelles, such as the Golgi apparatus and peroxisome, appear to be absent from at least some eukaryotes, such as the parasitic protists *Giardia* and *Entamoeba* (He 2007; Gabaldón 2010). In addition, the ER plays a key role in the biogenesis of other components of the endomembrane system, including the nucleus, during cell division (He 2007; Güttinger 2009; Hettema and Motley 2009; Gabaldón 2010).

While a structure homologous to the eukaryotic endomembrane system is not known from prokaryotes, some prokaryotes do possess lipid-bilayer-membrane-bound compartment(s) in their cytoplasm, albeit not to the level of complexity that a typical eukaryotic cell displays (Koops et al. 1976; Diekmann and Pereira-Leal 2013). For example, specialized cell compartments known as magnetosomes are found in magnetotactic bacteria; this organelle allows bacteria to orient in response to the geomagnetic field (Murat et al. 2010). Magnetosomes are structured in a linear array of membrane-bound vesicles that house magnetite crystals (Scheffel and Schüler 2006). Each magnetosome membrane layer is continuous with the inner cell membrane (IM; Murat et al. 2010). It is possible that some other compartments may be detached from the cytoplasmic membrane entirely. These include the anammoxosomes in anaerobic ammonium-oxidizing bacteria (Murat et al. 2010; Neumann et al. 2014), the chromatophores of purple photosynthetic bacteria such as *Rhodobacter sphaeroides* (Tucker et al. 2010; Scheuring et al. 2014), and the thylakoid membranes of cyanobacteria (Van De Meene et al. 2006; Ting et al. 2007; Liberton et al. 2011). In addition, a nucleus-like structure has been identified from members of the Plancomycetes–Verrucumicrobia–Chlamydiae (PVC) bacterial superphylum. This cell feature has drawn considerable attention due to its possible link to the origin of the eukaryotic nucleus (e.g., Fuerst and Sagulenko 2011; Fuerst 2013). A growing body of evidence, however, suggests that the PVC organisms have a gram-negative cell wall structure, or a modified version of it, with two membranes separated by a peptidoglycan layer (Devos 2014; Van Teeseling et al. 2015). This means that what was interpreted as the nuclear envelope in these bacteria corresponds to the inner (=cytoplasmic) membrane of the Gram-negative cell wall (Santarella-Mellwig et al. 2013). Thus their similarity to the nuclear envelope is by analogy, not by homology (McInerney et al. 2011; Devos 2014).

B THE CYTOSKELETON

The actin- and tubulin-based cytoskeleton plays a central structural role in eukaryotes. Both actin and tubulin can form linear polymers: actin filaments (F-actin) consist of two spirally wound protofilaments, and microtubules consist of a hollow tubular structure made up by protofilaments comprising repeating α/β-tubulin dimers (Wickstead and Gull 2011). Both play a key role in eukaryotic cell division. Actin filaments are essential for cytokinesis, while microtubules—through the formation of the spindle apparatus—enable chromosomal segregation during cell division (Wickstead and Gull 2011). Actin filaments and microtubules are also core elements of dynamic cytoplasmic extension mechanisms (e.g., in lamellipodia; Mitchison and Cramer 1996) and eukaryotic flagella/cilia (Moran et al. 2014), respectively.

It was once thought that the cytoskeleton was unique to eukaryotes and absent in prokaryotes, which typically have a rigid wall (e.g., the peptidoglycan layer in bacteria) that is borne outside of the cytoplasmic membrane. Thus, prokaryotes do not seem to be in need of intracellular scaffolding (Knoll 2003). However, a number of studies over the past two decades have shown that prokaryotes have homologs of actin (e.g., crenactins, FtsA, MamK, MreB, and ParM) and tubulin (e.g., FtsZ and TubZ), including those that play roles in cell shape, cell division (cytokinesis), and DNA segregation (Ettema et al. 2011; Wickstead and Gull 2011; Busiek and Margolin 2015). One example is the bacterial actin-like protein, MamK, which forms filaments required during the formation of the magnetosome chain in magnetotactic bacteria (Cornejo et al. 2016).

From the perspective of phylogeny, the most eukaryote-like tubulin homologs are found in the Verrucumicrobia bacterium *Prosthecobacter* (Jenkins et al. 2002). Some suggest that these bacterial tubulin genes—which do not show strong affinity with particular eukaryotic tubulin subfamilies—originated via an ancient gene transfer from a eukaryote, perhaps before the major diversification of the tubulin family took place (Pilhofer et al. 2011). The next most closely related prokaryotic homologs of eukaryotic tubulin are reported from members of the archaeal "Asgard" superphylum (specifically the Odinarchaeota), which is the archaeal group most closely related to the eukaryotes (Zaremba-Niedzwiedzka et al. 2017). Artubulins of the Thaumarchaeota (Archaea) species are next in line; like the Verrucomicrobia and "Asgard" proteins, they are more closely related to eukaryotic tubulins than other prokaryotic homologs like FtsZ (Yutin and Koonin 2012). For actin, closely related prokaryotic homologs—lokiactins and crenactins—are found among members of the archaea. Lokiactins, found in members of the "Asgard" archaea, are the prokaryotic homologs most closely related to the actins of eukaryotes (Zaremba-Niedzwiedzka et al. 2017). Crenactins, reported from some members of the archaeal "TACK" superphylum (=Proteoarchaeota; Bernander et al. 2011), are more distantly related to eukaryotic actins than lokiactins (Spang et al. 2015). Nonetheless, crenactins form filaments that are structurally very similar to actin filaments (Izoré et al. 2016).

Despite a growing number of studies that support the origin of the eukaryotic cytoskeleton in prokaryotes, there are still several synapomorphies of the cytoskeleton of eukaryotes. For instance, mechanical movement of actin filaments and microtubules in eukaryotes is facilitated by cytoskeletal motors (e.g., dyneins, kinesins, and myosin), which are not known in prokaryotes (Wickstead and Gull 2011). In addition, there is no comparative structure known in prokaryotes for eukaryotic microtubule-containing structures, including the centrioles/basal bodies, cilia/flagella, and spindle apparatus. All of these structures are thought to have been present in the last eukaryotic common ancestor (LECA) (Carvalho-Santos et al. 2011).

C OTHER FEATURES

Several additional eukaryote-specific traits are broadly distributed across the tree of eukaryotes and therefore, predicted to be present in the LECA. These include spliceosomal introns, spatial separation of transcription and translation, endocytosis (e.g., phagocytosis) (with the possible exception for the bacterial planctomycete *Gemmata obscuriglobus*), mitosis/meiosis, linear chromosomes with telomeres at the ends (prokaryotic chromosomes are, with few exceptions, circular), an RNA interference system, and the endomembrane system-related processes such as membrane trafficking and autophagy (i.e., degradation of intracellular components via lysosomal machinery) (Shabalina and Koonin 2008; Lonhienne et al. 2010; Starokadomskyy and Dmytruk 2013; Kuzminov 2014). In addition, all eukaryotic organisms have (or once had) mitochondria or mitochondrion-derived organelles, such as hydrogenosomes or mitosomes, which originated from an ancestral member of the α-proteobacteria (Stairs et al. 2015).

Furthermore, sterols and the ubiquitin protein modifier system are each pre-dominantly eukaryotic, although they are found in some prokaryotes. Sterols—a class of lipids with a multiple-ring structure—are integral to the eukaryotic cell, in particular, in the regulation of membrane fluidity and dynamics (Dufourc 2008). Most eukaryotes synthesize or are auxotrophic for sterols, although some low-oxygen-dwelling organisms, such as the excavate *Andalucia incarcerata*, instead, produce hopanoids or related molecules like tetrahymanol that fill in the role of sterols (Takishita et al. 2012). In contrast, sterols are absent from archaea and most bacteria. Corroborating this, only about 0.1% of surveyed bacterial genomes harbor genes for which sterol production has been experimentally confirmed in some corresponding bacteria (16 out of 18 tested strains) (Wei et al. 2016). In phylogenetic trees of oxidosqualene cyclase, the bacterial sequences are para-phyletic, and branch outside of a clade that includes eukaryotic homologs (plus some bacterial sequences that likely originated from eukaryotes via lateral gene transfer (LGT); Wei et al. 2016). These observations support the hypothesis that eukaryotic sterol synthesis had its origin in bacteria.

The ubiquitin protein modifier system, found in all eukaryotic organisms, has implications for an impressive array of processes including protein degradation and endocytosis (Hochstrasser 2009). While it was once considered a eukaryotic inno-vation, a homologous, but simplified, version has been identified from a number of prokaryotes, including the "Asgard" archaea (Nunoura et al. 2011; Maupin-Furlow 2014; Zaremba-Niedzwiedzka et al. 2017). Based on comparative genomic analyses, Grau-Bové et al. (2015) suggested that the ubiquitin system of eukaryotes originated from an archaeal ancestor.

5.3 SYMBIOTIC ORIGINS OF THE EUKARYOTIC CELL

A Composite Nature of the Eukaryotic Cell

The origin of eukaryotic cells is a topic of ongoing discussion. While numerous, often elaborate hypotheses have been put forward concerning the nature and pro-cess of eukaryogenesis (see reviews in Archibald 2015; Martin et al. 2015), all posit one or more symbiotic mergers as key steps toward the eukaryotic cell state. This stems from the observation that the eukaryotic cell displays a mixture of both bacte-rial and archaeal traits. Broadly speaking, the eukaryotic cell inherited, with some exceptions, genes related to information processing (e.g., replication, transcription, and translation) from its archaeal ancestor, and those related to metabolism (e.g., respiration) from one or more bacterial sources (Rochette et al. 2014). An alternative, but not necessarily conflicting view to this is that the archaeal ancestor of the eukaryotes provisioned more central and essential genes, in terms of the organism's survivability, whereas the genes inherited from the bacterial partner(s) tend to be less significant, albeit numerically dominant (McInerney et al. 2014).

As far as the archaeal ancestry of the eukaryotes is concerned, the "Asgard" archaea—a group of anaerobic archaea with no currently cultured representatives—are closest to the eukaryotes in phylogenetic trees based on ribosomal proteins and ribosomal RNA genes (Zaremba-Niedzwiedzka et al. 2017). In addition, "Asgard" archaea share a number of genetic features with the eukaryotes, including gelsolin-domain proteins (required for actin filament assembly in eukaryotes) and an expanded set of GTPases (Spang et al. 2015; Zaremba-Niedzwiedzka et al. 2017).

The bacterial heritage of the eukaryotic cell lies in the α-proteobacterial precursor of the mitochondrion (for all eukaryotes) and the cyanobacterial precursor of the plastid (for plastid-bearing eukaryotes, such as green algae) (Archibald 2015a; Ku et al. 2015). A number of eukaryotic genes show affinities to various other bacterial groups, but the evolutionary origins of these genes remain obscure and open to interpretation (Ku et al. 2015; Pittis and Gabaldón 2016).

It also needs to be noted that a large fraction of eukaryotic genes do not have clear homologs in archaea or bacteria (Pittis and Gabaldón 2016). These likely originated after the archaeal-bacterial merger, thereby, tentatively representing true eukaryotic innovations (Dacks et al. 2016).

B MITOCHONDRIAL EVOLUTION

It is widely accepted that the mitochondrion and plastid originated from an endosymbiotic α-proteobacterium and cyanobacterium, respectively. While symbiogenetic models have been proposed for other eukaryotic compartments, including the flagellum (Margulis et al. 2006) and peroxisome (Duhita et al. 2010), evidence supporting such hypotheses is not robust, and their origins are generally assumed to be autogenous in nature (Koumandou et al. 2013; Gabaldón and Pittis 2015).

With respect to mitochondrial evolution, one of the ongoing debates concerns whether the host cell that incorporated its α-proteobacterial precursor was a simple archaeon or a more complex, phagocytosing cell (Archibald 2015). Some models, including the syntrophy hypothesis, phagocytosing archaeon theory, and neomuran theory, postulate that the host cell was more complex than known extant prokaryotic cells and was capable of phagocytosis (i.e., internalization of large particulate matter), which enabled the uptake of the mitochondrial ancestor (Martijn and Ettema 2013; Cavalier-Smith 2014; López-García and Moreira 2015). In contrast, other models, including the hydrogen hypothesis and inside-out theory, posit that the merger between two simple prokaryotic cells triggered eukaryogenesis, including the evolution of the endomembrane system and phagocytosis (Lane 2011; Baum and Baum 2014; Gould et al. 2016). Under these scenarios, phagocytosis is not considered a prerequisite for the acquisition of the mitochondrion.

There are indeed examples of a nonphagocytotic cell harboring another cell, providing alternative mechanistic routes for intracellular associations (Corsaro and Venditti 2006). For instance, the bacterium *Burkholderia rhizoxinica* penetrates into the cytoplasm of the nonphagocytotic fungus *Rhizopus microsporus* via topical cell wall lysis (Moebius et al. 2014). The predatory bacterium *Bdellovibrio* and its close relatives are capable of entering through the outer cell membrane and peptidoglycan cell wall layer of target bacteria and reside within the periplasm (=space between the outer and inner membranes in gram-negative bacteria) (Sockett 2009). In addition, some bacteria such *Bacillus* are capable of forming a dormant, non-reproductive structure known as an endospore, via a process that includes engulfment of one cell by other. This engulfment process is unrelated to eukaryotic phagocytosis, as different molecular components are utilized (Tan and Ramamurthi 2014).

C PLASTID EVOLUTION

Unlike the case for the mitochondrion, not all eukaryotic organisms have or once had a plastid. Also, whereas the mitochondrion-generating symbiosis occurred only once, plastid acquisition took place on multiple occasions in taxonomically diverse host eukaryotes that already possessed mitochondria (Archibald 2015). Plastid acquisition played an important role as a major driver of eukaryotic diversification, spawning several major lineages, including the dinoflagellates, green algae plus land plants (together known as the Chloroplastida or Viridiplantae), haptophytes, and red algae (Graham et al. 2006).

The evolution of the first photosynthetic eukaryotes, which occurred >1 billion years ago, was the result of a cellular merger between a heterotrophic eukaryotic host and a cyanobacterium, the process known as primary plastid-generating endosymbiosis (Graham et al. 2016). This event is generally held to have led directly to the evolution of the common ancestor of green algae (plus their land plant descendants), red algae, and glaucophytes, together known as the Archaeplastida (or Plantae) (Adl et al. 2012; but see section 5.4 for alternative hypotheses). In addition, several unrelated eukaryotes acquired a plastid from photosynthetic eukaryotes, via processes known as secondary or tertiary endosymbiotic events (Graham et al. 2016). A series of these eukaryote-to-eukaryote amalgamations gave rise to several eukaryotic groups, including the apicomplexans (and the related chromerids, which include *Chromera* and *Vitrella*), chlorarachniophytes, cryptophytes, dinoflagellates, euglenophytes, haptophytes, and ochrophytes (i.e., plastid-bearing members of the stramenopiles) (Archibald 2009). More complicated cases of plastid acquisition are known from some dinoflagellates, which replaced or reacquired the plastid via tertiary or even possibly quaternary endosymbiotic associations (Morden and Sherwood 2002). Additionally, groups of uncultured plastid-bearing eukaryotes, including rappemonads have been identified over the past decade (Kim et al. 2011; Choi et al. 2017). While their plastids show affinity to haptophyte plastids, the nature of the host eukaryotes, or the endosymbiotic events that gave rise to these groups, remains unclear.

It is generally assumed that the heterotrophic ancestors of "eukaryotic algae" were phagotrophic, and utilized their feeding apparatuses in the internalization of either a cyanobacterial or eukaryotic algal precursor of a plastid, which bypassed host digestion and persisted in the host cytoplasm (Maruyama and Kim 2013). Besides the canonical phagocytotic mechanism of engulfing a whole cell, myzocytosis—a specialized feeding mechanism where the prey cell membrane is ruptured and its cellular content, including organelles, is sucked up—has been proposed to play a role in plastid acquisition, in particular in the evolution of dinoflagellate and euglenophyte plastids (Cavalier-Smith and Chao 2004; Yamaguchi et al. 2012).

In terms of the cyanobacterial ancestry of the plastid, a recent study by Ponce-Toledo et al. (2017) suggested that the plastid originated from an ancient cyanobacterium similar to the extant *Gloeomargarita*, which is unicellular, not capable of fixing atmospheric nitrogen (unlike many other cyanobacteria), and restricted to freshwater environments. In case of the host ancestry of eukaryotic algae, some plastid-bearing eukaryotes have known nonphotosynthetic, plastid-lacking relatives, which could serve as models for their heterotrophic ancestors (O'Kelleys 1993). For example, cryptophytes are closely related to obligatory phagotrophs such as *Goniomonas* and kathablepharids (Kim and Archibald 2013); euglenophytes are closely related to eukaryovores like *Anisonema* (Yamaguchi et al. 2012). In contrast, heterotrophic eukaryotes closely related to the primary plastid-bearing lineages remain somewhat obscure. Even so, plastid-less members of the Cryptista, such as *Palpitomonas* and *Goniomonas*, have been proposed as good models for the host eukaryote that gave rise to the first photosynthetic eukaryotes (Kim and Graham 2008; Yabuki et al. 2010).

D   OTHER SYMBIONT-DERIVED "ORGANELLES"

Besides the mitochondrion and plastid, there are other symbiont-derived, vertically-inherited structures[1] in eukaryotic cells (Bodył et al. 2007). One well-known example is found in the rhizarian amoeba *Paulinella chromatophora*, which harbors a photosynthetic compartment known as the chromatophore, which is of cyanobacterial origin (Nowack et al. 2016). Chromatophore acquisition, which is estimated to have happened 60–120 Myr ago (Delaye et al. 2016), occurred independently of primary plastid evolution in other eukaryotes and is supported by phylogenetic data showing that the two organelles are not closely related to each other (Marin et al. 2005).

Another example is a nitrogen-fixing compartment, known as the spheroid body, in rhopalodiacean diatoms (Nakayama et al. 2011). While the spheroid body originated from a *Cyanothece*-like cyanobacterium relatively recently (~12 Myr ago), it has lost the capacity to photosynthesize (Nakayama et al. 2014), which may at least be partially explained by the fact that the host diatoms did, at the time of the spheroid body acquisition, and still do, possess a functioning photosynthetic plastid. In addition, aphids and many other plant sap-feeding insects harbor obligatory and heritable "endosymbionts" of various bacterial origins, which provide nutrients such as amino acids to the host insects (Bennet and Moran 2015).

---

[1] Whether or not such entities should be considered *bona fide* organelles has been a topic of much discussion (Theissen and Martin 2006; Bodył et al. 2007). A growing body of evidence supports the notion that boundaries between endosymbionts and organelles (of symbiotic origins) are not sharp, and the two concepts are better viewed in the context of continuum.

Generally considered obligatory in nature, many of these insect–bacterial associations have persisted for millions of years, in some cases even longer than 100–200 Myr (Bennet and Moran 2015).

5.4 GENOMIC CONSEQUENCES OF ENDOSYMBIOSES

While various genetic changes are associated with endosymbioses, we focus here on those cellular mergers that involve cyanobacterial and eukaryotic algal endosymbionts.

A COMPOSITE NATURE OF THE PLASTID PROTEOME

In comparison to free-living cyanobacteria, whose genomes typically contain several thousand protein-coding genes, plastid genomes code for only up to ~230 proteins, and are much smaller in size (Raven et al. 2013; Ng et al. 2017). These observations suggest that, during the transition from a free-living cyanobacterium to an organelle, the majority of the original cyanobacterial genes were either transferred to the host nucleus or lost completely. Comparative genomic studies have shown that the nuclear genomes of plastid-bearing eukaryotes are indeed impacted by endosymbiotic gene transfer (EGT), a specialized type of lateral gene transfer (LGT; also known as horizontal gene transfer; Archibald 2015). While methods for identifying and enumerating nuclear genes arising from EGT vary from study to study, 5%–20% of the surveyed genes of some eukaryotic algae are estimated to be of cyanobacterial origins (Curtis et al. 2012; Price et al. 2012; Dagan et al. 2013).

Predictably, protein products of such endosymbiont-derived nuclear genes are targeted back to the plastid compartment, which is accomplished through a peptide domain "tag" located at the N-terminus (Patron and Waller 2007). This, however, applies to less than 50% of the transferred genes and, thus, the majority of them are now localized outside of the plastid (Martin et al. 2002; Curtis et al. 2012). This suggests neofunctionalization of the transferred genes and/or relocation of components of the endosymbiont's metabolic machinery to the nucleocytoplasm of the host.

Surprisingly, not all plastid-localized proteins originated from a cyanobacterial ancestor. This includes a number of plastid proteins that are host-derived. For example, many plastid solute transporters, including those that transport sugar, phosphate, and magnesium, are recruited from the host eukaryote (Karkar et al. 2015). Further, various other bacteria, as well as viruses, may have contributed to the plastid proteome, as has been shown for the (plastid-like) chromatophore of *P. chromatophora* (Nowack et al. 2016). As such, modern-day plastids operate through proteins of mixed origin.

In a photosynthetic eukaryote, there are usually several hundred or more nucleus-encoded, plastid-targeted proteins, which have a wide range of functional roles (Curtis et al. 2012; Dagan et al. 2013). Here, we provide some details on two relatively well-studied examples of plastid-targeted proteins. One is ribulose-1,5-bisphosphate carboxylase–oxygenase (rubisco), a core enzyme involved in CO_2 fixation, and considered the most abundant protein on the planet (Andersson and Backlund 2008; Raven 2013). The rubisco of eukaryotic algae is made of two differently sized subunits, which are encoded in the *rbcS* and *rbcL* genes (Tabita et al. 2007). These two genes are located in the plastid DNA in many eukaryotic algae. In green algae and land plants, however, *rbcL* is located in plastid DNA, whereas *rbcS* resides in nuclear DNA. This indicates that a transfer of *rbcS* from plastid to nuclear DNA occurred in the ancestry of this lineage (Andersson and Backlund 2008).

Another example is oxygen-evolving enhancer protein 1 (PsbO), a ubiquitous component of Photosystem II that plays a role in stabilizing the manganese (Mn) cluster where water oxidation takes place (De Las Rivas and Barber 2004). In all photosynthetic eukaryotes, PsbO is encoded in the nuclear DNA (Archibald et al. 2003). This suggests that plastid-to-nucleus transfer of the gene (*psbO*) occurred early during the establishment of primary plastids. It also follows that secondary or tertiary plastid-bearing eukaryotes did not acquire *psbO* in their nuclear DNA directly from plastid DNA, but rather from the nuclear DNA of their algal endosymbionts. Due to its conservative nature and restricted distribution among cyanobacteria and photosynthetic eukaryotes, PsbO is considered a useful marker in elucidating the phylogenetic history of plastids (Ishida and Green 2002). For instance, PsbO is one of the markers that showed that euglenophytes and chlorarachniophytes acquired their plastids independently from different endosymbiotic green algae (Rogers et al. 2007; Takahashi et al. 2007).

B Cases for Other Cyanobacterium-Derived "Organelles"

The chromatophore of *P. chromatophora* and the spheroid body of rhopalodiacean diatoms have more recent origins than plastids, and their genomes include more protein-coding genes than those of plastids—867 and 1,720, respectively (Nowack et al. 2008; Nakayama et al. 2014). Nonetheless, the gene number of these "organelles" represents <40% of that in their respective free-living relatives, which is suggestive of ongoing reductive genome evolution (Nowack et al. 2008; Nakayama et al. 2014). In the case of *P. chromatophora*, for which a transcriptome and a partial genome have been assembled, ~60 EGT-derived genes, including photosystem-I-related genes *psaE*, *psaK1*, and *psaK2*, have been identified thus far, and some of these were confirmed to be chromatophore-targeted (Nowack and Grossman 2012; Nowack et al. 2016).

C Gene Transfers Associated with Modern-Day Symbioses and Kleptoplastidy

There is a wealth of literature concerning photosynthetic endosymbionts in animals and other heterotrophic eukaryotic hosts (e.g., Buchner 1965; Paracer and Ahmadjian 2000). Unlike the plastid and other organelle-like entities (e.g., the chromatophore), these symbionts are generally characterized as being nonpermanent (not heritable over generations) and independent (capable of living freely), although no clear distinction can be made between the two concepts due to a number of intermediate forms that exist in nature (Bodył et al. 2007). In warm oligotrophic marine waters, the photosynthetic dinoflagellate *Symbiodinium* is the champion of endosymbioses, as it associates with a range of marine fauna and protists, including corals, sea anemones, jellyfish, giant clams, foraminifers, and radiolarians (Trench 1993; Pochon and Gates 2010). In intertidal habitats, some prasinophyte green algae, such as *Nephroselmis* and *Tetraselmis*, live in the cytoplasm of flatworms and heterotrophic protists (Paracer and Ahmadjian 2000; Okamoto and Inouye 2005). In freshwater environments, the green alga *Chlorella* excels at endosymbioses, as it associates with a number of invertebrates, including sponges and *Hydra*, as well as many ciliate species (Graham and Graham 1980). In a vernal pond, the green alga *Oophila* thrives inside the eggs of the salamander *Ambystoma maculatum*; this consortium also includes a surprising intracellular association, which is otherwise unknown amongst vertebrates (Kerney et al. 2011). While the list goes on, only a handful of them have been investigated, so far, from the perspective of EGT or of other genome-level impacts made by such associations.

One member of that handful is the sea anemone *Aiptasia*, which recruits the dinoflagellate *Symbiodinium* into its gastrodermal cells (Garrett et al. 2013). From the sea anemone's genome, 29 putative EGT-derived genes, representing ~0.1% of the total of *Aiptasia*'s genes, have been identified (Baumgarten et al. 2015). Of these, 17 genes are specific to *Aiptasia*, but, interestingly, 12 also occur in other cnidarians, including the nonsymbiotic *Nematostella vectensis* (Baumgarten et al. 2015). One such example is a gene that codes for a fusion protein containing a 3-dehydroquinate synthase domain and an O-methyltransferase domain, both of which are involved in the synthesis of mycosporine-like amino acids, compounds known for providing UV protection. This fused protein-coding gene is found only among some cnidarians and dinoflagellates. This fact, together with the gene's phylogeny, suggests that cnidarians acquired the gene from an ancient dinoflagellate, which may have been an endosymbiont of early cnidarian ancestors (Baumgarten et al. 2015).

Kleptoplastidy is another form of acquired phototrophy that has been reported from several eukaryotic groups, including ciliates and invertebrates (Graham et al. 2016). In this association, the host organism maintains the plastid compartment of algal prey. As this "stolen" organelle, known as the kleptoplastid, is not capable of multiplication, such an association is not generally considered a case of endosymbiosis (Graham et al. 2016). A well-known example of kleptoplastidy is found in the dinoflagellate *Dinophysis acuminata*, which displays an intricate form of cellular association that involves three organisms (Wisecaver and Hackett 2010). The dinoflagellate acquires a kleptoplastid not from an alga but from the ciliate *Myrionecta rubra*. The ciliate is a heterotrophic protist but performs acquired phototrophy by retaining the plastid (as well as other compartments, including the mitochondrion and nucleus) of a prey organism, a cryptophyte alga (Johnson 2011; Wisecaver and Hackett 2010). From the dinoflagellate transcriptome data, five putative nucleus-encoded, plastid-targeted genes (out of the total of 816 annotated contigs) were identified, of which only one—*psbM*—is of cryptophyte origin (Wisecaver and Hackett 2010). Similarly, only one kleptoplastidy-originated gene was identified from the transcriptome data of the foraminifer *Elphidium margaritaceum*, which sequesters the plastid compartment of its diatom prey and retains the organelle for up to several months (Pillet and Pawlowski 2013). Additional cases of gene transfer associated with kleptoplastidy may be uncovered when more complete transcriptomic or genomic data become available for appropriate taxa. Nevertheless, the extent to which kleptoplastidy-driven gene transfers has taken place does not seem numerically comparable to that associated with the full integration of plastids.

Solar-powered sacoglossan nudibranchs, such as *Elysia chlorotica*, represent another case of kleptoplastidy. These green colored sea slugs feed on eukaryotic algae, including the filamentous stramenopile *Vaucheria litorea*. The nudibranchs acquire phototrophy by retaining the plastid, but not other compartments, of the algal prey within their endodermal cells for as long as several months (Rumpho et al. 2000). As the majority of plastid proteins are encoded in nuclear DNA, long-term retention of isolated plastids within the context of an animal cell was surprising and prompted scientists to search for external factors, such as genes laterally transferred from algal prey to sea slug genomes, which may explain this seemingly extraordinary phenomenon. Unfortunately, the results from multiple research groups have not been congruent with each other on this question (Rauch et al. 2015). While some reported the presence of alga-derived genes from analyses of transcriptome data or by PCR-based experiments (Rumpho et al. 2008; Pierce et al. 2012), others reported no indication of algal gene transfer, either from the genomic data of the same host species or from transcriptome data of related taxa (Wägele et al. 2011; Bhattacharya et al. 2013). Such discrepancies may be attributed to analysis artifacts (e.g., non-removal of singleton reads), individual variability, sample contamination, the possible presence of extrachromosomal DNA, or any combination of these (Bhattacharya et al. 2013; Rauch et al. 2015). These ongoing developments illustrate that extreme caution is needed both in experimental design and data analysis, especially when searching for recent gene transfer events (Arakawa 2016; Koutsovoulos et al. 2016).

D   ALGAL GENES IN UNSUSPECTED HOST EUKARYOTES

Interestingly and unexpectedly, some heterotrophic organisms that are not known to associate with photosynthetic symbionts harbor alga-derived genes in their genomes. In *Monosiga brevicollis*, which is a colonial choanoflagellate that feeds on microorganisms via phagocytosis, >100 genes were found to have their closest homologs in photosynthetic organisms (Sun et al. 2010). Alga-derived genes have also been found in various other heterotrophs, including the tunicate *Ciona*, the oomycete *Phytophthora*, the heterolobosean amoeba *Naegleria*, and the ciliates *Tetrahymena* and *Paramecium* (Tyler et al. 2006; Reyes-Prieto et al. 2008; Maruyama et al. 2009; Ni et al. 2012). Some have suggested that such alga-derived genes may represent relics of ancient endosymbiotic events, while others have proposed instead that they originated via gene transfer from viruses or ingested prey organisms.

Another interesting and puzzling observation is the presence of both green-alga-derived genes in eukaryotes bearing a plastid of red algal origin, and red-alga-derived genes in eukaryotes with a plastid of green algal origin. For instance, the nuclear genome of the cryptophyte *Guillardia theta* includes 252 genes that have their closest homologs in green algae, which represent more than double the number of genes showing red algal affiliation. However, this ratio decreases to 1:1 when genes otherwise present in only red or green algae are excluded (Curtis et al. 2012).

A similar pattern has been noted for other eukaryotes with a red alga-derived plastid, such as diatoms and the photosynthetic alveolate *Chromera velia* (Moustafa et al. 2009; Woehle et al. 2011). In the chlorarachniophyte *Bigellowiella natans*, which possesses a plastid of green algal origin, 45 genes have their closest homologs in red algae, whereas 207 genes are most closely related to green algal sequences (Curtis et al. 2012). While some suggested that genes of unexpected algal origins may have stemmed from "cryptic" endosymbiotic events (Moustafa et al. 2009), others have attributed such mixed signals to factors like biased taxon sampling and random-phylogenetic error (Woehle et al. 2011; Burki et al. 2012; Deschamps and Moreira 2012).

E METHODOLOGICAL CONSIDERATION

EGT-derived (or other laterally transferred) genes are usually identified by comparatively analyzing individual gene phylogenies for congruency or other rare genomic features (e.g., gene fusion) (Huang and Yue 2013). If the phylogeny of a certain gene is in conflict with the expected organismal relationships (e.g., a generally accepted species tree), and such discord is not due to lineage sorting or gene duplication and subsequent extinction, the gene may be considered to be of a foreign origin (Maddison 1997). However, in practice, it is not always straightforward to distinguish between competing scenarios, especially when they pertain to very ancient events. Phylogenetic signals erode over time as noise accumulates and signals saturate, leading to unresolved trees. Many intermediate taxa may have gone extinct, which can cause interpretational artifacts. The assumed organismal (reference) tree may have been impacted by hidden paralogy, lineage sorting, or LGT (Maruyama and Archibald 2012). Data quality and methodological artifacts, such as long-branch attraction, are other factors to be considered (Huang and Yue 2013). For instance, the genome of a tardigrade, *Hypsibius dujardini*, was initially reported to have been greatly impacted by extensive LGT events (Boothby et al. 2015). However, it was later found that the majority of the LGT candidates came from contaminating reads (Arakawa 2016; Hashimoto et al. 2016; Koutsovoulos et al. 2016).

Further, given the past and ongoing genome shuffling by LGT, it is not easy to develop a robust baseline for the ancestral genomic landscape (particularly for prokaryotes), thereby, limiting our confidence in analyses that are based exclusively on modern-day taxa (Lorenz and Wackernagel 1994; Shi and Falkowski 2008). An example relevant to this point may be the proposed tripartite endosymbiotic event at the origin of eukaryotic photosynthesis. Based on the presence of Chlamydiales-like genes in photosynthetic eukaryotes, it has been proposed that a Chlamydiales endosymbiont existed in the ancestors of photosynthetic eukaryotes and, more importantly, played critical roles in the metabolic integration of the cyanobacterial precursor of the plastid (Ball et al. 2013; Cenci et al. 2017). However, this proposal, known as the *ménage-à-trois* hypothesis (MATH), has been a subject of vigorous debate (Domman et al. 2015; Ball et al. 2016a and b; Gould 2016). While skeptics of the hypothesis pointed out possible methodological artifacts in the studies, such as the choice of protein evolutionary models, and the lack of controlled experiments (Stiller 2011; Domman et al. 2015), a more serious problem may lie in the assumption that the phylogenetic relationships among extant species should reflect their >1 billion-year-old ancestors. Over such a long time span, prokaryotes have likely undergone extensive genomic rearrangements, thereby blurring their distant history (Rujan and Martin 2001; Maruyama and Archibald 2012).

5.5 CHANGING CLASSIFICATION OF PLASTID-BEARING EUKARYOTES

A DOMINANCE AND DEMISE OF PARSIMONY-BASED HYPOTHESES

All the plastid-bearing eukaryotes were once proposed to be classified into only three groups—namely, Archaeplastida, Chromalveolata, and Cabozoa—based on a simple, parsimony-based argument (Cavalier-Smith 1999, 2003). This "Big Three" super-group classification of eukaryotic phototrophs and their relatives, particularly that of Archaeplastida and Chromalveolata, proposed by Cavalier-Smith (1999) and others (with some modifications on details), was popular and influential in the field until the mid-2000s (Figure 5.1a) (Bhattacharya et al. 2004; Hjorth et al. 2004; Keeling 2004).

This hypothesis proposes that there were only three plastid-generating endosymbioses, each of which gave rise to one of the "Big Three" lineages. Whereas the plastid in the common ancestor of Archaeplastida is proposed to have originated directly from an endosymbiotic cyanobacterium (and is thus known as the primary plastid), those of Chromalveolata and Cabozoa derived from eukaryotic red algal or green algal symbionts, respectively (Cavalier-Smith 1999, 2003). This proposal was based on the principle of parsimony—in this case, finding the minimal number of symbioses required to explain the observed diversity of eukaryotic phototrophs. Parsimony, often referred to as "Ockham's Razor" (named after the fourteenth century philosopher William of Ockham), is a methodological principle that emphasizes economy in explaining a phenomenon. Although its utility as an overarching principle has been questioned, Ockam's Razor has nevertheless had a critical influence in studies concerning the origin and diversification of plastids (Stiller 2014). To give a pertinent example, since the plastids of euglenophytes and chlorarachniophytes are similar to each other (e.g., in pigment composition), it is most parsimonious to assume that the two algal groups share a common ancestor that underwent a single endosymbiotic event that gave rise to those plastids. This group, the Cabozoa, includes these two algal groups, as well as their nonplastid-bearing relatives (Cavalier-Smith 1999).

Notwithstanding their great appeal, especially, as a force for economy in taxonomy, parsimony-based cabozoan and chromalveolate hypotheses have been rejected by a series of analyses based on emerging cell biological, phylogenetic, and phylogenomic data (Bodył 2005; Rice and Palmer 2006; Burki et al. 2007; Rogers et al. 2007; Takahashi et al. 2007; Sanchez-Puerta and Delwiche 2008; Bodył et al. 2009; Lane and Archibald 2009; Brown et al. 2012; Stiller et al. 2014). A major counterargument to the cabozoan and chromalveolate hypotheses came from growing recognition of the SAR supergroup, which comprises the large-scale groups Stramenopila, Alveolata, and Rhizaria (Burki et al. 2007; Adl et al. 2012). Acceptance of the SAR (or Sar) clade necessarily refutes the monophyly of Cabozoa (=Rhiazaria plus Excavata), as well as Chromalveolata (=Cryptophyta plus Haptophyta plus Stramenopila plus Alveolata).

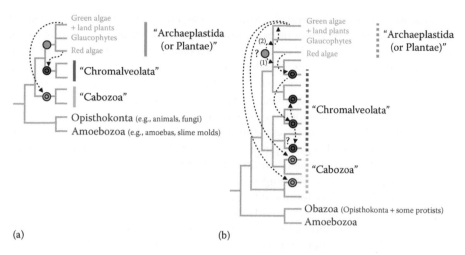

(a)   (b)

FIGURE 5.1 Changing views on the evolutionary history of plastids and their hosts. An earlier view (a), advocates of which include Cavalier-Smith (2003), was simple and driven by the principle of parsimony. Under this hypothesis, there were only three plastid-generating endosymbioses; three was the minimum number required to explain the plastid diversity known at that time. This scheme, however, posits a number of independent and apparently complete losses of plastids within the chromalveolates and cabozoans (but not within the archaeplastidan group). With additional data, this earlier, simpler scenario became no longer tenable, in particular concerning the chromalveolate and carbozoan hypotheses. One current model (b) instead suggests at least six independent plastid acquisition events, including two involving green algal endosymbionts. The red algal plastid spread into multiple eukaryotic lineages via serial transfer events. Whether the archaeplastidan groups originated from a single primary plastid-generating event or via multiple events (perhaps serially) is an ongoing topic of debate. The model depicted here (b) is mapped onto the "mega" phylogeny by Burki et al. (2016) and is based on hypotheses proposed by Sanchez-Puerta and Delwiche (2008), Stiller et al. (2014), and Kim and Maruyama (2014). Primary and secondary plastid-generating endosymbioses are represented as single and double circles, respectively. A green circle indicates an endosymbiosis involving a green algal endosymbiont whereas a magenta circle denotes an acquisition of red-alga-derived plastid. Arrows indicate the direction of plastid transfer. A pair of arrows with a numeric label (in b) shows two competing scenarios concerning the evolution of the archaeplastidan plastid. (Courtesy of Burki, F. et al., *Proc. Biol. Sci.*, 283, 20152802; Courtesy of Sanchez-Puerta, M. and Delwiche, C.F., *J. Phycol.*, 44, 1097–1107, 2008; Courtesy of Stiller, J.W., *J. Phycol.*, 50, 462–471, 2014; Courtesy of Kim, E. and Maruyama, S., *Acta Soc. Bot. Pol.*, 83, 331–336, 2014.)

Some plastid-related traits, including the phylogeny of PsbO, also refute the cabozoan hypothesis (Rogers et al. 2007; Takahashi et al. 2007). Another critique pointed out that the proposal by Cavalier-Smith, in particular, that of the Cabozoa, did not even have support from the then-current models of host relationships (Keeling 2010). In fact, the cabozoan and chromalveolate hypotheses imply multiple occasions of plastid loss, as both proposed groups include many plastid-less subgroups, including ciliates, oomycetes, bicosoecids, kinetoplastids, jakobids, and foraminifers. Thus, the cabozoan and chromalveolate hypotheses might be parsimonious insofar as only the number of plastid gains is concerned (Cavalier-Smith 1999), but not necessarily when one assumes that plastid loss is similarly "difficult" to plastid gain (Stiller 2014).

Baurain et al. (2010) provided additional compelling evidence that challenged the chromalveolate hypothesis by examining the separate phylogenies within a cell. The authors reasoned that, if all the chromalveolate sublineages originated from a single endosymbiotic event, then phylogenies deduced from all three genetic compartments—nucleus, mitochondrion, and plastid—should be more or less congruent. In other words, if (for instance) the plastid compartment shows discordant phylogenic signals and evolutionary rates from the other two, that would falsify the null hypothesis that all cellular compartments, including the plastid, share a common history. Based on this principle of "proof by contradiction," the authors refuted the chromalveolate hypothesis (Baurain et al. 2010).

Other studies have also challenged the chromalveolate hypothesis, and a series of more complex models for the evolution of red algal type plastids have been proposed (e.g., Sanchez-Puerta and Delwiche 2008; Petersen et al. 2014). One is the model of serial plastid endosymbioses, which posits that one red algal plastid spread serially into multiple eukaryotic groups (Figure 5.1b) (Sanchez-Puerta and Delwiche 2008; Stiller et al. 2014). According to a version of this model, a secondary plastid-generating event involving a red algal symbiont gave rise to the cryptophytes, a member of which subsequently formed an endosymbiosis with an ancient heterotrophic stramenopile, thereby, generating the Ocrhophyta (which is the only plastid-bearing group within Stramenopila). An ochrophyte was then associated with an unrelated eukaryote, which led to the rise of the haptophytes (Stiller et al. 2014). Another endosymbiotic event involving a different ancient ochrophyte may have been responsible for the generation of the plastid of chromerids (and perhaps other alveolates, including dinoflagellates) (Petersen et al. 2014).

One important notion, which has become more obvious through the course of the above-mentioned debates is that plastids are laterally transferrable, and are thus best thought of as portable evolutionary units. Therefore, neither plastids themselves nor other associated traits are necessarily indicative of the host-lineage relationships. Conversely, host-lineage relationships do not necessarily reflect the evolutionary trajectory of their plastids. In other words, while all the descendants of a single plastid-generating event form a clade, a group of plastid-bearing eukaryotes being monophyletic does not necessarily mean that their plastids arose via a single endosymbiotic event at the ancestry of the group (Figure 5.2). It is circular logic if one attempts to assert that plastid-bearing members of two related groups stemmed from a single plastid-generating event simply based on the presence of "similar" plastids. How strongly the host phylogeny affects our inference of their plastid evolution is dependent on other factors, such as the "portability" of the plastid via eukaryote-to-eukaryote symbioses (Maruyama and Archibald 2012; Stiller 2014). A relevant example for this would be the chromerid plastid (Moore et al. 2008), whose evolutionary origin remains debated. While some favor a simpler hypothesis saying that the chromerids, as well as other alveolates and stramenopiles, originated from a common algal ancestor (Figure 5.2a) (Janouskovec et al. 2010), others have proposed more complex scenarios—albeit simpler from the perspective of plastid loss, as fewer or no plastid loss events are assumed—that describe multiple plastid acquisition events within alveolate evolution (Figure 5.2b and c) (Petersen et al. 2014; Ševčíková et al. 2015).

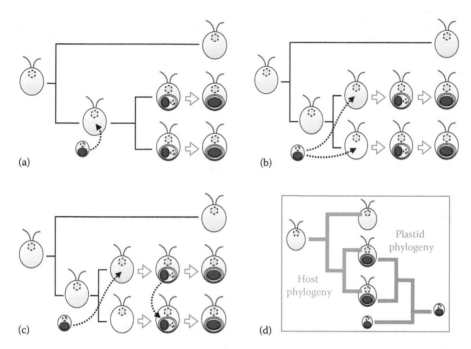

FIGURE 5.2 A cartoon illustrating multiple possible evolutionary paths by which related eukaryotes (as shown in blue and yellow colored flagellates) could harbor a similar plastid. Nuclei are indicated by blue circles bounded by broken lines, plastids by magenta ovals. The simplest scenario would be that two eukaryotic groups acquired a plastid from a common endosymbiotic event (a). In an alternative scenario, two related eukaryotes associated with the same or closely related algal species (b). It is also possible that the plastid of one eukaryote was obtained serially from a related eukaryotic alga (c). The inset shows relationships of the hosts versus plastids (d).

A solution to this conundrum may lie in the adoption of a probabilistic model that considers, among other things, the rate of plastid loss and gain, which has probably been variable throughout evolutionary time (Stiller 2014). Such an approach would be more realistic over an *a priori* adoption of the principle of parsimony.

B ARCHAEPLASTIDA: THE LAST UNEXPLORED CONTINENT OR AN ILLUSORY MONSTER?

Among Cavalier-Smith's "Big Three" large-scale lineages, only the Archaeplastida hypothesis has survived, and it is still the dominant phylogenetic framework, utilized by a majority of researchers (Adl et al. 2012). While some earlier phylogenetic studies based on the use of nucleus-encoded proteins provided moderate to strong support for a monophyletic Archaeplastida (Rodríguez-Ezpeleta et al. 2007; Burki et al. 2008), more recent studies do not. Instead, these show that archaeplastidan monophyly is interrupted by other eukaryotic groups such as cryptistans, partly due to improved taxon sampling (Brown et al. 2013; Burki et al. 2016). Even so, it is important to be reminded that the host phylogeny is not necessarily reflected in the evolutionary history of their plastids. For example, when monophyly is not recovered, Archaeplastida can still be paraphyletic, meaning that a single symbiotic event gave rise to the three archaeplastidan groups plus the "interrupting" groups, which could have lost the primary plastid secondarily. Alternatively, polyphyletic origins of Archaeplastida would be inferred if plastids of the three archaeplastidan groups were acquired via separate symbiotic events (Maruyama and Archibald 2012). Given the portable nature of the plastid, whether Archaeplastida is monophyletic or not should be a matter independently addressed from that for their plastids.

Similar to the patterns revealed by Baurain et al. (2010) regarding chromalveolate evolution, phylogenomic analyses focusing on the lineages comprising Archaeplastida demonstrated incongruent phylogenetic signals between nuclear and plastid genes, especially in terms of the internal relationships among the three primary plastid-bearing groups (i.e., green algae plus land plants, red algae, and glaucophytes) (Deschamps and Moreira 2009). Additional sampling of genes and taxa were suggested as ways to solve this "problem" (Deschamps and Moreira 2009). It is notable that data incongruence is considered a "problem," which may reflect a subjective bias for the concept of Archaeplastida. However, if one approaches the data without a prior assumption of the Archaeplastida hypothesis, the observed signal conflict could be a hint that so-called "primary plastids" followed a more complicated evolutionary trajectory than had previously been assumed. For instance, the conflicting signals may stem from multiple occasions of plastid acquisition such as via serial transfer of plastids among "primary plastid" bearing lineages (Figure 5.1b) (Kim and Maruyama 2014).

C On the "Sameness" of Plastids

Plastids form a robust clade to the exclusion of cyanobacteria in molecular phylogenies (Ponce-Toledo et al. 2017). In other words, as far as plastid evolution is concerned, there are no known extant cyanobacteria that break up the union of the primary plastids (plus their secondary and tertiary descendants) in multigene phylogenies. In addition, there are several plastid-specific traits that are not known from cyanobacteria, such as the presence of Toc34 (Kim and Archibald 2009). Together, these data suggest that all plastids trace back to a single cyanobacterial ancestor (Archibald 2015). Some notable plastid-specific features are associated with the plastid protein translocon (TIC–TOC) complex: in particular, Toc34 and Tic110 (note, though, that the role of Tic110 as a core component of TIC has recently been questioned, Nakai 2015); plastid solute transporters, including the sugar phosphate transporter UhpC and sugar phosphate/phosphate translocators (PTs); and light-harvesting complex (LHC) proteins, particularly those with three transmembrane helices (Tyra et al. 2007; Price et al. 2012; Sturm et al. 2013). It should be noted that neither multi-helix LHC proteins nor plastid PTs are known for glaucophytes (Price et al. 2012), which supports the hypothesis that these features arose after the common ancestor of plastids originated but prior to the divergence of green and red algal plastids (Engelken et al. 2010).

Even so, there are reasons to be cautious and not consider the single plastid origin hypothesis as "set in stone." For instance, it is possible that cyanobacteria having plastid-specific traits like UhpC and Toc34 existed at the time of plastid origin, but since went extinct, or have so far not been found (Rujan and Martin 2001; Maruyama and Archibald 2012). Howe et al. (2008) noted: "the combination of Ockham's Razor with limited taxon sampling and mass extinction will implicitly favor the conclusions that (i) evidence favors monophyly over polyphyly but (ii) we cannot be sure."

Howe et al. conclusion (ii) may seem to point out the incompleteness of the methodology in the inference of plastid origins, but this is a natural consequence of probabilistic inference. It is a great challenge to calculate and to test statistically which of any given hypotheses is correct, or (more formally) is more probable than the others (Sober and Steel 2002). An ancient event cannot be correctly reconstructed if information relevant to that event is partly or entirely lost. When inferring deep evolutionary history, such as that of the plastid, it is perhaps not necessary to "prove" an *a priori* hypothesis or to search for the "best" hypothesis; but rather the key is to be open-minded and continue updating evolutionary models as new data and methods become available.

Howe et al. conclusion (i) is more substantial. Parsimony is not always the best choice (Sober 1988), and this appears to be the case in deducing the evolutionary history of the plastid. For instance, the multiple plastid origin hypothesis would not be less likely than the single origin hypothesis if one assumes the following. Perhaps, there was a certain cyanobacterium prone to be predated on, and subsequently, retained as a symbiont by early eukaryotes (maybe akin to the modern day *Symbiodinium*, which associates with many unrelated host eukaryotes). If so, the probability of the "permanent" incorporation of that specific cyanobacterial type was much higher than often assumed. Meanwhile, one would be inclined towards the single origin hypothesis if it is assumed that a primary plastid-generating endosymbiosis is extremely difficult.

Beyond considering how easy or difficult plastid acquisition may have been (Howe et al. 2008), there is a need for reevaluating our perception of "sameness" between plastids. For example, unlike green algal and glaucophyte plastids, which have the cyanobacterium-derived rubisco form I-B, red algal plastids possess form I-D, which is also found in proteobacteria (Morden et al. 1992; Delwiche and Palmer 1996; Tabita et al. 2008). This means that, at least in terms of gene composition, the three types of primary plastids are not completely identical, and their variance cannot be simply explained by differential gene loss. Nevertheless, it is worth noting that this finding itself is compatible with both the single origin and multiple origins hypotheses. Within the framework of the single plastid origin hypothesis, the "unusual" red algal rubisco genes can be explained as a consequence of LGT that occurred prior to the last common ancestor of red algal plastids, but after the initial establishment of primary plastids (Delwiche and Palmer 1996). Alternatively, one could interpret that the red algal plastid originated—independently from the green algal or glaucophyte plastid—from an ancestral cyanobacterium having rubisco form I-D, which had been acquired via LGT prior to the red algal plastid acquisition event. By the same logic, "sameness" or difference of any plastid-related characters, including the conserved genes in plastid genomes, does not directly address whether plastid origin was singular or multiple (Kim and Archibald 2009). This further illustrates that the single plastid origin hypothesis owes much to the principle of parsimony. If simpler, parsimony-driven hypotheses are not preferred by default; one may as well infer that the multiple plastid origins hypotheses are no less likely. This is the very lesson learned from the rise and demise of the cabozoan and chromalveolate hypotheses.

In a review of phylogenetic analyses of the relationships between prokaryotes and eukaryotes, Brown and Doolittle (1997) remarked:

> "'We need to have more data....,' is the incessant refrain for most summaries of this sort. However, archaeal, bacterial, and eukaryotic genome sequencing projects are generating tremendous volumes of relevant data. Thus, the major challenge is the synthesis of a grander view of the prokaryote-eukaryote transition."

Twenty years later, the same kind of refrain is continuing in the field of phylogenomics.

Phylogenomic data, even including those based on over hundreds of proteins, do not seem to provide unequivocal answers concerning the validity of the Archaeplastida concept (Kim and Maruyama 2014). Notwithstanding a growing volume of genome sequence data, besides common possession of a primary plastid, there is still no nucleocytoplasmic trait that both stands up to rigorous investigation and firmly unites the three archaeplastidan groups. Also, many questions remain to be addressed regarding the structural, genomic, and physiological processes relevant to plastid acquisition (e.g., the mechanism of algal phagocytosis and composition of the components responsible for it) (Maruyama and Kim 2013; Burns et al. 2015). The synthesis of a grander view of eukaryotic algal evolution is still challenging but also very promising in the "post" genomic era.

5.6   CONCLUSION

This chapter provided an overview on the origin and early diversification of eukaryotes with emphasis on symbiosis-driven transformation of cell structures and genomes. Modern-day eukaryotic cells are descended from a symbiotic merger between an archaeon and one or more bacteria that presumably occurred >2 billion years ago. In other words, the eukaryotic cell is a chimera between multiple, distinct prokaryotic cells. The respiratory mitochondrion was acquired from an α-proteobacterial symbiont prior to the diversification of the extant major groups of eukaryotes. On the other hand, the plastid—an organelle best known for photosynthesis—arose from a cyanobacterial precursor, and its distribution across the eukaryotic tree of life is patchy. Both the mitochondrion and plastid have made a significant impact on the nuclear genomes of eukaryotes. These two organelles, however, differ notably in their mode of evolution: the mitochondrion-generating symbiosis occurred once; acquisition of the plastid took place via a series of endosymbioses, first with a cyanobacterial symbiont and later with plastid-bearing eukaryotic symbionts. Our understanding of plastid evolution has changed dramatically over the past decade with emerging new data, analytical tools, and insights. Notably, a growing number of phylogenetic studies refute the parsimony-driven chromalveolate and cabozoan hypotheses concerning plastid evolution. Further, the Archaeplastida hypothesis, which posits the monophyly of and a common endosymbiotic origin for three algal groups (the green algae plus their land plant descendants; red algae; and glaucophytes), is also being challenged. We suggest that our ability to decipher the evolutionary trajectory of plastid-bearing eukaryotes is limited not only by extrinsic factors like data availability but also by our own intrinsic interpretational biases. Adoption of a probabilistic model that considers factors like the rate of plastid loss and gain may provide a more realistic view of the plastid origins and evolution.

ACKNOWLEDGMENTS

The authors are indebted to Aaron Heiss, John A Burns, Ryan Kerney, Asako Goto, and Alexandros Pittis for helpful comments and editorial suggestions. This work was supported by the NSF CAREER (#1453639) and Simons Foundation (SF-382790) awards to E. Kim, and the JSPS KAKENHI (#JP15K18562) award to S. Maruyama.

REFERENCES

Adl, S. M., Simpson, A. G. B., Lane, C. E., Bass, D., Bowser, S. S., Brown, M. W., and Burki, F. 2012. The revised classification of eukaryotes. *J Eukaryot Microbiol* 59: 429–514.

Andersson, I. and Backlund, A. 2008. Structure and function of Rubisco. *Plant Physiol Biochem* 46: 275–291.

Arakawa, K. 2016. No evidence for extensive horizontal gene transfer from the draft genome of a tardigrade. *Proc Natl Acad Sci* 113: E3057.

Archibald, J. M. 2009. The puzzle of plastid evolution. *Curr Biol* 19: R81–R88.

Archibald, J. M. 2015. Endosymbiosis and eukaryotic cell evolution. *Curr Biol* 25: R911–R921.

Archibald, J. M., Rogers, M. B., Toop, M., Ishida, K., and Keeling, P. J. 2003. Lateral gene transfer and the evolution of plastid-targeted proteins in the secondary plastid-containing alga *Bigelowiella natans. Proc Natl Acad Sci* 100: 7678–7683.

Ball, S. G., Bhattacharya, D., and Weber, A. P. 2016b. Infection and the first eukaryotes—Response. *Science* 352: 1065–1066.

Ball, S. G., Bhattacharya, D., Qiu, H., and Weber, A. P. 2016a. Commentary: Plastid establishment did not require a chlamydial partner. *Front Cell Infect Microbiol* 6: 43.

Ball, S. G., Subtil, A., Bhattacharya, D., Moustafa, A., Weber, A. P., Gehre, L., Colleoni, C., Arias, M. C., Cenci, U., and Dauvillée, D. 2013. Metabolic effectors secreted by bacterial pathogens: Essential facilitators of plastid endosymbiosis? *Plant Cell* 25: 7–21.

Baum, D. A. and Baum, B. 2014. An inside-out origin for the eukaryotic cell. *BMC Biol* 12: 76.

Baumgarten, S., Simakov, O., Esherick, L. Y., Liew, Y. J., Lehnert, E. M., Michell, C. T., Li, Y. 2015. The genome of *Aiptasia*, a sea anemone model for coral symbiosis. *Proc Natl Acad Sci USA* 112: 11893–11898.

Baurain, D., Brinkmann, H., Petersen, J., Rodríguez-Ezpeleta, N., Stechmann, A., Demoulin, V., Roger, A. J., Burger, G., Lang, B. F., and Philippe, H. 2010. Phylogenomic evidence for separate acquisition of plastids in cryptophytes, haptophytes, and stramenopiles. *Mol Biol Evol* 27: 1698–1709.

Bennett, G. M. and Moran, N. A. 2015. Heritable symbiosis: The advantages and perils of an evolutionary rabbit hole. *Proc Natl Acad Sci USA* 112: 10169–10176.

Bernander, R., Lind, A. E., and Ettema, T. J. 2011. An archaeal origin for the actin cytoskeleton: Implications for eukaryogenesis. *Commun Integr Biol* 4: 664–667.

Bhattacharya, D., Pelletreau, K. N., Price, D. C., Sarver, K. E., and Rumpho, M. E. 2013. Genome analysis of *Elysia chlorotica* Egg DNA provides no evidence for horizontal gene transfer into the germ line of this Kleptoplastic Mollusc. *Mol Biol Evol* 30: 1843–1852.

Bhattacharya, D., Yoon, H. S., and Hackett, J. D. 2004. Photosynthetic eukaryotes unite: Endosymbiosis connects the dots. *Bioessays* 26: 50–60.

Bodył, A. 2005. Do plastid-related characters support the chromalveolate hypothesis? *J Phycol* 41: 712–719.

Bodył, A., Mackiewicz, P., and Stiller, J. W. 2007. The intracellular cyanobacteria of *Paulinella chromatophora*: Endosymbionts or organelles? *Trends Microbiol* 15: 295–296.

Bodył, A., Stiller, J. W., and Mackiewicz, P. 2009. Chromalveolate plastids: Direct descent or multiple endosymbioses? *Trends Ecol Evol* 24: 119–121; author reply 121–122.

Boothby, T. C., Tenlen, J. R., Smith, F. W., Wang, J. R., Patanella, K. A., Nishimura, E. O., Tintori, S. C. et al. 2015. Evidence for extensive horizontal gene transfer from the draft genome of a tardigrade. *Proc Natl Acad Sci USA* 112: 15976–15981.

Brown, J. R. and Doolittle, W. F. 1997. Archaea and the prokaryote-to-eukaryote transition. *Microbiol Mol Biol Rev* 61: 456–502.

Brown, M. W., Kolisko, M., Silberman, J. D., and Roger, A. J. 2012. Aggregative multicellularity evolved independently in the eukaryotic supergroup Rhizaria. *Curr Biol* 22: 1123–1127.

Brown, M. W., Sharpe, S. C., Silberman, J. D., Heiss, A. A., Lang, B. F., Simpson, A. G., and Roger, A. J. 2013. Phylogenomics demonstrates that breviate flagellates are related to opisthokonts and apusomonads. *Proc Biol Sci* 280: 20131755.

Buchner, P. 1965. *Endosymbiosis of Animals with Plant Microorganisms*. New York: John Wiley & Sons.

Burki, F., Flegontov, P., Obornik, M., Cihlář, J., Pain, A., Lukeš, J. and Keeling, P. J. 2012. Re-evaluating the green versus red signal in eukaryotes with secondary plastid of red algal origin. *Genome Biol Evol* 4: 626–635.

Burki, F., Kaplan, M., Tikhonenkov, D. V., Zlatogursky, V., Minh, B. Q., Radaykina, L. V., Smirnov, A., Mylnikov, A. P., and Keeling, P. J. 2016. Untangling the early diversification of eukaryotes: A phylogenomic study of the evolutionary origins of Centrohelida, Haptophyta and Cryptista. *Proc Biol Sci* 283: 20152802.

Burki, F., Shalchian-Tabrizi, K., and Pawlowski, J. 2008. Phylogenomics reveals a new "megagroup" including most photosynthetic eukaryotes. *Biol Lett* 4: 366–369.

Burki, F., Shalchian-Tabrizi, K., Minge, M., Skjæveland, Å., Nikolaev, S. I., Jakobsen, K. S., and Pawlowski, J. 2007. Phylogenomics reshuffles the eukaryotic supergroups. *PLoS One* 2: e790.

Burns, J. A., Paasch, A., Narechania, A., and Kim, E. 2015. Comparative genomics of a bacterivorous green alga reveals evolutionary causalities and consequences of phagomixotrophic mode of nutrition. *Genome Biol Evol* 7: 3047–3061.

Busiek, K. K. and Margolin, W. 2015. Bacterial actin and tubulin homologs in cell growth and division. *Curr Biol* 25: R243–R254.

Carvalho-Santos, Z., Azimzadeh, J., Pereira-Leal, J. B., and Bettencourt-Dias, M. 2011. Tracing the origins of centrioles, cilia, and flagella. *J Cell Biol* 194: 165–175.

Cavalier-Smith, T. 1999. Principles of protein and lipid targeting in secondary symbiogenesis: euglenoid, dinoflagellate, and sporozoan plastid origins and the eukaryote family tree. *J Eukaryot Microbiol* 46: 347–366.

Cavalier-Smith, T. 2003. Protist phylogeny and the high-level classification of Protozoa. *Euro J Protistol* 39: 338–348.

Cavalier-Smith, T. 2014. The neomuran revolution and phagotrophic origin of eukaryotes and cilia in the light of intracellular coevolution and a revised tree of life. *Cold Spring Harb Perspect Biol* 6: a016006.

Cavalier-Smith, T. and Chao, E. E. 2004. Protalveolate phylogeny and systematics and the origins of Sporozoa and dinoflagellates (phylum Myzozoa nom. nov.). *Eur J Protistol* 40: 185–212.

Cenci, U., Bhattacharya, D., Weber, A. P., Colleoni, C., Subtil, A., and Ball, S. G. 2017. Biotic host–pathogen interactions as major drivers of plastid endosymbiosis. *Trends Plant Sci* 22: 316–328.

Choi, C. J., Bachy, C., Jaeger, G. S., Poirier, C., Sudek, L., Sarma, V. V. S. S., Mahadevan, A., Giovannoni, S. J., and Worden, A. Z. 2017. Newly discovered deep-branching marine plastid lineages are numerically rare but globally distributed. *Curr Biol* 27: R15–R16.

Cornejo, E., Subramanian, P., Li, Z., Jensen, G. J., and Komeili, A. 2016. Dynamic remodeling of the magnetosome membrane is triggered by the initiation of biomineralization. *MBio* 7: e01898–15.

Corsaro, D. and Venditti, D. 2006. Bacterial endosymbionts in prokaryotes. In *Complex Intracellular Structures in Prokaryotes* (Ed.) Shively, J. M., pp. 359–371. Berlin, Germany: Springer-Verlag.

Curtis, B. A., Tanifuji, G., Burki, F., Gruber, A., Irimia, M., Maruyama, S., Arias, M. C. et al. 2012. Algal genomes reveal evolutionary mosaicism and the fate of nucleomorphs. *Nature* 492: 59–65.

Dacks, J. B., Field, M. C., Buick, R., Eme, L., Gribaldo, S., Roger, A. J., Brochier-Armanet, C., and Devos, D. P. 2016. The changing view of eukaryogenesis–fossils, cells, lineages and how they all come together. *J Cell Sci* 129: 3695–703.

Dagan, T., Roettger, M., Stucken, K., Landan, G., Koch, R., Major, P., Gould, S. B. et al. 2013. Genomes of Stigonematalean cyanobacteria (subsection V) and the evolution of oxygenic photosynthesis from prokaryotes to plastids. *Genome Biol Evol* 5: 31–44.

De Las Rivas, J. and Barber, J. 2004. Analysis of the structure of the PsbO protein and its implications. *Photosynth Res* 81: 329–343.

Delaye, L., Valadez-Cano, C., and Pérez-Zamorano, B. 2016. How really ancient is *Paulinella chromatophora*? *PLoS Curr Tree Life.* doi:10.1371/currents.tol. e68a099364bb1a1e129a17b4e06b0c6b.

Delwiche, C. F. and Palmer, J. D. 1996. Rampant horizontal transfer and duplication of rubisco genes in eubacteria and plastids. *Mol Biol Evol* 13: 873–882.

DeSalle, R., Perkins, S. L., and Wynne, P. J. 2015. *Welcome to the Microbiome: Getting to Know the Trillions of Bacteria and Other Microbes in, on, and Around You.* New Haven, CT: Yale University Press.

Deschamps, P. and Moreira, D. 2009. Signal conflicts in the phylogeny of the primary photosynthetic eukaryotes. *Mol Biol Evol* 26: 2745–2753.

Deschamps, P. and Moreira, D. 2012. Reevaluating the green contribution to diatom genomes. *Genome Biol Evol* 4: 683–688.

Devos, D. P. 2014. PVC bacteria: Variation of, but not exception to, the Gram-negative cell plan. *Trends Microbiol* 22: 14–20.

Diekmann, Y. and Pereira-Leal, J. B. 2013. Evolution of intracellular compartmentalization. *Biochem J* 449: 319–331.

Domman, D., Horn, M., Embley, T. M., and Williams, T. A. 2015. Plastid establishment did not require a chlamydial partner. *Nat Commun* 6: 642.

Dufourc, E. J. 2008. Sterols and membrane dynamics. *J Chem Biol* 1: 63–77.

Duhita, N., Satoshi, S., Kazuo, H., Daisuke, M., and Takao, S. 2010. The origin of peroxisomes: The possibility of an actinobacterial symbiosis. *Gene* 450: 18–24.

Engelken, J., Brinkmann, H., and Adamska, I. 2010. Taxonomic distribution and origins of the extended LHC (light-harvesting complex) antenna protein superfamily. *BMC Evol Biol* 10: 233.

Ettema, T. J., Lindås, A. C., and Bernander, R. 2011. An actin-based cytoskeleton in archaea. *Mol Microbiol* 80: 1052–1061.

Fuerst, J. A. 2013. The PVC superphylum: Exceptions to the bacterial definition? *Antonie Van Leeuwenhoek* 104: 451–466.

Fuerst, J. A. and Sagulenko, E. 2011. Beyond the bacterium: Planctomycetes challenge our concepts of microbial structure and function. *Nat Rev Microbiol* 9: 403–413.

Gabaldón, T. 2010. Peroxisome diversity and evolution. *Philos Trans R Soc Lond B Biol Sci* 365: 765–773.

Gabaldón, T. and Pittis, A. A. 2015. Origin and evolution of metabolic sub-cellular compartmentalization in eukaryotes. *Biochimie* 119: 262–268.

Garrett, T. A., Schmeitzel, J. L., Klein, J. L., Hwang, J. J., and Schwarz, J. A. 2013. Comparative lipid profiling of the cnidarian *Aiptasia pallida* and its dinoflagellate symbiont. *PLoS One* 8: e57975.

Gilbert, S. F., McDonald, E., Boyle, N., Buttino, N., Gyi, L., Mai, M., Prakash, N., and Robinson, J. 2010. Symbiosis as a source of selectable epigenetic variation: Taking the heat for the big guy. *Philos Trans R Soc Lond B Biol Sci* 365: 671–678.

Gould, S. B. 2016. Infection and the first eukaryotes. *Science*, 352: 1065.

Gould, S. B., Garg, S. G., and Martin, W. F. 2016. Bacterial vesicle secretion and the evolutionary origin of the eukaryotic endomembrane system. *Trends Microbiol* 24: 525–534.

Graham, L. E. and Graham, J. M. 1980. Endosymbiotic Chlorella (Chlorophyta) in a species of Vorticella (Ciliophora). *Trans Am Microsc Soc* 99: 160–166.

Graham, L. E., Graham, J. M., Wilcox, L. M., and Cook, M. E. 2016. *Algae*, 3rd ed. LJLM Press LLC.

Grau-Bové, X., Sebé-Pedrós, A., and Ruiz-Trillo, I. 2015. The eukaryotic ancestor had a complex ubiquitin signaling system of archaeal origin. *Mol Biol Evol* 32: 726–739.

Güttinger, S., Laurell, E., and Kutay, U. 2009. Orchestrating nuclear envelope disassembly and reassembly during mitosis. *Nature Rev Mol Cell Biol* 10: 178–191.

Hashimoto, T., Horikawa, D. D., Saito, Y., Kuwahara, H., Kozuka-Hata, H., Shin, T., Minakuchi, Y. et al. 2016. Extremotolerant tardigrade genome and improved radiotolerance of human cultured cells by tardigrade-unique protein. *Nat Commun* 7: 12808.

He, C. Y. 2007. Golgi biogenesis in simple eukaryotes. *Cell Microbiol* 9: 566–572.

Hettema, E. H. and Motley, A. M. 2009. How peroxisomes multiply. *J Cell Sci* 122: 2331–2336.

Hjorth, E., Hadfi, K., Gould, S. B., Kawach, O., Sommer, M. S., Zauner, S., and Maier, U. G. 2004. Zero, one, two, three, and perhaps four–Endosymbiosis and the gain and loss of plastids. *Endocytobiol Cell Res* 15: 459–468.

Hochstrasser, M. 2009. Origin and function of ubiquitin-like proteins. *Nature* 458: 422–429.

Howe, C. J., Barbrook, A. C., Nisbet, R. E. R., Lockhart, P. J., and Larkum, A. W. D. 2008. The origin of plastids. *Philos Trans R Soc Lond B Biol Sci* 363: 2675–2685.

Huang, J. and Yue, J. 2013. Horizontal gene transfer in the evolution of photosynthetic eukaryotes. *J Syst Evol* 51: 13–29.

Ishida, K. I. and Green, B. R. 2002. Second-and third-hand chloroplasts in dinoflagellates: Phylogeny of oxygen-evolving enhancer 1 (PsbO) protein reveals replacement of a nuclear-encoded plastid gene by that of a haptophyte tertiary endosymbiont. *Proc Natl Acad Sci USA* 99: 9294–9299.

Izoré, T., Kureisaite-Ciziene, D., McLaughlin, S. H., and Löwe, J. 2016. Crenactin forms actin-like double helical filaments regulated by arcadin-2. *eLife* 5: e21600.

Janouskovec, J., Horák, A., Oborn, M., Lukeš, J., and Keeling, P. J. 2010. A common red algal origin of the apicomplexan, dinoflagellate, and heterokont plastids. *Proc Natl Acad Sci* 107: 10949–10954.

Jenkins, C., Samudrala, R., Anderson, I., Hedlund, B. P., Petroni, G., Michailova, N., Pinel, N., Overbeek, R., Rosati, G., and Staley, J. T. 2002. Genes for the cytoskeletal protein tubulin in the bacterial genus Prosthecobacter. *Proc Natl Acad Sci USA* 99: 17049–17054.

Johnson, M. D. 2011. Acquired phototrophy in ciliates: A review of cellular interactions and structural adaptations. *J Euk Microbiol* 58: 185–195.

Karkar, S., Facchinelli, F., Price, D. C., Weber, A. P., and Bhattacharya, D. 2015. Metabolic connectivity as a driver of host and endosymbiont integration. *Proc Natl Acad Sci USA* 112: 10208–10215.

Keeling, P. J. 2004. Diversity and evolutionary history of plastids and their hosts. *Am J Bot* 91: 1481–1493.

Keeling, P. J. 2010. The endosymbiotic origin, diversification and fate of plastids. *Philos Trans R Soc Lond B Biol Sci* 365: 729–748.

Kerney, R., Kim, E., Hangarter, R. P., Heiss, A. A., Bishop, C. D., and Hall, B. K. 2011. Intracellular invasion of green algae in a salamander host. *Proc Natl Acad Sci USA* 108: 6497–6502.

Kim, E. and Archibald, J. M. 2009. Diversity and evolution of plastids and their genomes. In *The Chloroplast* (Eds.) Sandelius, A. S. and Aronsson, H., pp. 1–39. Berlin, Germany: Springer.

Kim, E. and Archibald, J. M. 2013. Ultrastructure and molecular phylogeny of the cryptomonad *Goniomonas avonlea* sp. nov. *Protist* 164: 160–182.

Kim, E. and Graham, L. E. 2008. EEF2 analysis challenges the monophyly of Archaeplastida and Chromalveolata. *PLoS One* 3: e2621.

Kim, E. and Maruyama, S. 2014. A contemplation on the secondary origin of green algal and plant plastids. *Acta Soc Bot Pol* 83: 331–336.

Kim, E., Harrison, J. W., Sudek, S., Jones, M. D., Wilcox, H. M., Richards, T. A., Worden, A. Z., and Archibald, J. M. 2011. Newly identified and diverse plastid-bearing branch on the eukaryotic tree of life. *Proc Natl Acad Sci USA* 108: 1496–1500.

Knoll, A. H. 2003. *Life on a Young Planet: The First Three Billion Years of Evolution on Earth*. Princeton, NJ: Princeton University Press.

Koops, H. P., Harms, H., and Wehrmann, H. 1976. Isolation of a moderate halophilic ammonia-oxidizing bacterium, *Nitrosococcus mobilis* nov. sp. *Arch Microbiol* 107: 277–282.

Koumandou, V. L., Wickstead, B., Ginger, M. L., van der Giezen, M., Dacks, J. B., and Field, M. C. 2013. Molecular paleontology and complexity in the last eukaryotic common ancestor. *Crit Rev Biochem Mol Biol* 48: 373–396.

Koutsovoulos, G., Kumar, S., Laetsch, D. R., Stevens, L., Daub, J., Conlon, C., Maroon, H., Thomas, F., Aboobaker, A. A., and Blaxter, M. 2016. No evidence for extensive horizontal gene transfer in the genome of the tardigrade Hypsibius Dujardini. *Proc Natl Acad Sci USA* 113: 5053–5058.

Ku, C., Nelson-Sathi, S., Roettger, M., Sousa, F. L., Lockhart, P. J., Bryant, D., Hazkani-Covo, E., McInerney, J. O., Landan, G., and Martin, W. F. 2015. Endosymbiotic origin and differential loss of eukaryotic genes. *Nature* 524: 427–432.

Kuzminov, A. 2014. The precarious prokaryotic chromosome. *J Bacteriol* 196: 1793–1806.

Lane, C. E. and Archibald, J. M. 2009. Reply to Bodył, Stiller and Mackiewicz: "Chromalveolate plastids: Direct descent or multiple endosymbioses?" *Trends Ecol Evol* 24: 121–122.

Lane, N. 2011. Energetics and genetics across the prokaryote-eukaryote divide. *Biol Direct* 6: 35.

Liberton, M., Austin, J. R., Berg, R. H., and Pakrasi, H. B. 2011. Unique thylakoid membrane architecture of a unicellular N_2-fixing cyanobacterium revealed by electron tomography. *Plant Physiol* 155: 1656–1566.

Lonhienne, T. G., Sagulenko, E., Webb, R. I., Lee, K. C., Franke, J., Devos, D. P., Nouwens, A., Carroll, B. J., and Fuerst, J. A. 2010. Endocytosis-like protein uptake in the bacterium Gemmata obscuriglobus. *Proc Natl Acad Sci USA* 107: 12883–12888.

López-García, P. and Moreira, D. 2015. Open questions on the origin of eukaryotes. *Trends Ecol Evol* 30: 697–708.

Lorenz, M. G. and Wackernagel, W. 1994. Bacterial gene transfer by natural genetic transformation in the environment. *Microbiol Rev* 58: 563–602.

Maddison, W. P. 1997. Gene trees in species trees. *Syst Biol* 46: 523–536.

Margulis, L., Chapman, M., Guerrero, R., and Hall, J. 2006. The last eukaryotic common ancestor (LECA): Acquisition of cytoskeletal motility from aerotolerant spirochetes in the Proterozoic Eon. *Proc Natl Acad Sci USA* 103: 13080–13085.

Marin, B., Nowack, E. C., and Melkonian, M. 2005. A plastid in the making: Evidence for a second primary endosymbiosis. *Protist* 156: 425–432.

Martijn, J. and Ettema, T. J. 2013. From archaeon to eukaryote: The evolutionary dark ages of the eukaryotic cell. *Biochem Soc Trans* 41: 451–457.

Martin, W. F., Garg, S., and Zimorski, V. 2015. Endosymbiotic theories for eukaryote origin. *Phil Trans R Soc B* 370: 20140330.

Martin, W., Rujan, T., Richly, E., Hansen, A., Cornelsen, S., Lins, T., Leister, D., Stoebe, B., Hasegawa, M., and Penny, D. 2002. Evolutionary analysis of Arabidopsis, cyanobacterial, and chloroplast genomes reveals plastid phylogeny and thousands of cyanobacterial genes in the nucleus. *Proc Natl Acad Sci USA* 99: 12246–12251.

Maruyama, S. and Archibald, J. M. 2012. Endosymbiosis, gene transfer and algal cell evolution. In *Advances in Algal Cell Biology* (Eds.) Heimann, K. and Katsaros, C., pp. 21–42. Berlin, Germany: De Gruyter.

Maruyama, S. and Kim, E. 2013. A modern descendant of early green algal phagotrophs. *Curr Biol* 23: 1081–1084.

Maruyama, S., Matsuzaki, M., Misawa, K., and Nozaki, H. 2009. Cyanobacterial contribution to the genomes of the plastid-lacking protists. *BMC Evol Biol* 9: 197.

Maupin-Furlow, J. A. 2014. Prokaryotic ubiquitin-like protein modification. *Annu Rev Microbiol* 68: 155–175.

McInerney, J. O., Martin, W. F., Koonin, E. V., Allen, J. F., Galperin, M. Y., Lane, N., Archibald, J. M., and Embley, T. M. 2011. Planctomycetes and eukaryotes: A case of analogy not homology. *Bioessays* 33: 810–817.

McInerney, J. O., O'connell, M. J., and Pisani, D. 2014. The hybrid nature of the Eukaryota and a consilient view of life on Earth. *Nat Rev Microbiol* 12: 449–455.

Mitchison, T. J. and Cramer, L. P. 1996. Actin-based cell motility and cell locomotion. *Cell* 84: 371–379.

Moebius, N., Üzüm, Z., Dijksterhuis, J., Lackner, G., and Hertweck, C. 2014. Active invasion of bacteria into living fungal cells. *eLife* 3: e03007.

Moore, R. B., Oborník, M., Janouskovec, J., Chrudimský, T., Vancová, M., Green, D. H., Wright, S. W. et al. 2008. A photosynthetic alveolate closely related to apicomplexan parasites. *Nature* 451: 959–963.

Moran, J., McKean, P. G., and Ginger, M. L. 2014. Eukaryotic flagella: Variations in form, function, and composition during evolution. *BioScience* 64: 1103–1114.

Morden, C. W. and Sherwood, A. R. 2002. Continued evolutionary surprises among dinoflagellates. *Proc Natl Acad Sci USA* 99: 11558–11560.

Morden, C., Delwiche, C. F., Kuhsel, M., and Palmer, J. D. 1992. Gene phylogenies and the endosymbiotic origin of plastids. *Biosystems* 28: 75–90.

Moustafa, A., Beszteri, B., Maier, U. G., Bowler, C., Valentin, K., and Bhattacharya, D. 2009. Genomic footprints of a cryptic plastid endosymbiosis in diatoms. *Science* 324: 1724–1726.

Murat, D., Byrne, M., and Komeili, A. 2010. Cell biology of prokaryotic organelles. *Cold Spring Harb Perspect Biol* 2: a000422.

Nakai, M. 2015. The TIC complex uncovered: The alternative view on the molecular mechanism of protein translocation across the inner envelope membrane of chloroplasts. *Biochim Biophys Acta* 1847: 957–967.

Nakayama, T., Ikegami, Y., Nakayama, T., Ishida, K. I., Inagaki, Y., and Inouye, I. 2011. Spheroid bodies in rhopalodiacean diatoms were derived from a single endosymbiotic cyanobacterium. *J Plant Res* 124: 93–97.

Nakayama, T., Kamikawa, R., Tanifuji, G., Kashiyama, Y., Ohkouchi, N., Archibald, J. M., and Inagaki, Y. 2014. Complete genome of a nonphotosynthetic cyanobacterium in a diatom reveals recent adaptations to an intracellular lifestyle. *Proc Natl Acad Sci USA* 111: 11407–114012.

Neumann, S., Wessels, H. J., Rijpstra, W. I. C., Sinninghe Damsté, J. S., Kartal, B., Jetten, M. S., and Niftrik, L. 2014. Isolation and characterization of a prokaryotic cell organelle from the anammox bacterium *Kuenenia stuttgartiensis*. *Mol Microbiol* 94: 794–802.

Ng, P. K., Lin, S. M., Lim, P. E., Liu, L. C., Chen, C. M., and Pai, T. W. 2017. Complete chloroplast genome of *Gracilaria firma* (Gracilariaceae, Rhodophyta), with discussion on the use of chloroplast phylogenomics in the subclass Rhodymeniophycidae. *BMC Genomics* 18: 40.

Ni, T., Yue, J., Sun, G., Zou, Y., Wen, J., and Huang, J. 2012. Ancient gene transfer from algae to animals: Mechanisms and evolutionary significance. *BMC Evol Biol* 12: 83.

Nowack, E. C. and Grossman, A. R. 2012. Trafficking of protein into the recently established photosynthetic organelles of Paulinella chromatophora. *Proc Natl Acad Sci USA* 109: 5340–5345.

Nowack, E. C., Melkonian, M., and Glöckner, G. 2008. Chromatophore genome sequence of *Paulinella* sheds light on acquisition of photosynthesis by eukaryotes. *Curr Biol* 18: 410–418.

Nowack, E. C., Price, D. C., Bhattacharya, D., Singer, A., Melkonian, M., and Grossman, A. R. 2016. Gene transfers from diverse bacteria compensate for reductive genome evolution in the chromatophore of *Paulinella chromatophora*. *Proc Natl Acad Sci USA* 113: 12214–12219.

Nunoura, T., Takaki, Y., Kakuta, J., Nishi, S., Sugahara, J., Kazama, H., Chee, G. J. et al. 2011. Insights into the evolution of Archaea and eukaryotic protein modifier systems revealed by the genome of a novel archaeal group. *Nucleic Acids Res* 39: 3204–3223.

O'Kelly, C. J. 1993. Relationships of eukaryotic algal groups to other protists. In *Ultrastructure of Microalgae* (Ed.) Berner, T., pp. 269–293. Boca Raton, FL: CRC Press.

Okamoto, N. and Inouye, I. 2005. A secondary symbiosis in progress? *Science* 310: 287.

Paracer, S. and Ahmadjian, V. 2000. *Symbiosis: An Introduction to Biological Associations*, 2nd ed. New York: Oxford University Press.

Patron, N. J. and Waller, R. F. 2007. Transit peptide diversity and divergence: A global analysis of plastid targeting signals. *Bioessays* 29: 1048–1058.

Petersen, J., Ludewig, A.-K., Michael, V., Bunk, B., Jarek, M., Baurain, D., and Brinkmann, H. 2014. *Chromera velia*, endosymbioses and the rhodoplex hypothesis–Plastid evolution in cryptophytes, alveolates, stramenopiles, and haptophytes (CASH lineages). *Genome Biol Evol* 6: 666–684.

Pierce, S. K., Fang, X., Schwartz, J. A., Jiang, X., Zhao, W., Curtis, N. E., Kocot, K. M., Yang, B., and Wang, J. 2012. Transcriptomic evidence for the expression of horizontally transferred algal nuclear genes in the photosynthetic sea slug, *Elysia chlorotica*. *Mol Biol Evol* 29: 1545–1556.

Pilhofer, M., Ladinsky, M. S., McDowall, A. W., Petroni, G., and Jensen, G. 2011. Microtubules in bacteria: Ancient tubulins build a five-protofilament homolog of the eukaryotic cytoskeleton. *PLoS Biol* 9: e1001213.

Pillet, L. and Pawlowski, J. 2013. Transcriptome analysis of foraminiferan *Elphidium margaritaceum* questions the role of gene transfer in kleptoplastidy. *Mol Biol Evol* 30: 66–69.

Pittis, A. A. and Gabaldón, T. 2016. Late acquisition of mitochondria by a host with chimaeric prokaryotic ancestry. *Nature* 531: 101–104.

Pochon, X. and Gates, R. D. 2010. A new *Symbiodinium* clade (Dinophyceae) from soritid foraminifera in Hawai'i. *Mol Phylogen Evol* 56: 492–497.

Ponce-Toledo, R. I., Deschamps, P., López-García, P., Zivanovic, Y., Benzerara, K., and Moreira, D. 2017. An early-branching freshwater cyanobacterium at the origin of plastids. *Curr Biol* 27: 386–391.

Price, D. C., Chan, C. X., Yoon, H. S., Yang, E. C., Qiu, H., Weber, A. P., Schwacke, R. et al. 2012. *Cyanophora paradoxa* genome elucidates origin of photosynthesis in algae and plants. *Science* 335: 843–847.

Rauch, C., de Vries, J., Rommel, S., Rose, L. E., Woehle, C., Christa, G., Laetz, E. M. et al. 2015. Why it is time to look beyond algal genes in photosynthetic slugs. *Genome Biol Evol* 7: 2602–2607.

Raven, J. A. 2013. Rubisco: Still the most abundant protein of Earth? *New Phytol* 198: 1–3.

Raven, J. A., Beardall, J., Larkum, A. W., and Sánchez-Baracaldo, P. 2013. Interactions of photosynthesis with genome size and function. *Philo Trans R Soc Lond B Biol Sci* 368: 20120264.

Reyes-Prieto, A., Moustafa, A., and Bhattacharya, D. 2008. Multiple genes of apparent algal origin suggest ciliates may once have been photosynthetic. *Curr Biol* 18: 956–962.

Rice, D. W. and Palmer, J. D. 2006. An exceptional horizontal gene transfer in plastids: Gene replacement by a distant bacterial paralog and evidence that haptophyte and cryptophyte plastids are sisters. *BMC Biol* 4: 31.

Rochette, N. C., Brochier-Armanet, C., and Gouy, M. 2014. Phylogenomic test of the hypotheses for the evolutionary origin of eukaryotes. *Mol Biol Evol* 31: 832–845.

Rodríguez-Ezpeleta, N., Brinkmann, H., Burger, G., Roger, A. J., Gray, M. W., Philippe, H. and Lang, B. F. 2007. Toward resolving the eukaryotic tree: The phylogenetic positions of jakobids and cercozoans. *Curr Biol* 17: 1420–1425.

Rogers, M. B., Gilson, P. R., Su, V., McFadden, G. I., and Keeling, P. J. 2007. The complete chloroplast genome of the chlorarachniophyte *Bigelowiella natans*: Evidence for independent origins of chlorarachniophyte and euglenid secondary endosymbionts. *Mol Biol Evol* 24: 54–62.

Rujan, T. and Martin, W. 2001. How many genes in *Arabidopsis* come from cyanobacteria? An estimate from 386 protein phylogenies. *Trends Genet* 17: 113–120.

Rumpho, M. E., Summer, E. J., and Manhart, J. R. 2000. Solar-powered sea slugs. Mollusc/algal chloroplast symbiosis. *Plant Physiol* 123: 29–38.

Rumpho, M. E., Worful, J. M., Lee, J., Kannan, K., Tyler, M. S., Bhattacharya, D., Moustafa, A., and Manhart, J. R. 2008. Horizontal gene transfer of the algal nuclear gene *psbO* to the photosynthetic sea slug *Elysia chlorotica*. *Proc Natl Acad Sci USA* 105: 17867–17871.

Sagan, L. 1967. On the origin of mitosing cells. *J Theor Biol* 14: 225–274.

Sanchez-Puerta, M. and Delwiche, C. F. 2008. A hypothesis for plastid evolution in chromalveolates. *J Phycol* 44: 1097–1107.

Santarella-Mellwig, R., Pruggnaller, S., Roos, N., Mattaj, I. W., and Devos, D. P. 2013. Three-dimensional reconstruction of bacteria with a complex endomembrane system. *PLoS Biol* 11: e1001565.

Scheffel, A. and Schüler, D. 2006. Magnetosomes in magnetotactic bacteria. In *Complex Intracellular Structures in Prokaryotes* (Ed.) Shively, J. M., pp. 167–191. Berlin, Germany: Springer-Verlag.

Scheuring, S., Nevo, R., Liu, L. N., Mangenot, S., Charuvi, D., Boudier, T., Prima, V., Hubert, P., Sturgis, J. N., and Reich, Z. 2014. The architecture of *Rhodobacter sphaeroides* chromatophores. *Biochimica et Biophysica Acta* 1837: 1263–1270.

Ševčíková, T., Horák, A., Klimeš, V., Zbránková, V., Demir-Hilton, E., Sudek, S., Jenkins, J. et al. 2015. Updating algal evolutionary relationships through plastid genome sequencing: Did alveolate plastids emerge through endosymbiosis of an ochrophyte. *Sci Rep* 5: 10134.

Shabalina, S. A. and Koonin, E. V. 2008. Origins and evolution of eukaryotic RNA interference. *Trends Ecol Evol* 23: 578–587.

Shi, T. and Falkowski, P. G. 2008. Genome evolution in cyanobacteria: The stable core and the variable shell. *Proc Natl Acad Sci USA* 105: 2510–2515.

Sober, E. 1988. *Reconstructing the Past: Parsimony, Evolution, and Inference.* Cambridge, MA: MIT Press.

Sober, E. and Steel, M. 2002. Testing the hypothesis of common ancestry. *J Theor Biol* 218: 395–408.

Sockett, R. E. 2009. Predatory lifestyle of *Bdellovibrio bacteriovorus. Annu Rev Microbiol* 63: 523–539.

Spang, A., Saw, J. H., Jørgensen, S. L., Zaremba-Niedzwiedzka, K., Martijn, J., Lind, A. E., van Eijk, R. et al. 2015. Complex archaea that bridge the gap between prokaryotes and eukaryotes. *Nature* 521: 173–179.

Stairs, C. W., Leger, M. M., and Roger, A. J. 2015. Diversity and origins of anaerobic metabolism in mitochondria and related organelles. *Phil Trans R Soc B* 370: 20140326.

Starokadomskyy, P. and Dmytruk, K. V. 2013. A bird's-eye view of autophagy. *Autophagy* 9: 1121–1126.

Stiller, J. W. 2011. Experimental design and statistical rigor in phylogenomics of horizontal and endosymbiotic gene transfer. *BMC Evol Biol* 11: 259.

Stiller, J. W. 2014. Toward an empirical framework for interpreting plastid evolution. *J Phycol* 50: 462–471.

Stiller, J. W., Schreiber, J., Yue, J., Guo, H., Ding, Q., and Huang, J. 2014. The evolution of photosynthesis in chromist algae through serial endosymbioses. *Nat Commun* 5: 5764.

Sturm, S., Engelken, J., Gruber, A., Vugrinec, S., Kroth, P. G., Adamska, I., and Lavaud, J. 2013. A novel type of light-harvesting antenna protein of red algal origin in algae with secondary plastids. *BMC Evol Biol* 13: 159.

Sun, G., Yang, Z., Ishwar, A., and Huang. J. 2010. Algal genes in the closest relatives of animals. *Mol Biol Evol* 27: 2879–2889.

Tabita, F. R., Hanson, T. E., Li, H., Satagopan, S., Singh, J., and Chan, S. 2007. Function, structure, and evolution of the RubisCO-like proteins and their RubisCO homologs. *Microbiol Mol Biol Rev* 71: 576–599.

Tabita, F. R., Satagopan, S., Hanson, T. E., Kreel, N. E., and Scott, S. S. 2008. Distinct form I, II, III, and IV Rubisco proteins from the three kingdoms of life provide clues about Rubisco evolution and structure/function relationships. *J Exp Bot* 59: 1515–1524.

Takahashi, F., Okabe, Y., Nakada, T., Sekimoto, H., Ito, M., Kataoka, H., and Nozaki, H. 2007. Origins of the secondary plastids of Euglenophyta and Chlorarachniophyta as revealed by an analysis of the plastid-targeting, nuclear-encoded gene *psbO. J Phycol* 43: 1302–1309.

Takishita, K., Chikaraishi, Y., Leger, M. M., Kim, E., Yabuki, A., Ohkouchi, N., and Roger, A. J. 2012. Lateral transfer of tetrahymanol-synthesizing genes has allowed multiple diverse eukaryote lineages to independently adapt to environments without oxygen. *Biol Direct* 7: 5.

Tan, I. S. and Ramamurthi, K. S. 2014. Spore formation in *Bacillus subtilis. Environ Microbiol Rep* 6: 212–225.

Theissen, U. and Martin, W. 2006. The difference between organelles and endosymbionts. *Curr Biol* 16: R1016–R1017.

Ting, C. S., Hsieh, C., Sundararaman, S., Mannella, C., and Marko, M. 2007. Cryo-electron tomography reveals the comparative three-dimensional architecture of *Prochlorococcus,* a globally important marine cyanobacterium. *J Bacteriol* 189: 4485–4493.

Trench, R. K. 1993. Microalgal-invertebrate symbioses: A review. *Endocyt Cell Res* 9: 135–175.

Tucker, J. D., Siebert, C. A., Escalante, M., Adams, P. G., Olsen, J. D., Otto, C., Stokes, D. L., and Hunter, C. N. 2010. Membrane invagination in *Rhodobacter sphaeroides* is initiated at curved regions of the cytoplasmic membrane, then forms both budded and fully detached spherical vesicles. *Mol Microbiol* 76: 833–847.

Tyler, B. M., Tripathy, S., Zhang, X., Dehal, P., Jiang, R. H., Aerts, A., Arredondo, F. D. et al. 2006. *Phytophthora* genome sequences uncover evolutionary origins and mechanisms of pathogenesis. *Science* 313: 1261–1266.

Tyra, H. M., Linka, M., Weber, A. P. M., and Bhattacharya, D. 2007. Host origin of plastid solute transporters in the first photosynthetic eukaryotes. *Genome Biol* 8: R212.

Van De Meene, A. M., Hohmann-Marriott, M. F., Vermaas, W. F., and Roberson, R. W. 2006. The three-dimensional structure of the cyanobacterium *Synechocystis* sp. PCC 6803. *Arch Microbiol* 184: 259–270.

Van Teeseling, M. C., Mesman, R. J., Kuru, E., Espaillat, A., Cava, F., Brun, Y. V., VanNieuwenhze, M. S., Kartal, B., and Van Niftrik, L. 2015. Anammox Planctomycetes have a peptidoglycan cell wall. *Nat Commun* 6: 6878.

Wägele, H., Deusch, O., Händeler, K., Martin, R., Schmitt, V., Christa, G., Pinzger, B. et al. (2011). Transcriptomic evidence that longevity of acquired plastids in the photosynthetic slugs *Elysia timida* and *Plakobranchus ocellatus* does not entail lateral transfer of algal nuclear genes. *Mol Biol Evol* 28: 699–706.

Wei, J. H., Yin, X., and Welander, P. V. 2016. Sterol synthesis in diverse bacteria. *Front Microbiol* 7: 990.

Wickstead, B. and Gull, K. 2011. The evolution of the cytoskeleton. *J Cell Biol* 194: 513–525.

Wisecaver, J. H. and Hackett, J. D. 2010. Transcriptome analysis reveals nuclear-encoded proteins for the maintenance of temporary plastids in the dinoflagellate *Dinophysis acuminata*. *BMC Genomics* 11: 366.

Woehle, C., Dagan, T., Martin, W. F., and Gould, S. B. 2011. Red and problematic green phylogenetic signals among thousands of nuclear genes from the photosynthetic and apicomplexa-related *Chromera velia*. *Genome Biol Evol* 3: 1220–1230.

Woese, C. R., Kandler, O., and Wheelis, M. L. 1990. Towards a natural system of organisms: Proposal for the domains Archaea, Bacteria, and Eucarya. *Proc Natl Acad Sci USA* 87: 4576–4579.

Yabuki, A., Inagaki, Y., and Ishida, K. I. 2010. *Palpitomonas bilix* gen. et sp. nov.: A novel deep-branching heterotroph possibly related to Archaeplastida or Hacrobia. *Protist* 161: 523–538.

Yamaguchi, A., Yubuki, N., and Leander, B. S. 2012. Morphostasis in a novel eukaryote illuminates the evolutionary transition from phagotrophy to phototrophy: Description of *Rapaza viridis* n. gen. et sp. (Euglenozoa, Euglenida). *BMC Evol Biol* 12: 29.

Yutin, N. and Koonin, E. V. 2012. Archaeal origin of tubulin. *Biol Direct* 7: 10.

Zaremba-Niedzwiedzka, K., Caceres, E. F., Saw, J. H., Bäckström, D., Juzokaite, L., Vancaester, E., Seitz, K. W. et al. 2017. Asgard archaea illuminate the origin of eukaryotic cellular complexity. *Nature* 541: 353–358.

6 Cellular Signaling Centers and the Maintenance and Evolution of Morphological Patterns in Vertebrates

Kathryn D. Kavanagh

CONTENTS

6.1 HISTORY OF THE TERM "SIGNALING CENTER"

Cellular signaling centers are defined as a cluster of nonproliferating cells that secrete molecular signals that regulate cell behaviors and the fate(s) of surrounding cells. This term has only been in use in the embryological literature for about 20 years, reflecting the timing of the first identification of molecular developmental signals directing morphogenesis.

Prior to the use of the term signaling center and the widespread use of molecular tools, localized tissue areas involved in regulating adjacent cell behaviors *were* recognized by experimental embryologists. The term "organizing center" or organizer was used for these areas, which were identified through experimental grafting or altering cell populations in developing embryos, primarily in chick, *Xenopus*, and *Ambystoma*, with the result of a repatterning of morphology in cells adjacent to the grafted cells (Spemann and Mangold 1924; Saunders and Gasseling 1968; Tickle et al. 1975). The knowledge that localized cell populations could organize—or control the fate of—the surrounding tissues was a fundamental conceptual advance in developmental biology made by these early twentieth century embryologists. The search for "morphogens," or signals expressed as a gradient through cell populations, as the source of regulatory control drove research for several decades in the nineteenth and twentieth century. Beginning in the last decade of the twentieth century, the search to identify the specific molecular signals involved in inducing morphogenesis has dominated the field.

This chapter focuses on cellular signaling centers identified in vertebrates. However, important concepts leading to the idea of cellular signaling centers began in the 1970s in *Drosophila*, especially; (a) the ideas of anterior and posterior axial organizing centers in the insect egg, followed by (b) molecular identification of potential morphogens in axis determination in *Drosophila* eggs (Nusslein-Volhard 1977; Roth et al. 1989; St. Johnston and Nusslein-Volhard 1992).

In the 1990s and early 2000s, the role of molecular signaling in vertebrate morphogenesis was dominated by models of interacting morphogenetic fields of signals, setting the stage for the idea of the existence of clusters of cells localized to release signals directing morphogenesis. These identified organizing centers in the limb such as the zone of proliferating activity (ZPA) were subsequently found to be the site of morphogen release, for example, cells of the ZPA are the source of the morphogen sonic hedgehog (SHH). Subsequently, the organizing center idea became synonymous with a signaling center, with the signal being the source location of a candidate morphogen, exciting researchers in vertebrate developmental biology with a new way to study morphogenesis (e.g., Slack 1987, entitled "We have a morphogen!"). Because inductive molecules were being identified, Gilbert and Saxén in 1993 wrote that, after 50 years seeking substances involved in embryonic induction, "we find ourselves poised at the beginning of a Renaissance in organizer studies."

A literature database search on Web of Science shows that the first use of the term signaling center in cellular developmental biology literature was in 1995 (Pankratz and Hoch 1995). In this case, hedgehog and wingless were found to control epithelial morphogenesis of the foregut. The first recorded uses of the term signaling center in vertebrates was the Nieuwkoop signaling center in the South African clawed toad *Xenopus* (Guger and Gumbiner 1995) and the enamel knot signaling center in molar teeth in mice (Jernvall 1996). The term came into use around the same time that the molecular biology field was rapidly accelerating and the importance of developmental signaling in morphogenesis was revealed.

As indicated above, a cellular signaling center is defined as a small, focused number of non-dividing cells that secrete signals to direct the behaviors of surrounding cells. Importantly, these cell clusters are temporary; the cells die when the maintenance factors diminish and eventually disappear (Jernvall et al. 1998). As more details of developmental signaling networks have become available, researchers have continued to describe more signaling centers in the developing embryo, revealing that these temporary signaling centers are a common mechanism for directing morphogenesis in the embryo and for enabling evolutionary change.

6.2 CELLULAR SIGNALING CENTERS AND MORPHOLOGICAL EVOLUTION

As more signaling centers are identified, it has become clear that signaling centers are prevalent in embryogenesis, used by many, if not most, organ-systems (Table 6.1). Here we present three of the earliest identified, and consequently most well-studied, signaling centers.

A THE NIEUWKOOP CENTER AND THE SPEMANN–MANGOLD ORGANIZER (THE NODE)

The Spemann–Mangold Organizer deserves special mention because it was the signaling center that was critical in bridging the transition from experimental embryology to molecular induction and developmental system modeling in vertebrates (e.g., Spemann and Mangold 1924; Gilbert and Saxén 1993; Meinhardt 2006, 2015).

In early frog zygotes, the Nieuwkoop Center is located on the opposite side from the point of sperm entry; its location is established by a gradient of nodal-related proteins, and it expresses Wnt proteins. Guger and Gumbiner (1995) showed that Wnt signaling mimics the Nieuwkoop organizing center. (This paper specifically shows the transition from the use of "organizing center" to "signaling center" as the molecular signal is identified.) These two points (sperm entry point and Nieuwkoop center) set up the dorsal–ventral (D–V) axis in the early embryo. The dorsal Nieuwkoop Center secretes signals to establish the Spemann–Mangold signaling center nearby at the dorsal lip of the blastopore. An antagonizing ventral center is set up on the opposite side of the embryo from the Spemann–Mangold Organizer; their interaction sets up D–V patterning.

The Spemann–Mangold signaling center was also called the *Primary Organizer* because of early experimental grafting that established the importance of these cells in directing morphogenesis of surrounding cells. The most famous of these experiments is the grafting of the dorsal lip of the blastopore to the ventral side of blastula embryos, resulting in a partial duplication of the embryo axis (Spemann and Mangold 1924; De Robertis 2006).

This primary induction was the focus of several influential papers in the early 1990s that established the definitive experimental basis for detecting induction in embryonic tissues. (De Robertis 2006, 2009). The early blastula organizers

TABLE 6.1
Signaling Centers Directing Morphogenesis

Signaling Center...	Directs Morphogenesis in...	References
Nieuwkoop center	Spemann's organizer	Guger and Gumbiner 1995
Node; Spemann's organizer	Primary axial patterning	Spemann and Mangold 1924; Beddington and Smith 1993; Gilbert and Saxén 1993; De Robertis 2006, 2009
Primitive streak	Germ layer formation	Mikawa et al. 2004
Rostral forebrain	Anterior neural plate or hypothalamus; Diencephalon	Kiecker and Lumsden 2004; Vieira and Martinez 2006; Pottin et al. 2011
Hindbrain roof plate	Rhombomere boundaries	Riley et al. 2001; Elsen et al. 2008
Forebrain	FEZ (Frontonasal ectodermal zone)	Hu and Marcucio 2009; Monuki 2007; Foppiano et al. 2007; Marcucio et al. 2007
Midline	Forebrain; Retinal differentiation	Martinez-Morales et al. 2005; Gupta and Sen 2015; Retaux and Kano 2010; Yao et al. 2016
Notochord	Neural tube D-V patterning	Yamada et al. 1993; Smith 1993
Ventral tail bud	Paraxial mesoderm	Liu et al. 2004
Early gill arch	Gill arches and pectoral fin	Gillis and Hall 2016
Early foregut	Gut epithelial folding	Pankratz and Hoch 1995
Dorsal aorta	Sympathic ganglia; adrenal medullary cells	Saito and Takahashi 2015
Adrenal gland	Glomerulosa regeneration	Vidal et al. 2016
Epicardium	Myocardial and coronary vasculature	Lavine and Ornitz 2008
Early tooth bud ectoderm	Enamel knot	Ahtiainen et al. 2016
Primary enamel knot	Tooth initiation	Jernvall et al. 1994; Vaahtokari et al. 1996
Secondary enamel knot	Tooth cusps	Jernvall and Thesleff 2000
Limb AER	Limb patterning	Fernandez-Teran and Ros 2008
Limb ZPA	A–P patterning in limb	Tickle 1975; Riddle et al. 1993; Tabin 1995; Dillon and Othmer 1999; Cohn and Tickle 1999; Capdevila and Izpisua Belmonte 2001; Grandel et al. 2000; Leal and Cohn 2016
Limb bud ectoderm	A–P limb patterning	Nissim et al. 2007; Harfe 2011
Limb skeleton condensations	Limb vasculature	Eshkar-Oren et al. 2009
Rugae growth zone	Palate	Welsh and O'Brien 2009
Skeletal condensations	Skeletal elements	Atchley and Hall 1991; Hall and Miyake 1992, 1995
Occipitocervical somite boundary	Skeletal muscle	Rowton et al. 2007
Cranial placodes	Sensory organs and ganglia	Baker and Bronner-Fraser 2001; Schlosser 2014; Schlosser et al. 2014
Blastemas	Regeneration of limbs	McCusker et al. 2015
Epithelial placodes	Epithelial organs (hair, sweat glands, mammary glands, reptile scales, turtle scutes, etc.)	Pispa and Thesleff 2003; Huh et al. 2013; DiPoi and Milinkovitch 2016; Moustakas-Verho et al. 2014; Voutilainen et al. 2012; Närhi et al. 2012
Pharyngeal pouch	Pharyngeal arch development	Edlund et al. 2014
Nasal placode	Lateral nasal skeleton	Szabo-Rogers et al. 2009

effectively create robust self-regulation of pattern formation and morphogenesis. The positional information network established by these dorsal and ventral signals has been conserved over evolution since the origin of bilateral animals (De Robertis 2009). The fish homolog of the Spemann organizer is the embryonic shield. The bird analog is Henson's node. In mammals, it is called the node.

The Spemann–Mangold signaling center dorsalizes ectoderm, and, during gastrulation, sets up the anterior–posterior axis and imprints, an axial Hox code, as the primary A–P axis is forming. The Spemann–Mangold signaling center secretes three signals: Noggin, Chordin, and Follistatin, which bind and inhibit the ventralizing transcription factors Bmp-4 and Frzb, thereby, "dorsalizing" the ectoderm. Goosecoid is the first growth factor expressed in the inductive tissue of the Spemann–Mangold signaling center. Many other growth factors are coexpressed and most are inhibitors of other growth factors, for example, Chordin, Noggin, and Follistatin are Bmp inhibitors, Frzb-1, Crescent, sFRP2, and Dkk are Wnt antagonists, and Cerberus inhibits Nodal (De Robertis 2006). These observations demonstrate that inhibitory molecules acting in feedback loops are an important network feature in the functioning of the signaling center.

B   LIMB BUD—COOPERATING SIGNALING CENTERS: APICAL ECTODERMAL RIDGE AND ZONE OF PROLIFERATING ACTIVITY

Similar to the Spemann's organizer, limb bud signaling centers were recognized as an organizing center on the basis of experimental grafting and induction experiments decades before the diffusible molecules were known (Saunders and Gasseling 1959; Tickle et al. 1975).

The *zone of polarizing activity* (ZPA) is a small cluster of cells in the posterior mesenchyme of the developing limb buds. In early experiments with chick embryos, when a second (donor) ZPA was grafted to the anterior margin of a host limb bud, mirror-image duplications of the digits along the A–P axis resulted. These experiments showed that the small cluster of cells had the ability to reorganize and orient surrounding limb tissues so that they developed into a sequence of digits. Riddle et al. (1992) described this polarizing activity as mediated by a morphogen gradient of SHH across the A–P axis. The ZPA stimulates proliferative activity in surrounding cells along a gradient. Decades of work was stimulated by this pivotal finding and the persistent questions surrounding the regulation of limb and digit skeletal patterning (e.g., Tickle 2017). Molecular work focused on the signaling networks of the ZPA, *apical ectodermal ridge* (AER), and the overlying ectoderm (Nissim et al. 2007; Dillon and Othmer 1999; Sheth et al. 2012; Harfe et al. 2004).

While the ZPA is a signaling center located on the posterior mesenchymal margin of the limb bud, the AER signaling center is a structurally distinct thick band of epithelium on the distal ridge. In marsupials, it appears to be less structurally distinct while still functional (Doroba and Sears 2010). The AER regulates outgrowth by FGF signaling and interacts with other patterning networks through feedback loops (Benazet and Zeller 2009). More recently, non-AER ectoderm on the dorsal–ventral edge was defined as a novel signaling center that actively moderates the concentration of the A–P SHH signal across the limb bud (Nissim et al. 2007). This is another example of separated signaling centers that coordinate complex morphogenesis of a single organ system.

The ZPA and AER signaling centers are found in all vertebrate fins and limbs, indicating that they are an evolutionarily basal organizing strategy for morphogenesis of paired appendages (Shubin et al. 2009; Grandel et al. 2000; Dahn et al. 2007; Johanson et al. 2000). In limbless forms, as expected, the limb bud may initiate, but they fail to maintain the AER, and the limb does not form (Leal and Cohn 2016). Experimentally, when the overlying AER was removed from a developing limb, the limb did not form or was arrested at the stage when the signaling center was removed (Cohn and Tickle 1999). These observations, in sum, suggest an essential and evolutionarily stable requirement for these coordinated signaling centers in the maintenance and evolution of paired appendages.

C   TOOTH ENAMEL KNOTS

Organizing centers in budding and branching epithelial organs such as the hair, sweat gland, feather, mammary gland, and tooth, were proposed but could not be identified by grafting experiments because the technique was too disruptive to the tissue (Vaahtokaari et al. 1996). In the tooth, the *enamel knot* was a historical term for the condensed cells in the developing tooth bud, to which no function could be assigned but which later was recognized as a cellular signaling center. Once candidate molecular signals were identified, the mouse molar tooth became an experimental model of organogenesis and led to a greater understanding of how signaling centers facilitate morphogenesis in general.

The cellular signaling center in developing molars was first described in the mid-nineteen-nineties (Jernvall et al. 1994; Vaahtokari et al. 1996). The non-proliferative enamel knot was revived, in name and in experimentation, by Jernvall and Thesleff in the more modern context of a molecular signaling center regulating tooth morphogenesis in mouse embryos. (A slightly earlier paper called the enamel knot a "control center" Jernvall et al. 1994, MacCord 2017). This enamel knot was later specified more precisely as a *primary enamel knot signaling center*, which controlled the initiation of the tooth bud, in contrast to *secondary enamel knot signaling centers*, which determined the location of individual cusps on the tooth crown in mouse molars (Figure 6.1). Earlier epithelial–mesenchymal interactions induce the formation of the epithelial placode that becomes the molar enamel knot signaling center (Pispa and Thesleff 2003).

The enamel knot signaling center has overlapping gene expression of several major families of signaling genes, for example, SHH, FGFs, BMP, Wnts, which include many of the same genes expressed in other vertebrate signaling centers mentioned above (Vaahtokari et al. 1996; Figure 6.2). The signaling center interacts with surrounding cells to regulate proliferation rates. If the enamel knot is larger, as identified by the area expressing SHH in the primary enamel knot, the size of the tooth is greater. Experiments have determined that inhibitors such as Ectodin around the developing tooth influence the tooth shape (Kassai et al. 2005).

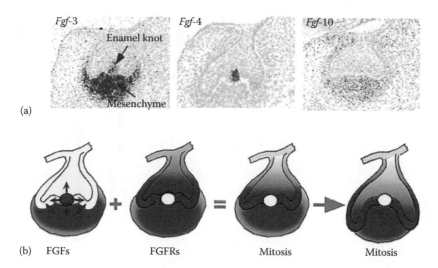

FIGURE 6.1 (a) Expression of different Fgfs (Fgf-3, -4, -10) in a cap stage mouse molar tooth. Only enamel knot cells express Fgfs in the dental epithelium. (Courtesy of Paivi Kettunen.) (b) The summed distributions of Fgfs (FGFs) and Fgf-receptors (FGFRs). (Courtesy of Jernvall, J. and Thesleff, I., *Mech. Dev.*, 92, 19–29, 2000.) Note how ligands and receptors both have inverted expression patterns in the epithelium and how prominent mitosis is in the cells surrounding the non-proliferative enamel knot (in red). The FGFs stimulate growth of the epithelia around the enamel knot and in the dental mesenchyme (arrows) and may cause the unequal growth resulting in down-growth of cervical loops the formation of the tooth crown base.

The tooth has a dense literature in many fields that intersect. Paleontological studies have given us a deep perspective on tooth shape evolution, since teeth are one of the best preserved structures in the fossil record, while the dental development literature has given deep insights into the molecular underpinnings of tooth morphogenesis in model species. This combination of details about the evolution and development of the mammalian molar has resulted in an unusually comprehensive understanding of this single organ. The tooth has served as an experimental and computational model for organogenesis; variations in enamel knot signaling have led to an amazing predictability in morphological simulations of tooth shape evolution (Salazar-Ciudad and Jernvall 2010).

Other epithelial organs may follow this same scenario as the tooth, where a cellular signaling center is established to direct the organization of an epithelial organ. Hair, feathers, mammary glands, sweat glands, lizard scales, and turtle scutes all follow a similar style of morphogenesis and some have been recognized as using signaling centers (Pispa and Thesleff 2003; Huh et al. 2013; Närhi et al. 2012; Moustakas-Verho et al. 2014; DiPoi and Milinkovitch 2016). Cellular signaling centers such as these are a common type of developmental strategy for organizing surrounding cells and initiating tissue and organ morphogenesis, but they are not the only localized signaling centers in the embryo.

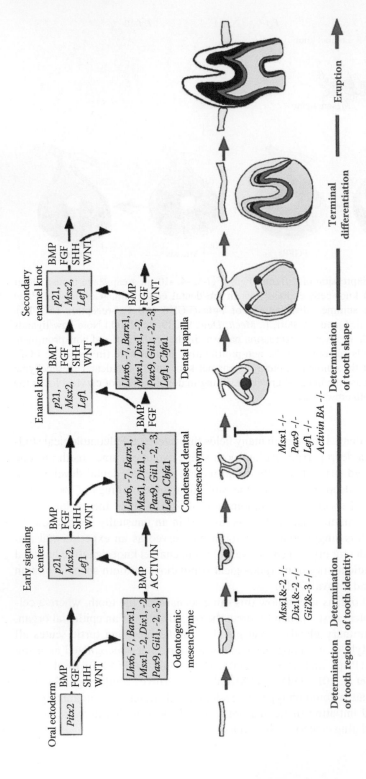

FIGURE 6.2 Schematic representation of the signals and transcription factors mediating the reciprocal signaling between epithelium and mesenchyme during advancing tooth development. The molecular cascades are shown above and the corresponding morphological stages below. The transcription factors and signals considered to be important for particular developmental stages are indicated in the squares and above the arrows, respectively. Note how the same signaling pathways are used reiteratively during advancing tooth development, and how tooth development arrests in the knockout mouse experiments to the early signaling center or the enamel knot stage. Yellow, tooth epithelium; red, enamel knots; blue, tooth mesenchyme. (Courtesy of Jernvall, J. and Thesleff, I., *Mech. Dev.*, 92, 19–29, 2000.)

6.3 ANLAGEN SIGNALING: CONDENSATIONS, PLACODES, AND BLASTEMAS

I have grouped another common type of localized signaling that directs morphogenesis under the term *Anlagen signaling centers* because they produce localized signaling within the earliest stages of a growing organ or structure (i.e., *anlage*).

Like the cellular signaling centers, anlagen signaling centers have signaling interactions with surrounding cells and tissues and affect morphogenesis (Figure 6.3). They differ from cellular signaling centers in that:

1. They are associated with a cell population that is fated to become a particular organ or structure in the adult, and
2. The cell population proliferates and grows.

These anlagen are also ephemeral in that the population of cells will eventually grow to a size, then, turn off proliferation in order to differentiate (Figure 6.2; Hall and Miyake 1995). At this stage in their development, they cease to be anlagen and are developing into organs with differentiating tissues or parts—a fundamentally different state. Skeletal condensations, sensory placodes, and blastemas fit this definition. All are defined cell clusters that are fated to become organs or specific anatomical

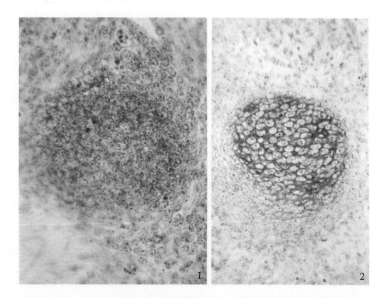

FIGURE 6.3 Panel 1. The condensation of a hyoid cartilage from a C57/BL6 mouse embryo 12 days 9 h post-fertilization, visualized with peanut agglutinin lectin conjugated to peroxidase (brown precipitate). Deposition of extracellular matrix has yet to begin; cf. panel 2. × 870. Panel 2. The hyoid cartilage from a C57/BL6 mouse embryo, 24-h later visualized with alcian blue staining to show deposition of extracellular matrix; × 545. Note that panel 1 is at a higher magnification than panel 2, reflecting extensive growth (condensation proliferation) between the two stages. (Courtesy of Hall, B.K. and Miyake, T., *Anat. Embryol.*, 186, 107–124, 1992.)

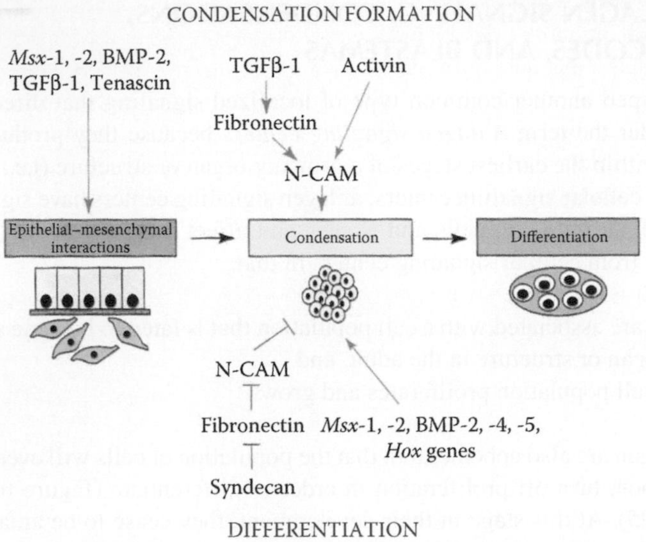

CONDENSATION FORMATION

FIGURE 6.4 A summary of the molecular pathways leading to condensation formation and to differentiation of prechondrogenic cells in the three major phases of chondrogenesis—epithelial–mesenchymal interaction; condensation, differentiation. Condensation is initiated by Msx-l, Msx-2, BMP-2, TGFβ-1, and Tenascin regulating epithelial–mesenchymal interactions that, in turn, initiate condensation. TGFβ-l, up-regulating fibronectin and activin by direct action stimulate accumulation of N-CAM and so promote condensation. Transition from condensation to overt self-differentiation is mediated negatively by suppression of further condensation proliferation and positively by direct enhancement of differentiation. Syndecan, by inhibiting fibronectin, breaks the link to N-CAM and so terminates condensation formation cessation of activin synthesis has the same effect. A number of Hox and Msx genes and BMP-2, -4, and -5 enhance differentiation directly by acting on condensed cells. See text for details. (Courtesy of Hall, B.K. and Miyake, T., *Int. J. Dev. Biol.*, 39, 881–893, 1995.)

structures (Hall and Miyake 1995; Schlosser 2014; McCusker et al. 2015, and see Chapter 7 by Schneider in this volume). The fact that these cell clusters both proliferate and secrete signals changes the dynamics of the interactions; the growing tissue changes shape and thus the signal concentration levels changes continually as the size and shape of the condensation changes (Figure 6.4).

While similar in their fate as progenitors of organs, these different types of anlagen signaling centers are diverse in origin and arise through different types of cellular signaling. For example, skeletal condensations, which are populations of mesenchymal cells (Figure 6.2a) that arise from mesoderm and neural crest, are induced to form by epithelial signaling (Figure 6.2c; Hall 2015), while sensory placodes are populations of ectodermal cells that are initiated by mesenchymal signals. Blastemas, which are populations of dedifferentiated cells involved in regeneration in adults, are induced to form by signals from nerves. This aspect, the formation of the anlagen, is generally well-studied. On the other hand, signaling from the anlagen to surrounding cell populations is poorly represented in the literature and in our thinking.

General features of these anlagen signaling centers are that they are developmentally modular and labile in development and evolution. For example, the skeletal condensations for endochondral ossification act as developmental and evolutionary modules. These condensations have been identified in vertebrate organogenesis as the fundamental unit of morphological change (Atchley and Hall 1991). Their modularity has been known for decades due to mutations causing modular phenotypic changes in skeletal development (Grüneberg 1963).

Other examples of early anlagen signaling centers include the sensory placodes. Neurogenic placodes form around the anterior edge of the head in the embryo. They are local epithelial condensations that have evolved with vertebrates (Baker and Bronner-Fraser 2001). Sensory placodes form as thick ectodermal tissue that goes on to grow and differentiate into the different sensory organs of the head. Sensory placodes interact with cranial neural crest, which joins the sensory organ to the brain with neural connections.

Given that phenotypic integration in the embryo essentially means that all developing organs *must* interact with surrounding cells, it may be that all developing organs have some level of a signaling role to coordinate integration of adjacent organs or structures. A few recent studies demonstrate this integration. The olfactory placode acts as a signaling center for lateral nasal skeletal development (Szabo-Rogers et al. 2009). Skeletal condensations signal to surrounding cells and coordinate vascularization (Eshkar-Oren et al. 2009), tendon, and muscle development (Blitz et al. 2009).

Interestingly, developing tendon condensations (Sox9/Scx progenitor populations) signal interactively with adjacent developing bone to form an attachment unit. This is a good example of signaling between different structures during early embryogenesis, where the structures must be integrated to make a fully functional unit later in ontogeny and in the adult (Zelzer et al. 2014). Furthermore, modularization of the tendon–bone attachment unit (Zelzer et al. 2014) may facilitate evolutionary lability of this feature.

Other systems also show signaling between adjacent structures. One recent study showed that the developing molar tooth interacts with surrounding jaw bone to affect the pattern of the cusps on the molar tooth crown (Renvoise et al. 2017). The developing first molar also interacts with the second molar and third in sequence to regulate size proportions (Kavanagh et al. 2007). Furthermore, predictable proportions of skeletal segmentation series also show, in effect, that there must be highly coordinated signaling between these sequentially developing, adjacent skeletal structures during morphogenesis (Young et al. 2015).

6.4 DISCUSSION

Signaling centers are integral to the development of many organs and body parts during vertebrate embryonic development (Table 6.1). They are employed for basal features such as the node, notochord, and limb buds, as well as for much more recently derived features such as the mammalian tooth crown and the secondary palate. Cellular signaling centers arise and degenerate as needed throughout ontogeny—in early developmental patterning in the gastrula, through organogenesis, and even in regeneration in adults. Cellular signaling centers are associated with all embryonic germ layers and with all major organ systems. In this chapter, I have highlighted the prevalence of cellular signaling centers as a means to coordinate development generally, rather than as a special case in a few model developmental systems.

The prevalence of signaling centers throughout the embryo and among all vertebrates indicates either that

1. Signaling centers are a basal and evolutionarily persistent mechanism to direct morphogenesis, or
2. Signaling centers arise convergently whenever needed.

We cannot distinguish between these two possibilities yet, but like most things in biology, it is probably some of both. Several lines of evidence indicate that cellular signaling centers self-organize, a phenomenon that would support either evolutionary scenario. Recent activator-inhibitor based modeling, supported by experimental observations, suggests that signaling centers can self-organize from near-homogeneous initial conditions (Meinhardt 2015; Nieuwkoop 1992). Even closely spaced opposing signaling centers can self-organize, for example, in a simulation where the Chordin—BMP signaling mimics the Spemann–Mangold center establishment (Meinhardt 2015). Signaling centers are not stable if they are transplanted too closely (Zwilling 1956), suggesting there is an optimal range of distances between opposing signaling centers—close but not too close; implant two ZPAs too close together into a limb bud and they initiate a single set of digits.

The essential inhibitory antagonism is thought to be an ancestral state (Meinhardt 2015). In computational experiments, modularization of signaling networks evolves inevitably under conditions of selection for optimality under changing adaptive conditions (Kashton and Alon 2005). Similarly, modularization of cellular signaling centers may be advantageous. The capacity for self-organization of cellular signaling centers in developing tissues, and of modularization of biological signaling networks, suggests that, structurally, cellular signaling centers are an evolutionarily and developmentally favored general mechanism for morphogenesis. This proposal needs to be explored.

Why are signaling centers a good system for directing morphogenesis? In other words, has evolution favored the emergence of signaling centers as a mechanism to provide transient, flexible, locally-responsive needs of embryonic organization during complex development? The alternative might be diffuse signals emanating from larger or dividing cell populations. It seems easier to utilize a point source (nonproliferating cluster of cells) rather than a dividing cell population with more diffuse, changing levels of signaling proteins. Also, nondividing cells have different properties than dividing cells—they stay together, have different cell-surface properties, and can be synchronized. These features are beneficial for the scale of morphogenesis. Localization is important because signals can be responsive to short distances. With a more diffuse source, the signals may not be able to tightly regulate the required signaling gradients as easily.

One question in developmental biology is the size scale/limits of morphogenetic interactions—it is obvious that all embryos and developing organs are small, so what limits the size at which morphogenesis can take place? In this context, it would be interesting to determine what might be the maximum distance possible for interacting signaling centers. Such boundaries would be a fascinating area to study, particularly in the context of developmental constraint on evolution.

It is clear that signaling centers can induce each other. In some cases, research has revealed that a chronological sequence, or cascade, of signaling centers, arise and lead to morphogenesis of a particular structure. That is, an early signaling center induces the formation of another signaling center (or multiple centers simultaneously) which then drive morphogenesis of the structure. The molar tooth is a well-characterized example, with primary and secondary enamel knots, inducing tooth morphogenesis and cusp (tooth crown) morphogenesis, respectively. Even further, later-developing *tertiary enamel knots* (Luukko 2003) and earlier-developing ectodermal signaling centers leading to the primary enamel knot of incisors (Ahtiainen et al. 2016) have been described. This demonstrated an extended sequence of signaling centers leading to a single anatomical structure, the tooth. Jernvall and Salazar-Ciudad (2007) suggest "the economy of tinkering," where, by iterative use of the same gene networks, the signaling center model facilitates efficient evolutionary variations and elaboration. Another example of a series of signaling centers is the Nieuwkoop center leading to the Spemann–Mangold center and to the notochord as inducer. Much more research into the spatial and temporal patterns of emergence of these transient signaling centers during embryogenesis would be insightful.

Understanding the logic and mechanisms of development at different levels of coordination in the embryo (Gilmour et al. 2017) may illuminate similarities among local cellular developmental processes within the bewildering complexity of embryogenesis. In particular, one can consider a three-part design of coordinating interactions that include:

1. Early tissue induction (epithelial–mesenchymal signaling) to establish placodes or condensations,
2. Nonproliferative cellular signaling centers to coordinate local morphogenesis of organs/structures, and
3. Anlagen signaling centers that integrate morphogenesis of adjacent organs/structures.

These interactions are common to all vertebrate embryos and organ systems. The self-organizing ability of signaling centers and their use in a wide variety of embryonic structures suggests that signaling centers are emergent properties and not uniquely selected situations for each developmental scenario. Further, the modularity of signaling centers suggests that they can vary independently and thus provide a way for evolution to modify embryogenesis while maintaining essential phenotypic coordination and gene interactions.

REFERENCES

Ahtiainen, L., Uski, I., Thesleff, I., and Mikkola, M. L. 2016. Early epithelial signaling center governs tooth budding morphogenesis. *J Cell Biol* 214: 753–767.

Atchley, W. R. and Hall, B. K. 1991. A model for development and evolution of complex morphological structures. *Bio Revs Camb Philos Soc* 66: 101–157.

Baker, C. V. and Bronner-Fraser, M. 2001. Vertebrate cranial placodes I. Embryonic induction. *Dev Biol* 232: 1–61.

Beddington, R. S. and Smith J. C. 1993. Control of vertebrate gastrulation: Inducing signals and responding genes. *Curr Opin Genet Devel* 3: 655–661.

Benazet, J. D. and Zeller, R. 2009. Vertebrate limb development: Moving from classical morphogen gradients to an integrated 4-dimensional patterning system. *Cold Spring Harbor Persp Biol* 1: a001339.

Blitz, E., Viukov, S., Sharir, A., Shwartz, Y., Galloway, J. L., Pryce, B. A., Johnson, R. L., Tabin, C. J., Schweitzer, R., and Zelzer, E. 2009. Bone ridge patterning during musculoskeletal assembly is mediated through SCX regulation of Bmp4 at the tendon-skeleton junction. *Dev Cell* 17: 861–873.

Capdevila, J. and Izpisua Belmonte, J. C. 2001. Patterning mechanisms controlling vertebrate limb development. *Annu Rev Cell Dev Biol* 17: 87–132.

Cohn, M. J. and Tickle, C. 1999. Developmental basis of limblessness and axial patterning in snakes. *Nature* 399: 474–479.

Dahn, R. D., Davis, M. C., Pappano, W. N., and Shubin, N. H. 2007. Sonic hedgehog function in chondrichthyan fins and the evolution of appendage patterning. *Nature* 445: 311–314.

De Robertis, E. M. 2006. Spemann's organizer and self-regulation in amphibian embryos. *Nat Rev Mol Cell Biol* 7: 296–302.

De Robertis, E. M. 2009. Spemann's organizer and the self-regulation of embryonic fields. *Mech Devel* 126: 925–941.

Dillon, R. and Othmer, H. G. 1999. A mathematical model for outgrowth and spatial patterning of the vertebrate limb bud. *J Theor Biol* 197: 295–330.

DiPoi, N. and Milinkovitch, M. 2016. The anatomical placode in reptile scale morphogenesis indicates shared ancestry among skin appendages in amniotes. *Sci Adv* 2: e1600708.

Doroba, C. K. and Sears, K. E. 2010. The divergent development of the apical ectodermal ridge in the marsupial *Monodelphis domestica*. *Anat Rec (Hoboken)* 293: 1325–1332.

Edlund, R. K., Ohyama, T., Kantarci, H., Riley, B. B., and Groves, A. K. 2014. Foxi transcription factors promote pharyngeal arch development by regulating formation of FGF signaling centers. *Dev Biol* 390: 1–13.

Elsen, G. E., Choi, L., Millen, K., Grinblat, Y., and Prince, V. E. 2008. Zic1 and Zic4 regulate zebrafish roof plate specification and hindbrain ventricle morphogenesis. *Dev Biol* 314: 376–392.

Eshkar-Oren, I., Viukov, S. V., Salameh, S., Krief, S., Oh, C. D., Akiyama, H., Gerber, H. P. Ferrara, N., and Zelzer, E. 2009. The forming limb skeleton serves as a signaling center for limb vasculature patterning via regulation of Vegf. *Development* 136: 1263–1272.

Fernandez–Teran, M. and Ros, M. A. 2008. The apical ectodermal ridge: Morphological aspects and signaling pathways. *Int J Devel Biol* 52: 857–871.

Foppiano, S., Hu, D., and Marcucio, R. S. 2007. Signaling by bone morphogenetic proteins directs formation of an ectodermal signaling center that regulates craniofacial development. *Dev Biol* 312: 103–114.

Gilbert, S. F. and Saxén, L. 1993. Spemann's organizer: Models and molecules. *Mech Dev* 41: 73–89.

Gillis, J. A. and Hall, B. K. 2016. A shared role for sonic hedgehog signalling in patterning chondrichthyan gill arch appendages and tetrapod limbs. *Development* 143: 1313–1317.

Gilmour, D., Rembold, M., and Leptin, M. 2017. From morphogen to morphogenesis and back. *Nature* 541: 311–320.

Grandel, H., Draper, B. W., and Schulte-Merker, S. 2000. Dackel acts in the ectoderm of the zebrafish pectoral fin bud to maintain AER signaling. *Development* 127: 4169–4178.

Grüneberg, H. 1963. *The Pathology of Development: A Study of Inherited Skeletal Disorders in Animals*. Oxford, UK: Blackwell.

Guger, K. A. and Gumbiner, B. M. 1995. beta-Catenin has Wnt-like activity and mimics the Nieuwkoop signaling center in *Xenopus* dorsal-ventral patterning. *Dev Biol* 172: 115–125.

Gupta, S. and Sen, J. 2015. Roof plate mediated morphogenesis of the forebrain: New players join the game. *Dev Biol* 413: 145–152.

Hall, B. K. 2015. *Bones and Cartilage: Developmental and Evolutionary Skeletal Biology, 2nd ed.* London, UK: Elsevier Academic Press.

Hall, B. K. and Miyake, T. 1992. The membranous skeleton: The role of cell condensations in vertebrate skeletogenesis. *Anat Embryol* 186: 107–124.

Hall, B. K. and Miyake, T. 1995. Divide, accumulate, differentiation: Cell condensation in skeletal development revisited. *Int J Dev Biol* 39: 881–893.

Harfe, B. D., Scherz, P. J., Nissim, S., Tian, H., McMahon, A. P., and Tabin, C. J. 2004. Evidence for an expansion-based temporal Shh gradient in specifying vertebrate digit identities. *Cell* 118: 517–528.

Harfe, B. D. 2011. Keeping up with the zone of polarizing activity: New roles for an old signaling center. *Dev Dyn* 240: 915–919.

Hu, D. and Marcucio, R. S. 2009. A SHH-responsive signaling center in the forebrain regulates craniofacial morphogenesis via the facial ectoderm. *Development* 136: 107–116.

Huh, S. H., Naerhi, K., Lindfors, P. H., Haara, O., Yang, L., Ornitz, D. M., Mikkola, M. L. 2013. Fgf20 governs formation of primary and secondary dermal condensations in developing hair follicles. *Genes Dev* 27: 450–458.

Jernvall, J. and Salazar-Ciudad, I. 2007. The economy of tinkering mammalian teeth. *Novartis Found Symp* 284: 207–216.

Jernvall, J. and Thesleff, I. 2000. Reiterative signaling and patterning during mammalian tooth morphogenesis. *Mech Dev* 92: 19–29.

Jernvall, J., Aberg, T., Kettunen, P., Keranen, S., and Thesleff, I. 1998. The life history of an embryonic signaling center: BMP-4 induces p21 and is associated with apoptosis in the mouse tooth enamel knot. *Development* 125: 161–169.

Jernvall, J., Kettunen, P., Karavanova, I., Martin, L. B., and Thesleff, I. 1994. Evidence for the role of the enamel knot as a control center in mammalian tooth cusp formation: Non-dividing cells express growth stimulating Fgf-4 gene. *Int J Dev Biol* 38: 463–469.

Johanson, Z., Joss, J., Boisvert, C. A., Ericsson, R., Sutija, M., and Ahlberg, P. E. 2000. Fish fingers: Digit homologues in sarcopterygian fish fins. *J Exp Zool Mol Dev Evol* 308B: 757–768.

Kashton, N. and Alon, U. 2005. Spontaneous evolution of modularity and network motifs. *Proc Natl Acad Sci USA* 102: 13773–13778.

Kassai, Y., Munne, P., Hotta, Y., Penttila, E., Kavanagh, K., Ohbayashi, N., Takada, S., Thesleff, I., Jernvall, J., and Itoh, N. 2005. Regulation of mammalian tooth cusp patterning by ectodin. *Science* 309: 2067–2070.

Kavanagh, K. D., Evans, A. R., and Jernvall, J. 2007. Predicting evolutionary patterns of mammalian teeth from development. *Nature* 449: 427–432.

Kiecker, C. and Lumsden, A. 2004. Hedgehog signaling from the ZLI regulates diencephalic regional identity. *Nat Neurosci* 7: 1242–1249.

Lavine, K. J. and Ornitz, D. M. 2008. Fibroblast growth factors and Hedgehogs: At the heart of the epicardial signaling center. *Trends Genet* 24: 33–40.

Leal, F. and Cohn, M. J. 2016. Loss and Re-emergence of legs in snakes by modular evolution of Sonic hedgehog and HOXD enhancers. *Curr Biol* 26: 2966–2973.

Liu, C., Knezevic, V., and Mackem, S. 2004. Ventral tail bud mesenchyme is a signaling center for tail paraxial mesoderm induction. *Dev Dyn* 229: 600–606.

Luukko, K., Loes, S., Furmanek, T., Fjeld, K., Kvinnsland, I. H., and Kettunen, P. 2003. Identification of a novel putative signaling center, the tertiary enamel knot in the postnatal mouse molar tooth. *Mech Dev* 120: 270–276.

MacCord, K. 2017. *Development, evolution, and teeth: How we came to explain the morphological evolution of the mammalian dentition*. PhD thesis, Arizona State University, Tempe, AZ.

Marcucio, R. S., Foppiano, S., and Hu, D. 2007. Bone morphogenetic proteins (BMPs) direct formation of a signaling center that regulates facial development. *Dev Biol* 306: 401–402.

Martinez-Morales, J. R., Del Bene, F., Nica, G., Hammerschmidt, M., Bovolenta, P., and Wittbrodt, J. 2005. Differentiation of the vertebrate retina is coordinated by an FGF signaling center. *Dev Cell* 8: 565–574.

McCusker, C., Bryant, S. V., and Gardiner, D. M. 2015. The axolotl limb blastema: Cellular and molecular mechanisms driving blastema formation and limb regeneration in tetrapods. *Regeneration* 11: 54–71.

Meinhardt, H. 2006. Primary body axes of vertebrates: Generation of a near-cartesian coordinate system and the role of spemann-type organizer. *Dev Dyn* 235: 2907–2919.

Meinhardt, H. 2015. Dorsoventral patterning by the Chordin-BMP pathway: A unified model from a pattern-formation perspective for drosophila, vertebrates, sea urchins and *Nematostella*. *Dev Biol* 405: 137–148.

Mikawa, T., Poh, A. M., Kelly, K. A., Ishii, Y., and Reese, D. E. 2004. Induction and patterning of the primitive streak, an organizing center of gastrulation in the amniote. *Dev Dyn* 229: 422–432.

Monuki, E. S. 2007. The morphogen signaling network in forebrain development and holoprosencephaly. *J Neuropathol Exp Neurol* 66: 566–575.

Moustakas-Verho, J. E., Zimm, R., Cebra-Thomas, J., Lempiainen, N. K., Kallonen, A., Mitchess, K. L., Hamalainen, K., Salazar-Ciudad, I., Jernvall, J., and Gilbert, S. F. 2014. The origin and loss of periodic patterning in the turtle shell. *Development* 141: 3033–3039.

Närhi, K., Tummers, M., Ahtiainen, L., Itoh, N., Thesleff, I., and Mikkola, M. L. 2012. Sostdc1 defines the size and number of skin appendage placodes. *Dev Biol* 364: 149–161.

Nieuwkoop, P. D. 1992. The formation of the mesoderm in urodelean amphibians. VI. The self-organizing capacity of the induced meso-endoderm. *Roux Arch Dev Biol* 201: 18–29.

Nissim, S., Allard, P., Bandyopadhyay, A., Harfe, B. D., and Tabin, C. J. 2007. Characterization of a novel ectodermal signaling center regulation Tbx2 and Shh in the vertebrate limb. *Devel Biol* 304: 9–21.

Nusslein-Volhard, C. 1977. Genetic analysis of pattern-formation in the embryo of Drosophila melanogaster: Characterization of the maternal-effect mutant Bicaudal. *Wilhelm Roux Arch Dev Biol* 183: 249–268.

Pankratz, M. J. and Hoch, M. 1995. Control of epithelial morphogenesis by cell signaling and integrin molecules in the *Drosophila* foregut. *Development* 121: 1885–1898.

Pispa, J. and Thesleff, I. 2003. Mechanisms of ectodermal organogenesis. *Dev Biol* 262: 195–205.

Pottin, K., Hinaux, H., and Retaux, S. 2011. Restoring eye size in *Astyanax mexicanus* blind cavefish embryos through modulation of the Shh and Fgf8 forebrain organising centres. *Development* 138: 2467–2476.

Renvoise, E., Kavanagh, K. D., Lazzari, V., Hakkinen, T. J., Rice, R., Pantalacci, S., Salazar-Ciudad, I., and Jernvall, J. 2017. Mechanical constraint from growing jaw facilitates mammalian dental diversity. *Proc Natl Acad Sci USA* 114: 9403–9408.

Retaux, S. and Kano, S. 2010. Midline signaling and evolution of the forebrain in chordates: A focus on the lamprey hedgehog case. *Integ Comp Biol* 50: 98–109.

Riddle, R. D., Johnson, R. L., Laufer, E., and Tabin, C. 1993. Sonic Hedgehog mediates the polarizing activity of the ZPA. *Cell* 75: 1401–1416.

Riley, B. B., Chiang, M. Y., Lekven, A. C., and Moon, R. T. 2001. A signaling network required to maintain rhombomere boundaries as organizing centers in the zebrafish hindbrain. *FASEB J* 15: A742.

Roth, S., Stein, D., and Nusslein-Volhard, C. 1989. A gradient of nuclear localization of the dorsal protein determines dorsoventral pattern in the Drosophila embryo. *Cell* 59: 1189–1202.

Rowton, M., Anderson, D., Huber, B., and Rawls, A. 2007. Regulation of a novel skeletal muscle signaling center at the occipitocervical somite boundary. *Dev Biol* 306: 401–401.

Saito, D. and Takahashi, Y. 2015. Sympatho-adrenal morphogenesis regulated by the dorsal aorta. *Mech Dev* 138: 2–7.

Salazar-Ciudad, I. and Jernvall, J. 2010. A computational model of teeth and the developmental origins of morphological variation. *Nature* 464: 583–586.

Saunders, J. and Gasseling, M. T. 1959. Effects of reorienting the wing-bud apex in the chick embryo. *J Exp Zool* 142: 553–569.

Saunders, J. W., Jr. and Gasseling, M. T. 1968. Ectodermal-mesodermal interactions in the origin of limb symmetry. In R. Fleischmajer and R. E. Billingham (Eds.), *Epithelial-Mesenchymal Interactions*, pp. 78–97. Williams & Wilkins, Baltimore, MD.

Schlosser, G. 2014. Early embryonic specification of vertebrate cranial placodes. *Wiley Interdiscip Rev Dev Biol* 3: 349–363.

Schlosser, G., Patthey, C., and Shimeld, S. M. 2014. The evolutionary history of vertebrate cranial placodes II. Evolution of ectodermal patterning. *Dev Biol* 389: 98–119.

Sheth, R., Marcon, L., Bastida, M. F., Junco, M., Quintana, L., Dahn, R., Kmita, M., Sharpe, J., and Ros, M. A. 2012. Hox genes regulate digit patterning by controlling the wavelength of a turing-type mechanism. *Science* 338: 1476–1480.

Shubin, N., Tabin, C., and Carroll, S. 2009. Deep homology and the origins of evolutionary novelty. *Nature* 457: 818–823.

Slack, J. M. W. 1987. We have a morphogen! *Nature* 327: 553–554.

Smith, J. C. 1993. Dorso-ventral patterning in the neural tube. *Curr Biol* 3: 582–585.

Spemann, H. and Mangold, H. 1924. Über Induktion von Embryonalanlagen durch Implantation artfremder Organisatoren. *Archiv für Mikroskopische Anatomie und Entwicklungsmechanik* 100: 599–638.

St. Johnston, R. and Nüsslein-Volhard, C. 1992. The origin of pattern and polarity in the *Drosophila* embryo. *Cell* 68: 201–219.

Szabo-Rogers, H. L., Geetha-Loganathan, P., Whiting, C. J., Nimmagadda, S., Fu, K., and Richman, J. M. 2009. Novel skeletogenic patterning roles for the olfactory pit. *Development* 136: 219–229.

Tabin, C. 1995. The initiation of the limb bud: Growth factors, Hox genes, and retinoids. *Cell* 80: 671–674.

Tickle, C. 2017. An historical perspective on the pioneering experiments of John Saunders. *Dev Biol* 429: 374–381 (and other articles in this Special Issue: Forming and Shaping the Field of Limb Development: A Tribute to Dr. John Saunders; ed. Lee Niswander).

Tickle, C., Summerbell, D., and Wolpert, L. 1975. Positional signaling and specification of digits in chick limb morphogenesis. *Nature* 254: 199–202.

Vaahtokari, A., Aberg, T., Jernvall, J., Keränen, S., and Thesleff, I. 1996. The enamel knot as a signaling center in the developing mouse tooth. *Mech Dev* 54: 39–43.

Vidal, V., Sacco, S., Rocha, A. S., da Silva, F., Panzolini, C., Dumontet, T., Doan, T. M. P. et al. 2016. The adrenal capsule is a signaling center controlling cell renewal and zonation through Rspo3. *Genes Dev* 30: 1389–1394.

Vieira, C. and Martinez, S. 2006. Sonic hedgehog from the basal plate and the zona limitans intrathalamica exhibits differential activity on diencephalic molecular regionalization and nuclear structure. *Neuroscience* 143: 129–140.

Voutilainen, M., Lindfors, P., Lefebvre, S., Ahtiainen, L., Fliniaux, I., Rysti, E., Murtoniemi, M., Schneider, P., Schmidt-Ullrich, R., and Mikkola, M. L. 2012. Ectodysplasin regulates hormone-independent mammary ductal morphogenesis via NF-kB. *Proc Natl Acad Sci USA* 109: 5744–5749.

Welsh, I. C. and O'Brien, T. P. 2009. Signaling integration in the rugae growth zone directs sequential SHH signaling during the rostral outgrowth of the palate. *Dev Biol* 336: 53–67.

Yamada, T., Pfaff, S. L., Edlund, T., and Jessell, T. M. 1993. Control of cell pattern in the neural tube: Motor neuron induction by diffusible factors from notochord and floor plate. *Cell* 73: 673–686.

Yao, Y., Minor, P. J., Zhao, Y. T., Jeong, Y., Pani, A. M., King, A. N., Symmons, O. et al. 2016. *Cis*-regulatory architecture of a brain-signaling center predates the origin of chordates. *Nat Genet* 48: 575–580.

Young, N. M., Winslow, B., Takkellapati, S., and Kavanagh, K. 2015. Shared rules of development predict patterns of evolution in vertebrate segmentation. *Nat Commun* 6: 6690.

Zelzer, E., Blitz, E., Killian, M. L., and Thomopoulos, S. 2014. Tendon-to-bone attachment: From development to maturity. *Birth Defects Res C* 102: 101–112.

Zwilling, E. 1956. Interaction between limb bud ectoderm and mesoderm in the chick embryo. II. Experimental limb duplication. *J Exp Zool A (Exp Integ Physiol)* 132: 173–187.

7 Cellular Control of Time, Size, and Shape in Development and Evolution

Richard A. Schneider

CONTENTS

7.1 A BRIEF HISTORY OF TIME, SIZE, AND SHAPE

The rules by which anatomical size and shape are generated have intrigued scientists for centuries. In 1638, Galileo suggested a mathematical relationship between proportional changes in the shape of bones as animals increase in size, which he argued was a functional necessity for weight bearing (1914). The formalism of Galileo, whereby, physical forces and mathematical laws became integrated with studies of size and shape in biology, was most conspicuously encapsulated over a hundred years ago in the 1917 monumental tome by D'Arcy Thompson entitled, *On Growth and Form* (Thompson 1917). In a breathtakingly comprehensive manner, Thompson synthesized the observations of numerous predecessors and contemporaries, and through countless examples built a theoretical and experimental framework for describing changes in morphology that persists to this day (Stern and Emlen 1999; Arthur 2006).

An essential component of Thompson's treatise was his system of Cartesian coordinates he employed to map the geometrical transformation of organs and organisms. Many other biologists in the 1920s and 1930s were motivated to address similar questions on size and shape in both the scholarly and popular literature (Gayon 2000). In 1926, John Haldane wrote a topical essay entitled, *On Being the Right Size*, in which he stated that, "The most obvious differences between different animals are differences of size, but for some reason, the zoologists have paid singularly little attention to them....For every type of animal, there is a most convenient size, and a large change in size inevitably carries with it a change of form" (p. 1) (Haldane 1926). This was just one of many topics during Haldane's career for which he showed remarkable prescience; by pointing out how little attention had been given to size and shape previously, he in effect anticipated a whole discipline.

A HUXLEY AND ALLOMETRY

Soon thereafter, Haldane's close friend Julian Huxley in his *Problems of Relative Growth* (which he dedicated to Thompson) expanded the discussion of size and shape to include mathematical representations of morphological transformations that arise over time, specifically during ontogeny and phylogeny (Huxley 1932). Along with Georges Teissier, Huxley symbolized relative growth with an algebraic power formula and introduced the term allometry to explain how changes in shape relate to changes in size (Huxley and Teissier 1936; Gayon 2000). A major motivation of Huxley, as well as many others who followed was to gain insight into the developmental (e.g., genetic and cellular) mechanisms generating allometric changes in proportion during evolution (Hersh 1934; von Bonin 1937; Lumer 1940; Needham and Lerner 1940; Anderson and Busch 1941; Lumer et al. 1942; Clark and Medawar 1945; Rensch 1948; Huxley 1950; Reeve 1950; Kermack and Haldane 1950; Bertalanffy and Pirozynski 1952; Gould 1966, 1971; Lande 1979; Alberch et al. 1979; Atchley 1981; Gould 1981; Coppinger and Coppinger 1982; Shea 1983; Atchley et al. 1984; Riska and Atchley 1985; Shea 1985; Coppinger et al. 1987; Deacon 1990; Godfrey and Sutherland 1995a; Coppinger and Schneider 1995; Stern and Emlen 1999; Smith et al. 2015).

Haldane and Huxley viewed size and shape predominantly through the prism of genetics, which during that period was surpassing embryology as the arbiter of acceptable explanations for mechanisms controlling the evolution of morphology. This grew from the seeding of Mendel alongside Darwin, and the paradigm being cultivated vis-à-vis genes, mechanisms of inheritance, and mutations that affect morphology from geneticists such as William Bateson, Richard Goldschmidt, and others (Bateson 1894; Bateson and Mendel 1902; Robb 1935; Goldschmidt 1938, 1940). Ten years after *Problems of Relative Growth*, Huxley published *Evolution, the Modern Synthesis* (Huxley 1942) in which he somewhat unintentionally helped push embryology out of the field of evolution for almost thirty years. This does not mean that embryologists were not thinking about evolution at the time or thereafter, but genetics ruled the roost due primarily to the robust and highly visible efforts of some former embryologists like Thomas Hunt Morgan (Morgan et al. 1915; Morgan 1919) and mathematical geneticists such as Ronald Fisher, Haldane, Sewall Wright, and Theodosius Dobzhansky (Fisher 1930; Wright 1931; Sinnott and Dunn 1932; Haldane 1932; Dobzhansky 1937).

Huxley was well-versed in embryology and evolution, given that his close colleague, Gavin de Beer had written *Embryology and Evolution* in 1930, and Huxley coauthored *Elements of Experimental Embryology* with de Beer in 1934 (de Beer 1930; Huxley and de Beer 1934). Even though Huxley's *Modern Synthesis* has been viewed as a nail in the coffin for evolutionary embryology, Huxley left some openings for development to play a role. He stated, "The course of Darwinian evolution is thus seen as determined (in varying degrees in different forms) not only by the type of selection, not only by the frequency of mutation, not only by the past history of the species, but also by the nature of the developmental effects of genes and of the ontogenetic process in general" (p. 555) (Huxley 1942). Likely, this inclusion of developmental growth was influenced by his own earlier studies (Huxley 1924, 1932) and by those who continued to work contemporaneously on allometry (Hersh 1934; Gregory 1934; Lumer 1940; Anderson and Busch 1941; Lumer and Schultz 1941; von Bonin 1937; Needham and Lerner 1940; Lumer et al. 1942), as well as other mechanisms of evolutionary embryology, especially those advanced by Walter Garstang (1922) and de Beer (1930).

B   DE BEER AND HETEROCHRONY

Gavin de Beer not only recognized the importance of allometry but also devised a series of definitions and schemas relating time to size and shape that is arguably one of the most important contributions in the history of the field of evolutionary developmental biology. First and foremost, de Beer was an evolutionary embryologist (Ridley 1985; Hall 2000a; Brigandt 2006). His work emphasized the significance of changes in the timing of developmental events, or heterochrony, in transforming the morphology of a descendant relative to an ancestor. Heterochrony was initially conceived by Ernst Haeckel and has been applied in various scenarios to link development and evolution (Kollmann 1885; Russell 1916; Bolk 1926; Garstang 1928; de Beer 1930; Dechambre 1949; Gould 1977; Hall 1984; McKinney 1988b; Klingenberg 1998; Smith 2003; Keyte and Smith 2014). de Beer classified eight modes of evolution through which ancestral and descendant ontogenies can differ (de Beer 1930). He provided examples for each type of heterochrony but argued that *neoteny*, defined as the retention of juvenile features in the adult form, was the one that truly allowed for large, rapid phenotypic change and morphological diversification. Other significant evolutionary concepts that de Beer advanced in the context of embryology include clandestine evolution, homology, and evolutionary plasticity. Notably, de Beer dropped the word *Evolution* that was in his first edition book title and instead adopted *Embryos and Ancestors* for the 1940 and subsequent editions (de Beer 1940, 1954, 1958), which can be seen as a reflection of how much the field of population genetics, and not embryology, laid claim to the study of evolution during that era.

C   THE HOLY TRINITY OF TIME, SIZE, AND SHAPE

Nonetheless, the theory that changes to the rate of growth and/or timing of events during ontogeny could alter the course of phylogeny continued as a subplot to the main story of evolution until becoming more generally accepted during the rebirth of *evo-devo* in the 1970s. Even Darwin in his *Origin of Species* was vexed and tantalized by the correlations of growth observed in embryos, which he acknowledged were a potential source of evolutionary variation (Darwin 1859). In *Chapter I, Variation Under Domestication*, Darwin wrote, "There are many laws regulating variation, some few of which can be dimly seen...I will here only allude to what may be called correlation of growth. Any change in the embryo or larva will almost certainly entail changes in the mature animal" (p. 11). He also stated: "If man goes on selecting, and thus augmenting, any peculiarity, he will almost certainly unconsciously modify other parts of the structure, owing to the mysterious laws of the correlation of growth. The result of the various, quite unknown, or dimly seen laws of variation is infinitely complex and diversified" (p. 12). Then again in *Chapter V, Laws of Variation*, Darwin explained that: "Changes of structure at an early age will generally affect parts subsequently developed; and there are very many other correlations of growth, the nature of which we are utterly unable to understand" (p. 168). Clearly, such *correlations of growth* were exactly on what Thompson focused, and his efforts helped lay the groundwork for a broad range of studies comparing changes in size and shape during development.

All the more so, about a decade before Thompson's seminal work, Charles Minot provided a complimentary and in many ways equally important embryological perspective that connected the size of animals and/or their organs with the regulation of cell number, differentiation, and rates of growth as a function of age (Minot 1908). Borrowing from this idea, In Chapter III, *The Rate of Growth*, Thompson equated age with time and stated that "the *form* of an organism is determined by its rate of *growth* in various directions; hence rate of growth deserves to be studied as a necessary preliminary to the theoretical study of form, and organic form itself is found, mathematically speaking, to be a *function of time*" (p. 79) (Thompson 1952). Similarly, Huxley latched on to the importance of time when contemplating evolutionary changes in relative size. He proposed potential genetic mechanisms involving "(a) mutations affecting the primary gradient of the early embryo, on which the time-relations of antero-posterior differentiation depend; (b) mutations affecting specific rate-genes; (c) mutations affecting specific 'time-genes'—genes controlling time of onset and not rate of processes" (p. 242) (Huxley 1932). Accordingly, changes to these *rate-genes* and *time-genes* can affect growth gradients and alter morphology at multiple levels in a coordinated way. Such theories were supported by Goldschmidt's discovery of genes that alter rates of development (Goldschmidt 1938, 1940; Dietrich 2000), something which was also integrated into evolutionary embryology, and more specifically heterochrony, by de Beer (Hall 2000a). On this point, de Beer stated, "By acting at different rates, the genes can alter the time at which certain structures appear" (p. 20) (de Beer 1954).

Therefore, primarily through the critical contributions of Minot, Thompson, Huxley, and de Beer during the first half of the twentieth century, the three parameters of time, size, and shape became unified in essence as the *holy trinity* of evolutionary morphology. But while some of these authors and others strived to integrate findings from the emerging field of developmental genetics led by classically trained embryologists and morphologists such as Goldschmidt (1938, 1940, 1953); Conrad Waddington (1939, 1940, 1957b, 1962) and Ivan Schmalhausen (1949), a deeper understanding of the molecular and cellular mechanisms that unite time, size, and shape during ontogeny and phylogeny would have to wait for almost fifty years. Moreover, the neo-Darwinians remained very skeptical that developmental genetics could contribute to evolutionary theory, and thought-leaders such as Dobzhansky (1937, 1951) and Ernst Mayr (1963, 1983) argued most vociferously that all evolution was microevolution arising from "the continuous adjustment of an integrated gene complex to a changing environment" (p. 332) (Mayr 1963). In other words, this was the prevailing synthetic theory that embraced natural selection and survival of the fittest, distribution of alleles at the level of populations, and gradual adaptive evolution as the sole agent of change. In this regard, the neo-Darwinians thoroughly rejected and even mocked the ideas of Goldschmidt (Gould 1982b), especially that small genetic changes affecting developmental time or rates could rapidly generate large phenotypic transformations in size and shape.

So, by the 1950s, evolutionary studies predicated on allometry and heterochrony were either vastly overshadowed by the neo-Darwinian paradigm, or more pointedly they were viewed as gross oversimplifications of embryonic growth by developmental biologists. Waddington, for example in a paper discussing how to measure size and shape in a meaningful and biologically relevant way stated that, "The validity of any biological conclusions which may be drawn from measurements of size and form depends far more on the adequacy of the physiological insight on which they are based than on the precision of the mathematical techniques used to summarize and compare them" (p. 515) (Waddington 1950). This sentiment begged the question of what governs growth over time and demanded a more in-depth probing of developmental mechanisms regulating size and shape.

7.2   TIME, SIZE, AND SHAPE REDUX

Despite Waddington's emphasis on acquiring a deeper understanding of developmental processes and his admonishment of expending too much energy on generating more sophisticated and precise methods for measuring size and shape (Waddington 1950), studies on allometry continued unabated for decades (Stern and Emlen 1999; Gayon 2000). Moreover, a whole field of morphometrics burgeoned based on multivariate methods and ultimately computer-based algorithms for quantifying and visualizing complex changes in size and shape (Bookstein 1978, 1990; Benson et al. 1982; Siegel and Benson 1982; Marcus 1996; Zelditch 2004; Hallgrimsson et al. 2015). Granted, the technical ability to analyze size and shape became more refined over time, but results generally remained phenomenological. Therefore, many morphometricians endeavored to frame their studies within the context of quantitative genetics and/or heterochrony, in order to make predictions about mechanisms through which size and shape can change during ontogeny and phylogeny (Gould 1966, 1981; Lande 1979; Alberch et al. 1979; Atchley 1981; Cheverud 1982; Benson et al. 1982; Riska 1986; McKinney 1988a; Atchley and Hall 1991; Coppinger and Schneider 1995; Klingenberg 1998; Roth and Mercer 2000; Drake 2011; Smith et al. 2015; Lord et al. 2016).

A CLOCKS FOR TIME, SIZE, AND SHAPE

Some of the most prominent work, applying numerical methods to characterize growth-related changes in size and shape came at the end of the 1960s and in the 1970s from Stephen Jay Gould, who almost single-handedly made allometry and heterochrony fashionable again and also acceptable as alternatives to the adaptationist program for studying evolution offered by the neo-Darwinians (Gould 1966, 1971, 1977; Gould and Lewontin 1979; Gould 1981, 1982a; Gayon 2000; De Renzi 2009).

Through a series of monographs and major papers on evolutionary allometry, Gould began to put developmental mechanisms front and center. Then, in a landmark book, *Ontogeny, and Phylogeny*, Gould (1977) traced the history of conceptual advances in understanding the relationship of development to evolution. Gould opened with the *Great Chain of Being* from the Greeks; continued to theories that ontogeny parallels or recapitulates phylogeny from various French and German tran-scendentalists such as Johann Meckel, Etienne Serres, Lorenz Oken, Louis Agassiz, and Ernst Haeckel; and finally described the outright rejection of recapitulation by Karl Ernst von Baer, Garstang, de Beer, and others. In the second half of his book, Gould revisited and expanded upon de Beer's schema for heterochrony and pre-sented his own semi-quantitative *clock model* in which the hands for size and shape depicted the morphology of a species relative to its ancestor. Each clock allowed for size, shape, and age (i.e., time) to be altered separately during evolution and accord-ingly could be adjusted to represent the many manifestations of heterochrony such as neoteny, progenesis, pedomorphosis, proportional dwarfism, and proportional gigantism.

In his impressive treatise and throughout his career, Gould confronted the neo-Darwinian view of morphological evolution head-on and argued forcefully for the role of development in macroevolutionary change (Gould 1966, 1971, 1977, 1982a, 2002; Eldredge and Gould 1972; Gould and Lewontin 1979; Gould and Vrba 1982). But Gould was a paleontologist, not an embryologist, and one critical issue was that his clock model was essentially qualitative and static (like the models of de Beer), and defined simple evolutionary patterns or end states rather than capture the complex and dynamic nature of developmental processes (Etxeberria and De la Rosa 2009). Shortly thereafter, the embryologist David Wake and his 24-year-old graduate student Pere Alberch invited Gould to collaborate on what was to become an especially celebrated paper that effectively launched the modern field of evolutionary developmental biology (Wake 1998; De Renzi 2009).

B   QUANTITATIVE METHODS FOR TIME, SIZE, AND SHAPE

Alberch thought Gould's clock models were a good start in theory but insufficient in actuality (Reiss et al. 2008), and so in a paper entitled *Size and shape in ontogeny and phylogeny*, Alberch, Gould, Oster, and Wake (1979) presented a tangible quantitative method to describe the relationship between heterochrony and evolution. Their intention was to "clothe" Gould's clock model in mathematics (Wake 1998) and in so doing build a better graphical framework for integrating changes in time with changes in size and shape during development and evolution. They formulated differential equations as a way to encapsulate a more dynamic view of heterochrony, which they described as shifts in the onset, cessation, or rate of growth, rather than as an end result (Etxeberria and De la Rosa 2009).

This highly cited work became an instant classic that helped spawn a decade of conferences and books on how to measure size, shape, and time in the context of heterochrony (Bonner 1982; Maderson et al. 1982; Raff and Kaufman 1983; McKinney 1988b; Wake and Roth 1989; De Renzi et al. 1999; Reiss et al. 2008). Moreover, as part of the re-birth of evo-devo as a discipline, heterochrony became the lens through which all kinds of biology was viewed (Alberch 1980a; Alberch and Alberch 1981; Balon 1981; Coppinger and Coppinger 1982; Gould 1982a; Haluska and Alberch 1983; Shea 1983; Bemis 1984; Hanken and Hall 1984; Roth 1984; Coppinger et al. 1987; Geist 1987; Hoberg 1987; Slatkin 1987; Foster and Kaesler 1988; Hafner and Hafner 1988; Coppinger and Smith 1989; Roth and Wake 1989; Shea 1989; Coppinger and Feinstein 1991; Blanco and Alberch 1992; Zelditch et al. 1992; Blackstone and Buss 1993; Klingenberg and Spence 1993; Allmon 1994; Duboule 1994; Coppinger and Schneider 1995; Godfrey and Sutherland 1995a, b; Richardson 1995; Gilbert et al. 1996; Maunz and German 1996; Richardson et al. 1997; Smith 1997; Nunn and Smith 1998; MacDonald and Hall 2001; Vaglia and Smith 2003; Crumly and Sanchez-Villagra 2004; Tokita et al. 2007; Drake 2011; Nagai et al. 2011; Mitgutsch et al. 2011).

C   CONSTRUCTION RULES FOR TIME, SIZE, AND SHAPE

While such morphometric approaches helped elucidate critical developmental stages and events whereby changes in size and shape occur, by necessity they often reduced the complex dynamic nature of development into something much more simplistic and static, their framework was typically applied globally at the level of organisms rather than in relation to individual systems or structures, they tended to divide continuous development into artificially discrete steps in order to compare ontogenetic trajectories, and also, they largely left much to be understood in terms of underlying molecular and cellular mechanisms.

Seemingly anticipating these points and echoing Waddington's sentiments, Alberch and his colleagues (1979) challenged the field when they expressed that: "We hope that our attempts to construct a quantitative theory will stimulate others to delve more deeply below the level of pure phenomenology and come to grips with the central issue underlying evolutionary diversification of size and shape—that is, the morphogenetic unfolding of genetic programs in ontogeny and their alteration in the course of phyletic evolution" (p. 297).

Such an emphasis on the mechanistic and more dynamic aspects of development grew directly out of Waddington's epigenetic landscapes and concepts like canalization, which basically served as metaphors for how gene regulation could alter the course of ontogeny and phylogeny (Waddington 1957a), and guided the remainder of Alberch's remarkably influential but tragically foreshortened career (De Renzi et al. 1999; Wake 1998; Reiss et al. 2008). To this very point, in his elegant first solo paper on the role of ontogeny in morphological diversification, Alberch (1980b) argued that "epigenetic interactions drastically constrain the universe of possible morphological novelties and impose directionality in morphological transformations through phylogeny" (p. 654). In other words, even if a genetic mutation is random, the morphological outcome is not. Why? Because developmental systems are highly integrated, iterative, accommodative, hierarchical, and ultimately defined by an "internal structure" that limits "the realm of possible morphologies" (Alberch 1982a:319).

In his subsequent and quite a formidable body of work, Alberch addressed the role of development in the evolution of size, shape, and other aspects of morphology on multiple levels in a wide range of organisms and organs. A critical concept that he advanced pertained to *construction rules* through which developmental systems are built and become altered from one morphological state to another during evolution (Alberch 1982a, 1985; Oster and Alberch 1982; Oster et al. 1988). Accordingly, development consists of interwoven "dynamical systems, where a small set of simple rules of cellular and physicochemical interactions" lead to complex morphology (Oster and Alberch 1982:455). Evolutionary changes in organ size, for example, can be achieved by varying the quantitative parameters of cells, including the number of progenitors, the rate of proliferation, length of the cell cycle, and timing of differentiation. Other parameter values that can potentially be modulated pertain to "biochemical, cell–cell, or tissue interactions" (Alberch 1985:50), which, in turn, can affect developmental processes such as "rates of diffusion, mitotic rate, cell adhesion, etc." (Alberch 1989:27).

Throughout his research program, Alberch combined insights from comparative morphology, experimental embryology, and teratology, to generate models and other mathematical tools that helped define morphogenesis as an emergent property of physical and biochemical interactions, as well as cyclical, multidimensional, and nonlinear feedback schemes operating at the level of molecules, genes, and proteins, and extending up through tissues. In stark contrast to the neo-Darwinian view of the relationship between genotype and phenotype, Alberch argued that "genes are just one step in the chain of interactions; gene expression is both the cause and the effect of a morphogenetic process" (Alberch 1991:6). Using amphibian limb buds as a model system for studying the relationship between construction rules and morphological outcomes, Alberch and his colleagues experimentally manipulated parameters such as cell number (using the mitotic inhibitor colchicine, for example) and showed that changes in size and shape of the limb, and number of the digits became altered in a non-random way once a critical threshold was reached (Alberch and Gale 1983, 1985; Shubin and Alberch 1986).

While these studies predated the technical ability to link such outcomes with molecular biology (specifically, underlying changes in gene expression), their results were completely consistent with predictions made in their mathematical models and pattern-generating algorithms (Oster et al. 1988), and showed that the phenotypes arising from perturbations to developmental programs were not stochastic. Because of such findings, Alberch argued that "even if the parameters of the system are randomly perturbed, by either genetic mutation or environmental variance or experimental manipulation during development, the system will generate a limited and discrete subset of phenotypes. Thus the realm of possible forms is a property of the internal structure of the system" (Alberch 1989:27). Analyses of genetic mutations and experimental manipulations in a range of model organisms by many subsequent workers in the field provided critical information on the internal structure of developmental systems. In particular, these types of approaches have been especially productive with regard to understanding how parameter changes in construction rules on the molecular and cellular levels have likely played a generative role during the evolution of size and shape in the vertebrate skull.

7.3   TIME, SIZE, AND SHAPE IN THE VERTEBRATE SKULL

For numerous reasons, including its inimitable paleontological record, its measurable geometry, its evolutionary adaptability, its functional significance, and its easily visualized embryogenesis, the vertebrate skull has long been the subject of intensive research on size and shape (de Beer 1937; Hanken and Hall 1993). This has occurred chiefly with regard to; (a) genes that affect skeletal element identity (Balling et al. 1989; Lufkin et al. 1992; Gendron-Maguire et al. 1993; Rijli et al. 1993; Schilling 1997; Qiu et al. 1997; Hunt et al. 1998; Smith and Schneider 1998; Pasqualetti et al. 2000; Grammatopoulos et al. 2000; Depew et al. 2002; Creuzet et al. 2002; Kimmel et al. 2005); (b) tissue interactions required for mesenchymal differentiation into cartilage and bone (Schowing 1968; Tyler 1978; Bee and Thorogood 1980; Hall 1980, 1982b; Tyler 1983; Thorogood et al. 1986; Thorogood 1987; Hall 1987; Richman and Tickle 1989, 1992; Dunlop and Hall 1995; Shigetani et al. 2000; Ferguson et al. 2000; Couly et al. 2002; Francis-West et al. 2003; Merrill et al. 2008); (c) secreted molecules that regulate axial polarity and skeletal outgrowth (Barlow and Francis-West 1997; Francis-West et al. 1998; Schneider et al. 2001; Hu et al. 2003; Abzhanov and Tabin 2004; Crump et al. 2004; Wilson and Tucker 2004; Wu et al. 2004; Abzhanov et al. 2004; Liu et al. 2005; Marcucio et al. 2005; Wu et al. 2006); and (d) mesenchymal control of species-specific pattern (Andres 1949; Wagner 1959; Noden 1983; Schneider and Helms 2003; Tucker and Lumsden 2004; Mitsiadis et al. 2006).

The special ability of mesenchyme to transmit species-specific information on size and shape has been recognized primarily through interspecific grafting experiments of mesenchymal cells destined to form the jaw skeleton (Noden and Schneider 2006; Lwigale and Schneider 2008). The exact molecular mechanisms through which mesenchyme performs this complicated function appear to involve the ability of mesenchyme to determine the timing of its own gene expression and differentiation, as well as that of adjacent tissues such as epithelia (Schneider and Helms 2003; Eames and Schneider 2005; Schneider 2005; Merrill et al. 2008). Taken together, results from genetic, molecular, and cellular studies lead to the conclusion that the regulation of skeletal size and shape by mesenchyme involves multiple gene regulatory networks, reciprocal signaling interactions with adjacent tissues, and hierarchical levels of control.

A BIRD BEAKS

Studies on the beaks of birds have been particularly helpful in identifying factors that influence skeletal size and shape (Helms and Schneider 2003; Schneider 2005, 2007; Fish and Schneider 2014c; Schneider 2015). For example, differential domains of *Bmp4* expression in beak progenitor cells underlie variation in beak depth and width among birds including Darwin's finches, cockatiels, chicks, and ducks (Abzhanov et al. 2004; Wu et al. 2004, 2006).

Beak length seems to be managed separately through a calmodulin-dependent pathway (Abzhanov et al. 2006; Schneider 2007). Similarly, factors including SHH, FGFs, WNTs, and BMPs, which are secreted from adjacent epithelial tissues also appear to affect the shape and outgrowth of the beak skeleton (MacDonald et al. 2004; Young et al. 2014; Hu and Marcucio 2009; Foppiano et al. 2007; Hu et al. 2015a, b; Hu and Marcucio 2012; Brugmann et al. 2007; Brugmann et al. 2010; Abzhanov and Tabin 2004; Bhullar et al. 2015; Grant et al. 2006; Wu et al. 2006; Ashique et al. 2002a; Richman et al. 1997; Rowe et al. 1992; Szabo-Rogers et al. 2008; Mina et al. 2002; Doufexi and Mina 2008; Havens et al. 2008; Schneider et al. 1999, 2001). A clearer picture of how these signaling pathways are regulated and how changes to their regulation affect skeletal size and shape has begun to emerge.

B QUAIL–DUCK CHIMERAS

In particular, additional details on molecular and cellular mechanisms through which the craniofacial skeleton acquires its proper size and shape have come from our studies, using a unique avian chimeric transplantation system that exploits species-specific differences between Japanese quail and white Pekin duck (Schneider and Helms 2003, 2005, 2007; Lwigale and Schneider 2008; Jheon and Schneider 2009; Ealba and Schneider 2013; Fish and Schneider 2014b).

As a proxy for studying the orchestration of morphogenesis more generally, we have been posing the question of how do skeletal elements in the jaw skeleton of quail and duck achieve their distinct size and shape? Quail have short and narrow jaws in comparison to those of duck, which are relatively long and broad (Figure 7.1a, b). We have focused on the lower jaw (Figure 7.1c), which forms embryonically within the paired mandibular primordia. Neural crest mesenchyme (NCM) that migrates from the caudal midbrain and rostral hindbrain is the only source of precursor cells that give rise to cartilage and bone within the skeleton of the face and jaws (Figure 7.1d)

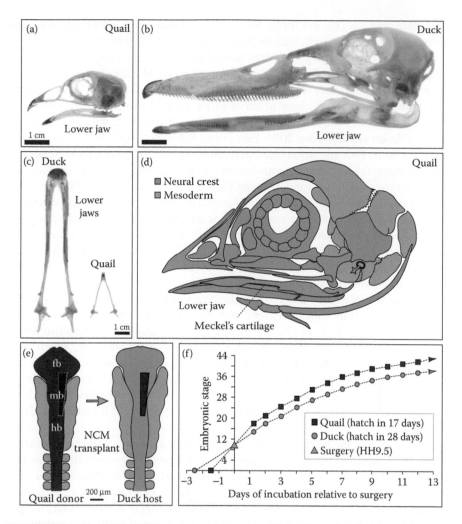

FIGURE 7.1 The quail–duck chimeric system for studying time, size, and shape in the head skeleton. (a) Head skeletons of adult Japanese quail (*Coturnix coturnix japonica*) and (b) white Pekin duck (*Anas platyrhyncos*) showing species-specific differences in size and shape. (c) Lower jaws of adult duck and quail. (d) Neural crest cells generate the facial and jaw skeletons (blue) whereas mesoderm forms the caudal cranial vault and skull base (orange). (e) Schematic of an embryonic rostral neural tube (dorsal view) depicting the origin of neural crest mesenchyme (NCM) from the forebrain (fb), midbrain (mb), and hindbrain (hb). NCM destined for the jaw primordia are grafted (green arrow) from a quail donor (red) to a duck host (blue). (f) Embryonic quail (red squares) and duck (blue circles) have distinct rates of maturation but can be stage-matched for surgery (green triangle on Y-axis) by setting eggs in the incubator at separate times. Approximately three embryonic stages distinguish faster-developing quail from duck embryos within two days following surgery, and this three-stage difference remains relatively constant during the period of jaw morphogenesis. (Panels [a, b] modified from Fish, J.L. et al., *Development*, 141, 674–684, 2014; Panels [c, e, f] modified from Eames, B.F., and Schneider, R.A., *Development*, 135, 3947–3958, 2008; Panel [d] modified from Schneider, R.A., *J. Anat.*, 207, 563–573, 2005, and based on a drawing from D. Noden.)

(Le Lièvre and Le Douarin 1975; Noden 1978; Couly et al. 1993; Köntges and Lumsden 1996; Helms and Schneider 2003; Noden and Schneider 2006).

Our experimental strategy involves transplanting pre-migratory NCM between quail and duck embryos (Figure 7.1e). We transplant NCM either bilaterally, so that donor cells fill both sides of the host jaw skeleton, or unilaterally, which allows the nonsurgical side of the host to serve as an internal control. Unilateral transplants enable us to compare donor- and host-derived tissues in the same chimeric embryo (Tucker and Lumsden 2004; Eames and Schneider 2005, 2008; Lwigale and Schneider 2008; Fish and Schneider 2014b; Solem et al. 2011; Tokita and Schneider 2009). A powerful and serendipitous feature of this chimeric system is the fact that quail embryos develop at a quicker rate than duck embryos (17 vs. 28 days from fertilization to hatching), which causes faster-developing quail cells and relatively slower-maturing duck cells to interact with one another over time while they become progressively asynchronous (Figure 7.1f). Having such divergent developmental trajectories conveniently offers a way to screen for the effects of donor cells on the host by looking for species-specific changes to the timing of gene expression, cell differentiation, and tissue formation. Consequently, and especially in the context of the aforementioned holy trinity of evolutionary developmental morphology, this system also affords us with the unique opportunity to evaluate directly and in the same embryo, the effects of changes in growth rates and the timing of developmental events on size and shape. We can use an anti-quail antibody, which does not recognize duck cells, to distinguish the contributions of donor versus host and we can quantify the proportion of quail versus duck cells at the molecular level, using a PCR-based strategy (Schneider 1999; Lwigale and Schneider 2008; Ealba and Schneider 2013; Fish and Schneider 2014b; Fish et al. 2014; Hall et al. 2014; Ealba et al. 2015).

Once quail and duck cells become mixed within chimeras, they become challenged to assimilate two separate morphogenetic programs controlling species-specific size and shape. Chimeric "*quck*" contain quail donor NCM inside of a duck host whereas "*duail*" have duck NCM in a quail host. As a result, we can discover mechanisms directing jaw patterning by (1) characterizing donor-mediated changes to jaw size and shape; (2) assaying for temporal and spatial shifts in developmental events underlying cartilage and bone formation such as mesenchymal condensation and differentiation; (3) analyzing the effects of donor NCM on non-NCM host derivatives that participate in skeletal patterning and growth such as epithelia, muscles, blood vessels, and osteoclasts; (4) looking for genes that become differentially expressed in chimeras; and (5) modulating the expression of these genes to test if they account for the chimeric phenotype and affect skeletal size and shape (Eames and Schneider 2005, 2008; Noden and Schneider 2006; Merrill et al. 2008; Tokita and Schneider 2009; Solem et al. 2011; Hall et al. 2014; Ealba et al. 2015).

An important point to emphasize is that this chimeric system can reveal in a more or less "normal" developmental context those molecular and cellular interactions between the donor and host that are divergent and ultimately generative of species-specific size and shape. In this context, the quail–duck chimeric system offers a unique opportunity to observe what Alberch et al. (1979) called "the morphogenetic unfolding of genetic programs in ontogeny and their alteration in the course of phyletic evolution" (p. 297). We can also probe for construction rules and identify those parameter changes that may account for evolutionary differences between each species. Along similar lines, Shubin and Alberch (1986) argued that while the basic morphogenetic rules have remained the same; what have changed during vertebrate evolution are the parameters through which these interactions occur (Etxeberria and De la Rosa 2009).

By combining a classical comparative method (Sanford et al. 2002) with experimental embryology (i.e., the quail–duck chimeric transplant system), we have found that NCM relies upon multiple mechanisms to exert cellular control over time, size, and shape, primarily through three phases of development:

- At the onset of NCM migration, quail and duck embryos allocate different numbers of progenitors to the presumptive jaw region, with duck having significantly more cells (Fish et al. 2014).
- Thereafter, when these populations of NCM expand, there is species-specific regulation of, and response to, critical signaling pathways in a manner that is dependent on their own rates of maturation (Eames and Schneider 2008; Merrill et al. 2008; Hall et al. 2014).
- Lastly, as these progenitors start to form cartilage and bone, they execute autonomous molecular and cellular programs for matrix deposition and resorption through patterns and processes that are inherent to each species and deeply rooted in the timing of developmental events (Eames and Schneider 2008; Merrill et al. 2008; Mitgutsch et al. 2011; Hall et al. 2014; Ealba et al. 2015; Schneider 2015).

Moreover, on the molecular level, the SHH, FGF, BMP, and TGFβ signaling pathways all seem to be clearly but not unexpectedly involved since many members and targets show species-specific expression and they become altered in quail–duck chimeras.

Thus, the ability of NCM to regulate the timing, levels, and spatial patterns of gene expression and to do so in a species-specific manner, likely modulates the proliferation, differentiation, and growth of skeletal progenitors, and determines the size and shape of cartilage and bone. Such work offers insight into the many ways NCM exerts its regulatory abilities during ontogeny and phylogeny, which has been a long-standing question in the field (Gans and Northcutt 1983; Noden 1983; Maderson 1987; Hall and Hörstadius 1988; Hanken 1989; Baker and Bronner-Fraser 1997; Hall 1999, 2000b; Graham 2003; Santagati and Rijli 2003; Trainor et al. 2003; Graham et al. 2004; Le Douarin et al. 2004; Noden and Schneider 2006; Jheon and Schneider 2009; Fish and Schneider 2014c; Sanchez-Villagra et al. 2016). Specific examples of the multiple ways NCM exercises control over skeletal size and shape, especially by keeping track of time, are detailed in the sections below.

7.4 EARLY CELLULAR DETERMINANTS OF JAW SIZE AND SHAPE

The genesis of NCM involves several sequential embryonic events, including induction at the boundary between neural and non-neural ectoderm, specification and regionalization along the dorsal margins of the neural folds, regulation of cell cycle and maintenance of multipotency, transition from epithelium to mesenchyme, and migration throughout the head and trunk (Betancur et al. 2010; Nikitina et al. 2008). NCM that arises from the midbrain through the first and second rhombomeres of the hindbrain migrates into the mandibular primordia (Le Lièvre and Le Douarin 1975; Noden 1978; Couly et al. 1993; Köntges and Lumsden 1996; Schneider et al. 2001). While the gene regulatory networks and developmental programs that govern these morphogenetic events are extremely conserved across vertebrates (Nikitina et al. 2008; Depew and Olsson 2008; Bronner-Fraser 2008; Northcutt 2005), much remains to be understood about when and where changes can lead to the evolution of species-specific morphology.

A   QUANTIFYING JAW PRECURSOR CELLS

For this reason, we have been concentrating on exactly when and where ducks assemble their long bills, compared to quail, who make relatively short beaks. Using a simple analogy that constructing a taller building might involve adding more bricks, as opposed to bigger bricks (Fish and Schneider 2014a), and following the spirit of Alberch and his construction rules, we set out to determine the number of jaw precursors that migrate into the mandibular primordia in duck versus quail (Figure 7.2a, b). We started by counting NCM at key embryonic stages (Fish et al. 2014).

At an initial embryonic stage, when NCM is specified along the neural folds, duck and quail appear to have the same total amount of cranial NCM. But soon afterwards, when NCM coalesces on the dorsal margins of the neural tube, duck has about 15% more NCM at the midbrain and rostral hindbrain levels, which is where the population that migrates into the presumptive jaw region originates. Remarkably only several stages later, the mandibular primordia of duck contain twice as many cells as do those of quail (Figure 7.2c). To understand how a 15% difference could quickly lead to a doubling in size, we looked for species-specific variation in cell proliferation and cell cycle length. Our results show that although duck has a longer cell cycle (13.5 versus 11 hours in quail), if the total duration of each embryonic stage during this period is taken into account in terms of absolute time (45 versus 32 hours), then duck cells, in fact, proliferate more than those of quail. By maintaining their intrinsic rates of maturation, duck deploy a cellular mechanism that increases jaw size progressively throughout development (Fish and Schneider 2014c; Fish et al. 2014; Schneider 2015). In so doing, they directly link developmental time with size.

B   GENES AND BRAIN REGIONALIZATION

But how might duck initially generate more midbrain NCM that can then migrate into the presumptive jaw region? To address this question, we assayed for species-specific variation in the expression of genes known to affect brain regionalization. We examined the expression of *Pax6* in the forebrain, *Otx2* in the forebrain and midbrain, *Fgf8* at the midbrain–hindbrain boundary, and *Krox20* in rhombomeres 3 and 5 of the hindbrain (Figure 7.2d). Landmark-based morphometrics was used to compare brain shape in duck and quail embryos after neurulation, and we pinpointed species-specific differences that were correlated with changes in domains

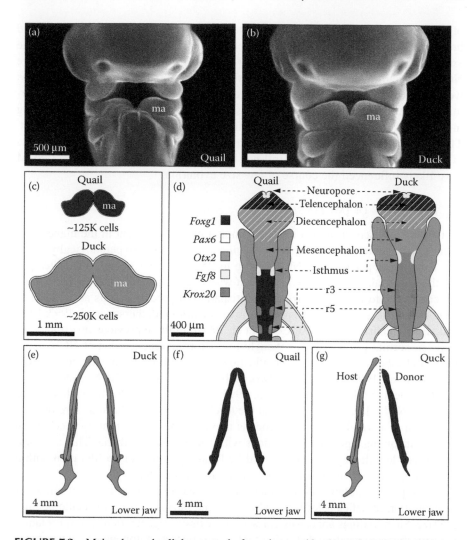

FIGURE 7.2   Molecular and cellular control of species-specific size and shape. (a, b) Frontal views of the heads of quail and duck embryos showing differences in the size of the mandibular primordia (ma), from which the lower jaw skeleton develops. (c) At this stage, the mandibular primordia (ma) in quail embryos (red) is approximately half the size of that of duck (blue) in terms of total number of cells. (d) Quail and duck embryos have distinct shapes and region-alization of the rostral neural tube. The duck midbrain (mesencephalon) is foreshortened and broader mediolaterally. Genes including *Foxg1, Pax6, Otx2, Fgf8,* and *Krox20* are expressed in specific domains, each domain being shifted more anteriorly in duck than in quail. The lower jaw skeletons of duck (e) and (f) quail show stage-specific and species-specific differ-ences in size and shape with duck being longer and more curved. (g) In quck mandibles, the quail donor-derived jaw skeleton (red) is shorter and straighter than the contralateral duck host-derived jaw skeleton (blue), which is longer and curved. (Panels [a, b, e, f, g] modi-fied from Fish, J.L. et al., *Development*, 141, 674–684, 2014; Panel c modified from Merrill, A.E. et al., *Development*, 135, 1223–1234, 2008; Panel [d] modified from Schneider, R.A., *Curr. Top Dev. Biol.*, 115, 271–298, 2015.)

of gene expression. Most strikingly, we found that the duck midbrain is shorter and broader and has a correspondingly restricted domain of *Otx2* expression along the anterior to posterior axis. Presumably, a broader midbrain in duck allows more NCM to accumulate and migrate into the mandibular primordia. Importantly, we observed these differences in *Otx2* expression even before neural tube formation or the genesis of NCM, demonstrating that species-specific patterning mechanisms affecting jaw size may function at the earliest stages of development. Thus, this work reveals how small spatial and temporal modifications to aspects of developmental programs controlling the allocation and proliferation of NCM have likely influenced the course of jaw size evolution.

C REGULATION OF JAW LENGTH

In addition to discovering that the total amount of NCM present in the mandibular primordia is a determinant of species-specific jaw size, we also paradoxically found that reducing or augmenting NCM by up to 25% does not significantly alter jaw length (Fish et al. 2014). This is consistent with other experiments showing that the jaw can return to its normal size after extirpation of precursor cells at the level of the neural folds, a process often referred to as *regulation* (Scherson et al. 1993; Hunt et al. 1995; Sechrist et al. 1995; Couly et al. 1996). In these previous investigations, however, normal jaw length was thought to arise from regeneration of NCM along the neural tube, either by a re-specification of remaining dorsal neuroepithelium (Sechrist et al. 1995; Hunt et al. 1995), or by an expansion of NCM generated by adjoining neural folds (Scherson et al. 1993; Couly et al. 1996). Instead, we conclude that NCM does not regenerate at the level of the neural tube and therefore, the restoration of normal jaw length depends upon another compensatory mechanism possibly involving signaling interactions with surrounding epithelia. That is to say, normal jaw length may also be affected by local regulation of proliferation within the postmigratory environment of the mandibular primordia.

This type of regulative development in the local environment would allow for compensation of deficiencies in NCM up to some intrinsic species-specific population size. Such findings are consistent with prior tissue regeneration and transplantation experiments revealing that individual organs possess autonomous determinants of size and can regulate growth appropriately in various contexts (Stern and Emlen 1999; Leevers and McNeill 2005). Moreover, that a strong correlation exists between innate rates of growth and overall size is well established in birds (Starck 1989; Ricklefs and Starck 1998; Starck and Ricklefs 1998).

7.5 CELLULAR CONTROL OF JAW SIZE AND SHAPE DURING SKELETAL DIFFERENTIATION

Once appropriate amounts of NCM are allocated to the mandibular primordia of quail versus duck, the next question is how these differences are integrated into the programs for skeletal differentiation that eventually produces species-specific size and shape? To answer this question, we examined the formation of Meckel's cartilage in the lower jaw skeleton (Eames and Schneider 2008).

A   MECKEL'S CARTILAGE AND SPECIES-SPECIFIC SIZE AND SHAPE

Meckel's cartilage is more-or-less a cylindrical rod that is derived exclusively from NCM (Helms and Schneider 2003; Noden and Schneider 2006; Noden 1978, 1982; Noden and Trainor 2005) and rarely goes on to ossify (Kavumpurath and Hall 1990; Ekanayake and Hall 1994; Eames et al. 2004; de Beer 1937). To identify molecular and cellular mechanisms through which Meckel's cartilage acquires its species-specific size and shape, we unilaterally transplanted NCM from quail embryos into a stage-matched duck. These quail donor NCM filled the right half of the duck host mandible, which allowed for an unambiguous comparison of donor quail-derived versus host duck-derived Meckel's cartilage development in the same chimeric mandible.

During normal growth of Meckel's cartilage, conspicuous stage-specific and species-specific differences in size and shape emerge in quail and duck. At early embryonic stages in both quail and duck, Meckel's cartilage goes from being slightly curved to more S-shaped. Shortly afterward, however, Meckel's cartilage in duck remains curved (Figure 7.2e), whereas Meckel's cartilage in quail becomes straightened (Figure 7.2f). Meckel's cartilage grows in each successive stage thereafter, but steadily gets larger in duck. In quck chimeras, quail donor NCM maintained its faster rate of growth within the relatively slower duck host environment, and Meckel's cartilage on the donor side was always accelerated by approximately three stages. Moreover, the size and shape of the donor side was consistently more quail-like compared to that observed on the contralateral duck host side (Figure 7.2g). Using landmark-based morphometrics and a Procrustes analysis (Chapman 1990; Rohlf and Bookstein 1990; Coppinger and Schneider 1995; Marcus 1996; Roth and Mercer 2000; Schneider and Helms 2003; Zelditch 2004), we quantified changes in Meckel's cartilage and found that NCM controls both stage-specific and species-specific size and shape.

To clarify the molecular and cellular mechanisms through which NCM accomplishes this complex task, we assayed for changes in the program of cartilage differentiation that might presage the genesis of size and shape. Cartilage differentiation involves the condensation of pre-chondrogenic mesenchyme, followed by overt chondrification where an abundant extracellular matrix is secreted by chondrocytes (Eames et al. 2003; Hall 2005). In quck chimeras, NCM on the donor side differentiated into chondrocytes and formed cartilage, following the timeframe of quail. Donor-dependent acceleration to the timing of cartilage differentiation was evident from the beginning of mesenchymal condensation. The transcription factor *Sox9*, which is the earliest known molecular marker of chondrogenic condensations (Healy et al. 1996; Zhao et al. 1997; Eames et al. 2003, 2004), and *Col2a1*, which is directly regulated by *Sox9* (Bell et al. 1997), were expressed prematurely by quail donor NCM relative to the duck host. We also observed that FGF signaling, which operates upstream of *Sox9* and chondrogenesis (Bobick et al. 2007; Govindarajan and Overbeek 2006; Murakami et al. 2000; Petiot et al. 2002; Healy et al. 1999; de Crombrugghe et al. 2000; Eames et al. 2004) was similarly regulated by donor NCM in temporal and spatial patterns like those observed in quail. For example, while the secreted ligands *Fgf4* and *Fgf8* were expressed continuously by duck host epithelium prior to and during formation of Meckel's cartilage, the receptor *Fgfr2* was prematurely expressed just by quail donor NCM. When we inhibited FGF signaling during this discrete temporal window of receptor activation, we blocked chondrogenesis. Thus by exerting control over the timing of FGF signaling and the expression of downstream targets such as *Sox9* and *Col2a1*, NCM likely transmits information establishing stage-specific and species-specific size and shape to Meckel's cartilage.

B Epithelia and Cartilage Patterning

While these experiments demonstrate that NCM dictates the size and shape of cartilage, other studies have shown that adjacent epithelia also play essential roles during cartilage pattern formation. For instance, in the 1980s Peter Thorogood advanced a *"flypaper model"* in which he proposed that interactions between epithelia and mesenchyme drive the production of extracellular matrix, which adhesively "traps" migrating NCM at their site of differentiation and leads to the induction of cartilage (Garrod 1986; Thorogood 1988, 1993). In the head, such epithelia are associated with the surface ectoderm and pharyngeal endoderm around the facial primordia, as well as the brain and sensory capsules, all of which are known to initiate and maintain chondrogenesis at one stage or another (Thorogood et al. 1986; Hall 1980, 1981).

While some data suggest that epithelia can play an inhibitory role during chondrogenesis (Mina et al. 1994), additional studies demonstrate that epithelia impart axial polarity and regional identity to the underlying NCM-derived skeletal tissues. More specifically, epithelia around the developing jaws and face (e.g., frontonasal, maxillary, mandibular primordia) seem to provide positional cues and maintenance factors necessary for patterned outgrowth of individual skeletal elements along the proximodistal, mediolateral, and dorsoventral axes (Hu et al. 2003; Foppiano et al. 2007; Hu and Marcucio 2009; Schneider et al. 1999; Young et al. 2000; Cordero et al. 2002; Helms and Schneider 2003; Young et al. 2010; Chong et al. 2012; Hu et al. 2015a). Experiments that rotate epithelium in the mid- and upper face, for example, cause mirror image duplications of distal upper beak structures along the dorsoventral axis (Hu et al. 2003; Marcucio et al. 2005). Similarly, transplantation studies and genetic analyses demonstrate that endodermal epithelium lining the pharynx transmits region-specific polarity and segmental identity to NCM, which is critical for the proper growth and orientation of bone and cartilage in the jaw skeleton (Couly et al. 2002; Haworth et al. 2007; Kikuchi et al. 2001; Kimmel et al. 1998; Veitch et al. 1999; Piotrowski and Nusslein-Volhard 2000; Miller et al. 2000; David et al. 2002; Crump et al. 2004). When either ectodermal or endodermal epithelia are rotated surgically, the underlying NCM-derived skeleton follows accordingly. Taken together, such studies indicate that the primary role for epithelia is to contribute local signals for generalized anatomical pattern, which in turn induce and/or maintain programmatic responses from underlying NCM (Richman and Tickle 1989; Langille and Hall 1993; Tucker et al. 1999; Ferguson et al. 2000; Mitsiadis et al. 2003; Santagati and Rijli 2003; Le Douarin et al. 2004; Wilson and Tucker 2004; Fish and Schneider 2014c; Foppiano et al. 2007; Hu and Marcucio 2009, 2012; Marcucio et al. 2011;

Hu et al. 2015b). Importantly, the timing of expression of these epithelial signals is under the regulatory control of NCM (Schneider and Helms 2003; Eames and Schneider 2005; Merrill et al. 2008).

The finding that NCM executes autonomous molecular and histological programs for cartilage size and shape can be combined with other experimental results about the role of epithelia in the following way. If the steps of skeletal patterning involve mesenchymal migration, proliferation, condensation, overt chondrocyte differentiation, and ultimately the morphogenesis of cartilage as a three-dimensional structure, then the interactions with pharyngeal endoderm and facial ectoderm, for example, would dictate cartilage orientation and regional identity along the oral cavity. Such interactions could happen before mesenchymal condensation and promote and align the spatial distribution of pre-chondrogenic mesenchyme. In this context, these epithelia would be acting instructively initially but then assume a more permissive role that facilitates the execution of NCM-dependent programs and enables chondrogenesis to proceed in a time-independent manner.

So, while epithelia derived from ectoderm and endoderm may define where chondrogenic condensations occur along an axis, which is presumably quite similar between quail and duck, our transplants reveal that NCM consequently responds via intrinsic, stage-specific and species-specific programs that determine cartilage size and shape. Equivalent roles have also been postulated for epithelia during osteogenesis of the mandible and other bones (Tyler and Hall 1977; Hall 1978, 1987; Bradamante and Hall 1980; Hall 1980, 1981, 1982a, b; Hall and Van Exan 1982; Hall et al. 1983; Van Exan and Hall 1984; Merrill et al. 2008). Further support is lent by the finding that several chondrogenic signaling pathways including FGFs and BMPs are expressed continuously by epithelia prior to and during the arrival of NCM in the mandible (Francis-West et al. 1994; Wall and Hogan 1995; Shigetani et al. 2000; Mina et al. 2002; Ashique et al. 2002b; Havens et al. 2006; Eames and Schneider 2008; Merrill et al. 2008). By controlling the timing of receptor activation, in this case, for $Fgfr2$, NCM allows the signal transduction required for chondrogenesis to proceed, and by doing so, initiates the program for cartilage size and shape.

Overall, this integrated perspective on the roles of mesenchyme and epithelium in the establishment of size and shape is also consistent with classic embryological work from the lab of Hans Spemann who first discovered the origins of species-specific pattern in the 1920s and 1930s through interspecific grafting experiments and especially by exchanging mouth-forming tissues between frogs and newts (Spemann and Mangold 1924; Spemann and Schotté 1932; Spemann 1938; Fassler 1996; Noden and Schneider 2006). These remarkable experiments showed that general anatomical features of the mouth are guided by local signals, but that species-specific pattern is dictated by information in the responding cells. Evidently, Spemann interpreted his finding to mean that, "The ectoderm says to the inducer, 'you tell me to make a mouth; all right, I'll do so, but I can't make your kind of mouth; I can make my own, and I'll do that"' (Harrison 1933). Ensuing transplant experiments between salamanders and frogs, between mice and chicks (in this instance for jaws and teeth), as well as divergent species of birds, including quail, chick, duck, and emu have also supported the conclusion that species-specific pattern is largely driven by NCM (Andres 1949; Wagner 1959; Lumsden 1988; Mitsiadis et al. 2003;

Lwigale and Schneider 2008; Sohal 1976; Yamashita and Sohal 1986; Schneider and Helms 2003; Tucker and Lumsden 2004; Eames and Schneider 2005; Schneider 2005; Noden and Schneider 2006; Eames and Schneider 2008; Jheon and Schneider 2009; Tokita and Schneider 2009; Fish and Schneider 2014c; Fish et al. 2014; Hall et al. 2014; Ealba et al. 2015; Schneider 2015).

C   TIME AS A DEVELOPMENTAL MODULE FOR CHONDROGENESIS

Our results suggest that the program for chondrogenesis through which NCM implements species-specific size and shape is integrated at multiple levels and through time as a developmental module. This is equivalent to the identification of developmental modules and the role proposed for mesenchyme in other embryonic systems such as epidermal appendages (Eames and Schneider 2005; Schneider 2005).

In a similar vein as that described by Alberch (Alberch 1982a), modularity is predicated on the observation that many developmental programs appear to function as semi-autonomous, self-directing, and hierarchical units that can be continuously iterated during development and rapidly diversified during evolution as a consequence of the inductive relationships among their constituent parts (Raff 1996; Bolker 2000; West-Eberhard 2003; Schlosser and Wagner 2004). The notion that NCM engineers size and shape by presiding over a highly integrated developmental module is supported by the fact that NCM executes autonomous molecular and cellular programs for both the formation of cartilage as a tissue and as a three-dimensional organ. Importantly, these programs include many of the same gene regulatory networks and signaling molecules that operate during NCM specification, proliferation, and differentiation such as members and targets of the BMP and FGF pathways, and that affect the size and shape of cartilage in the avian jaw and facial skeletons (Francis-West et al. 1994; Wall and Hogan 1995; Mina et al. 1995; Ekanayake and Hall 1997; Barlow and Francis-West 1997; Richman et al. 1997; Wang et al. 1999; Tucker et al. 1999; Barlow et al. 1999; Shigetani et al. 2000; Mina et al. 2002; Ashique et al. 2002b; Abzhanov et al. 2004; Wilson and Tucker 2004; Havens et al. 2006; Schneider 2007; Abzhanov and Tabin 2004; Wu et al. 2004, 2006; Foppiano et al. 2007).

Further evidence for modularity as a principal mechanism through which NCM exerts control over skeletal size and shape relates to the way NCM accounts for time. In fact, this is one of the most striking revelations to emerge from the quail–duck chimeric system: NCM keeps track of stage-specific and species-specific size and shape concurrently. In other words, as quail donor NCM shifts the timing of cartilage differentiation and morphogenesis in the duck to something like that found in the quail, these cells generate stage-appropriate and species-appropriate size and shape simultaneously. This result offers a novel mechanism that connects skeletal development with skeletal evolution vis-à-vis a single population of cells, the cranial NCM. Moreover, this melding of time links ontogeny with phylogeny in a manner completely consistent with previous theories of heterochrony as a means to understand transformations in size and shape.

While historically, heterochrony has been used to describe changes in the timing of developmental events between ancestors and descendants (Russell 1916;

de Beer 1930), the concept can also concern comparisons of closely related taxa (such as quail and duck) and be employed to assess the effects of changes in rates of growth on size and shape (Gould 1977; Alberch et al. 1979; Hall 1984; Roth 1984; McKinney 1988a; Foster and Kaesler 1988; Klingenberg and Spence 1993; Raff 1996). This type of growth heterochrony is probably one of many variables introduced by the faster-developing quail donor NCM in the relatively slower-growing duck host. While such an effect would largely arise from intrinsic species-specific differences in maturation rates, another variable could be any experimentally induced shifts in relative onsets, cessations, and/or durations of molecular and cellular events during chondrogenesis. Under normal circumstances, these types of changes would be considered instances of sequence heterochrony, which is another way changes in time can relate to changes in size and shape (Smith 2001, 2002, 2003), particularly with regard to reciprocal epithelial–mesenchymal interactions underlying skeletal evolution (Smith and Hall 1990).

The predisposition of quail NCM to follow its endogenous rate and time for cartilage development likely arises from cell-autonomous mechanisms that limit the cycling and proliferation of cells to a quail-specific timetable. As a result, chondrogenesis advances three embryonic stages ahead of schedule, and Meckel's cartilage attains species-specific size and shape. To be clear, this scenario was not the only theoretically possible outcome of our transplant experiments. Quail donor NCM could have acted naively, followed the timetable of the duck host, and made cartilage that was duck-like in morphology; or they could have become confused and created some novel anatomy that was either a combination of, or unlike what is normally observed in quail or duck. But instead, within a duck host environment and all that entails in terms of duck-specific signaling, quail donor NCM altered the relative timing and rates of differentiation, executed an innate program of cartilage morphogenesis that replaced and/or superseded the duck program, and in so doing, made something like that normally observed in quail. Thus, not only does NCM coordinate the developmental timing of its own derivatives, but host epithelium also responds to this premature induction and expresses secreted molecules on the donor timetable as well (Schneider and Helms 2003; Eames and Schneider 2005; Merrill et al. 2008).

Overall, our work supports the conclusion that heterochrony can underlie the species-specific evolution of size and shape, but in the case of the quail–duck chimeric system, such heterochrony does so with at least two important caveats. First, since quail donor NCM followed their own timetable and acted as they would normally do, the heterochrony we created is not heterochrony in the true sense of the word. In this chimeric system, absolute time remained the same, and the heterochrony we constructed can only be contemplated in terms of relative timing of developmental events (i.e., to that of the duck host). Second, timing is not the only thing that was altered in these transplants. Once inside the duck host, quail NCM seemingly and progressively implemented a quail-specific genome. Likely, donor NCM does so in response to shared common signals present in duck host epithelium (e.g., FGF and BMP) that appear to be expressed continuously during a broad developmental window, and which might be able to accommodate any difference in stage between the donor and the host. Even transplants in avian species with much wider disparities in maturation rates like quail and emu (i.e., 17 vs. 58 days from fertilization to hatching), which are separated by approximately seven embryonic stages during chondrogenesis, demonstrate that apparently there are few limits in the ability of the host to support the deployment of NCM-mediated programs for cartilage and bone at any given time during development (Hall et al. 2014). Similarly, duail chimeras, in which slower-growing duck NCM act out their programs on a delayed timetable relative to faster-developing quail host and consequently make duck-like structures, reveal that the same phenomenon is true in reverse (Schneider and Helms 2003; Eames and Schneider 2005; Merrill et al. 2008).

D   TIME AS A DEVELOPMENTAL MODULE FOR OSTEOGENESIS

In a similar manner to what we have observed for cartilage, NCM also appears to provide species-specific information on size and shape to the bone in the craniofacial skeleton by setting the timing of key events during osteogenesis. Following transplants of NCM destined to form the lower jaw, quail NCM maintains its faster timetable for development and autonomously executes molecular and cellular programs that initiate and synchronize each discrete step of osteogenesis including induction, proliferation, differentiation, osteoid deposition, mineralization, and matrix remodeling (Merrill et al. 2008; Hall et al. 2014; Ealba et al. 2015).

Again, this role as a developmental timekeeper holds true both in reverse and in the extreme, as evidenced by transplants of duck NCM into quail (i.e., *duail*) and quail NCM into emu (i.e., *qumu*), respectively. In accordance with one of Alberch's theoretical predictions concerning parameter changes to a construction rule, we find that NCM determines the timing of bone formation in the jaw skeleton by controlling cell cycle progression. In particular, we observed that NCM regulates the cell cycle through stage- and species-specific expression of cyclin and cyclin-dependent kinase inhibitors (CKI), including p27 (*Cdkn1b*), which is a CKI that decreases proliferation in cell types such as differentiating osteoblasts; cyclin E (*Ccne1*), which is required for G1/S phase transition; and cyclin B1 (*Ccnb1*), which is required for G2/M phase transition (Zavitz and Zipursky 1997; Coats et al. 1996; Drissi et al. 1999). Our data suggest that differences in the expression or post-translational processing of these cell cycle regulators may enable species such as quail to lessen mesenchymal proliferation and form a faster-developing and smaller beak. For example, in quail and quck we observed an up-regulation of p27 relative to that observed in duck. Previous studies have shown that p27 is correlated with size, including p27-deficient mice, which are substantially larger than wild-type littermates and have no apparent defects in skeletal development (Drissi et al. 1999). Additionally, the developing duck frontonasal process has a lower p27 level than in chick (Powder et al. 2012), and also the mandibular primordia shows tissue-specific post-translational regulation of p27, like what has been reported in other systems (Hirano et al. 2001; Zhang et al. 2005). Thus, modulating p27 may be a means to influence tissue- and species-specific size and/or overall growth. Ultimately, such a direct mechanistic link between the regulation of cell cycle progression and the sequence of developmental events during osteogenesis likely endows NCM with the capacity to generate changes in skeletal size and shape during evolution.

Another mechanism through which NCM appears to determine the size and shape of bones in the jaw skeleton is through members and targets of TGFβ and BMP pathways, especially, osteogenic transcription factors such as *Runx2*, which become expressed prematurely and at higher levels in quck chimeras (Merrill et al. 2008; Ealba and Schneider 2013; Hall et al. 2014; Ealba et al. 2015). *Runx2* is a master regulator of bone formation since its expression is sufficient to direct osteoblast differentiation, initiate the timing of mineralization, and affect skeletal size (Eames et al. 2004; Maeno et al. 2011; Ducy et al. 1997; Komori et al. 1997; Otto et al. 1997; Ducy et al. 1999; Pratap et al. 2003; Galindo et al. 2005; Thomas et al. 2004). Moreover, we observe premature expression of bone matrix-producing genes such as *Col1a1*.

Consistent with what we observed in our cartilage studies, the systemic environment of the duck host seems to be more or less permissive and supports osteogenesis independently by supplying circulating minerals and blood vessels. NCM controls precisely where and when bone forms by dictating the timing of cell cycle progression and by mediating the transition from cell proliferation to cell differentiation. If we experimentally induce premature cell cycle exit, we can mimic chimeras by accelerating and elevating expression of *Runx2* and *Col1a1* (Hall et al. 2014). Experimentally increasing and accelerating the timing of *Runx2* expression leads to a decrease in the size of the beak skeleton like that observed in quail. In effect, this mirrors the relationship between endogenous *Runx2* levels and species-specific beak size, since we also observed higher endogenous expression of *Runx2* in quail concomitant with their smaller beak skeletons. In fact, by the time the jaw becomes mineralized, *Runx2* levels in quail are more than double those of duck. This supports other studies, which have predicted a mechanistic connection between expected *Runx2* expression levels (based on ratios of tandem repeats in DNA) and facial length in dogs and other mammals (Fondon and Garner 2004; Sears et al. 2007; Pointer et al. 2012).

That the timing and levels of *Runx2* directly affect the size of the craniofacial skeleton, is a finding fulfilling predictions made around 75 years earlier by Huxley, Goldschmidt, and de Beer concerning genes that alter the time and rate of development (Huxley 1932; Goldschmidt 1938, 1940; de Beer 1954). Insight into how *Runx2* might play this role comes from *in vitro* studies in which *Runx2* both responds to and modulates cell cycle progression through direct and indirect mechanisms, including repressing rRNA synthesis, and up-regulating p27 expression (Young et al. 2007; Galindo et al. 2005; Thomas et al. 2004; Pratap et al. 2003).

These studies suggest that NCM controls jaw size by maintaining precise species-specific levels of essential transcription factors such as *Runx2* and by regulating the timing of skeletal cell differentiation. Duck NCM seemingly proliferates more slowly and expands in size for longer periods of time before differentiating, which then leads to larger skeletal elements. In contrast, quail embryos suppress proliferative signals more quickly, exit the cell cycle sooner, and achieve a smaller overall beak size. This scenario invokes possible changes to the balance between mesenchymal proliferation and differentiation during a critical phase of osteogenesis, which is condensation (Ettinger and Doljanski 1992; Hall 1980; Hall and Miyake 1992, 1995). Such changes would likely affect the size, shape, and location of these condensations, which can generate morphological variation in development and evolution (Atchley and Hall 1991; Dunlop and Hall 1995; Smith and Hall 1990; Hall and Miyake 2000; Smith and Schneider 1998). So, in terms of absolute time, earlier osteogenic condensations can lead to smaller skeletal elements and ultimately affect their shape through allometric growth. With regard to the differentiation of both cartilage and bone, the astonishing ability of NCM to transmit information on size and shape across embryonic stages and between species in parallel, lends strong support to the notion that development has played a generative role during the course of skeletal evolution (Alberch 1982b; Maderson et al. 1982).

7.6 CELLULAR CONTROL OF JAW SIZE AND SHAPE DURING LATE-STAGE GROWTH

Whereas most of our studies reveal that NCM imparts species-specific size and shape to the jaw skeleton by controlling the molecular and cellular programs that underlie the induction and deposition of cartilage and bone, we have also found that a previously unrecognized but perhaps equally important mechanism affecting size and shape is the ability of NCM to mediate the process of bone resorption (Ealba et al. 2015). Bone resorption is typically associated with bone deposition and as a metabolic process helps maintain homeostasis in the adult skeleton (Filvaroff and Derynck 1998; Buckwalter et al. 1996; Hall 2005; Teitelbaum 2000; Teitelbaum et al. 1997; Nguyen et al. 2013; O'Brien et al. 2008). The extent to which bone resorption affects the embryonic skeleton has not been studied extensively except for some hypotheses proposing differential fields of resorption to account for changes in size and shape that arise during the development of the human jaw skeleton (Enlow et al. 1975; Moore 1981; Radlanski et al. 2004; Radlanski and Klarkowski 2001).

A BONE RESORPTION

Bone resorption relies on the activities of two cell types, which can be distinguished by their distinct embryological lineages and morphology:

> *Osteoclasts*, which come from the mesodermal hematopoietic lineage (Jotereau and Le Douarin 1978; Kahn et al. 2009), have traditionally been thought of as the principal population of bone-resorbing cells (Hancox 1949; Martin and Ng 1994; Teitelbaum et al. 1997; Filvaroff and Derynck 1998; Teitelbaum 2000; Boyle et al. 2003). In our quail–duck chimeric system, all osteoclasts arise entirely from host mesoderm.
>
> *Osteocytes* are the second cell type that resorb bone (Belanger 1969; Qing et al. 2012; Tang et al. 2012; Xiong et al. 2014; O'Brien et al. 2008; Xiong and O'Brien 2012; Akil et al. 2014; Fowler et al. 2017; Jauregui et al. 2016), and in the skeleton of the jaws and face form solely from NCM (Helms and Schneider 2003; Noden 1978; Le Lièvre 1978). Hence in quail–duck chimeras, osteocytes are derived exclusively from donor NCM.

Both osteoclasts and osteocytes secrete an enzyme called tartrate-resistant acid phosphatase (TRAP) when they resorb bone (Minkin 1982; Qing et al. 2012; Tang et al. 2012). Also, osteoclasts and osteocytes express different molecular markers such as *Mmp9*, which is found in osteoclasts (Reponen et al. 1994; Engsig et al. 2000), and *Mmp13*, which is detected in osteocytes (Johansson et al. 1997; Behonick et al. 2007; Sasano et al. 2002). Therefore, following the transplant of NCM into the lower jaw of chimeric quck, *Mmp9* expression would be coming from duck host-derived osteoclasts whereas *Mmp13* would be expressed by quail donor-derived osteocytes.

When we examine the initiation of bone resorption in short-beaked quail versus long-billed duck we detect significantly higher levels and different spatial domains of TRAP, *Mmp9*, and *Mmp13* in quail, signifying that quail undergo more bone resorption than duck, and indicating that elevated resorption may relate to their shorter beaks. Correspondingly, chimeric quck have elevated quail-like levels of TRAP, *Mmp9*, and *Mmp13* coincident with their quail-like jaw skeletons. This means that in chimeric quck, quail donor NCM executes an autonomous species-specific program for bone resorption by way of higher *Mmp13* expression and TRAP activity, and also through upregulation of *Mmp9* expression in duck host osteoclasts. This reveals an unexpected NCM-mediated mechanism that potentially contributes to the shorter jaws of quail and chimeric quck. In other words, levels of bone resorption in bird beaks seem to be inversely proportional to jaw size.

To test if bone resorption is a determinant of jaw size, we used a biochemical approach to activate or inhibit resorption by osteocytes and osteoclasts. We administered treatments systemically when bone deposition is just starting and resorption has not yet begun. Inhibiting resorption causes the quail lower jaw to elongate, whereas activating resorption significantly shortens the jaw (Ealba et al. 2015). Thus, quail and duck express species-specific developmental programs for bone resorption that are distinct in terms of levels and spatial domains, these programs are governed by NCM, and bone resorption appears to be a contributing mechanism establishing beak length. Such experiments point to a novel function for NCM-mediated bone resorption, which is to help control species-specific jaw size, and they extend previous studies on Darwin's finches and other species, which argue that an important regulator of beak length is the calcium binding protein, *calmodulin* (Abzhanov et al. 2006; Schneider 2007; Gunter et al. 2014).

This connection is particularly intriguing because *calmodulin* has been shown to regulate osteocytes and osteoclasts in the local environment (Seales et al. 2006; Zayzafoon 2006; Choi et al. 2013a, b). Since calcium signaling is known to affect bone resorption (Hwang and Putney 2011; Kajiya 2012; Xia and Ferrier 1996; Xiong et al. 2014), NCM-mediated bone resorption may serve as another developmental mechanism that drives the evolvability of the avian beak more generally (Kirschner and Gerhart 1998), and determines jaw size more specifically (Gunter et al. 2014; Parsons and Albertson 2009). Additionally, this work suggests that bone resorption may act like a rheostat during jaw size evolution and one that is particularly sensitive to the availability of dietary calcium in the environment, the endocrine effects of calcium-dependent hormones, and gradients of calcium signaling within the jaw primordia (Schneider 2007).

The spatial and temporal regulation by the NCM of expression domains for genes including *Mmp9* and *Mmp13* are likely to affect shape as well, by establishing local zones of resorption that in effect sculpt the bone and promote or inhibit directional growth. Genetic disruptions to these genes and others that affect bone resorption are known to alter the morphology of the craniofacial skeleton. For example, mice with mutations in *Mmp2* have abnormal snouts (Egeblad et al. 2007), and defects in jaw morphology are observed in humans with clinical conditions such as Spondyloepimetaphyseal dysplasia (i.e., *Mmp13*), Juvenile Paget's disease (i.e., *Opg*), and following treatments with high doses of bisphosphonates, such as zoledronic acid, which inhibit bone resorption (Gorlin et al. 1990; Lezot et al. 2014).

Therefore, the remarkable ability of NCM to exert spatiotemporal control over not only the induction, differentiation, deposition, and mineralization of bone (Eames and Schneider 2008; Hall et al. 2014; Merrill et al. 2008; Schneider and Helms 2003), but also the resorption of bone, seamlessly integrates the molecular and cellular determinants of jaw size and shape, and confers NCM with its unique capacity to generate species-specific variation during development and evolution.

7.7   CONCLUSIONS

In the beginning of his collected works, Waddington (1975) expressed his "deeply ingrained conviction that the evolution of organisms must really be regarded as the evolution of developmental systems" (p. 7). A long line of evolutionary developmental biologists would certainly concur with Waddington's viewpoint particularly those researchers who have focused on allometry and/or heterochrony as mechanisms to explain species-specific transformations in size and shape. A major factor behind such transformations clearly involves modifications to the fundamental parameter of time, mainly in terms of total developmental time, differential rates of growth, and/or the timing of developmental events. Thus as in good comedy, timing is everything.

While in many respects Minot, Thompson, Huxley, de Beer, Goldschmidt, Waddington, Gould, and Alberch were way ahead of their own time, over the past 25 years, technological and conceptual advances in genetics, genomics, and molecular biology have revolutionized the study of pattern formation during development. Many of the genes that regulate basic developmental phenomena have been identified, and the processes they guide have been redefined in mechanistic terms. Notably, we have come to recognize that the construction rules of embryonic development and the genetic and epigenetic architecture required to enforce those rules are shared broadly across disparate taxa. This has led to the spread of a common language for evolutionary developmental biologists studying the embryos of seemingly diverse organisms, including but certainly not limited to mice, chicks, frogs, fish, flies, and worms. As a direct result, the pace of research in the field has greatly accelerated because discoveries in one species swiftly lead to progress in understanding the development of other species. This progress has transformed developmental biology from a descriptive science into one that can now explain the complexities of organ and tissue development as consequences of known signal transduction pathways and transcriptional programs.

Whether changes to the temporal and spatial programs for development become propagated at the level of genomic organization, at the *cis*-regulatory level of individual genes, at the transcriptional and post-transcriptional level through the epigenetic activities of non-coding RNA, at the level of nodes within gene regulatory networks, at the level of biochemical interactions among gene products such as enzymes and other proteins, at the level of post-translation modification of proteins, at the level of diffusion-reaction gradients and thresholds that establish the inductive abilities of cells, at the level of cell properties and cell movements, or at the level of physical and signaling interactions between tissues, the downstream effects on morphological phenotypes can range from subtle to profound. Internal modifications to developmental programs at any of these hierarchical levels of organization can generate the variation necessary for evolution, but they would also be buffered by the robustness and stability of the internal networks and nested interactions that ultimately work together to generate individual morphological units and ensure fidelity for structural and functional integration.

This feature of developmental systems has allowed the vertebrate craniofacial complex to be both highly conserved in its basic anatomical organization and extraordinarily diversified in its size and shape. Individual morphological units within the craniofacial complex can become modified rapidly over time, yet still, maintain connections and keep relationships that are required for meeting structural and functional demands. By focusing on the molecular and cellular regulation of species-specific pattern in the craniofacial complex we hope our work has helped pinpoint precisely where and when changes to developmental programs can affect the course of morphological evolution. Our experiments in quail and duck embryos reveal that NCM plays a special role in generating species-specific pattern in the craniofacial complex, by dominating its own signaling interactions with surrounding tissues and by way of autonomous morphogenetic programs that can span and accommodate fluctuations in time. Simply because of these virtues, cranial NCM has likely endured as a key effectuator of skeletal size and shape during development and evolution.

ACKNOWLEDGMENTS

This chapter is dedicated to Professor Raymond P. Coppinger who first inspired me to ponder evolutionary developmental biology while I was a young student at Hampshire College in 1987. If Ray had not made me drink deep from his old dog-eared copy of de Beer's *Embryos and Ancestors*, I do not know where I would be today. Ray was an intellectual giant who shepherded my career; his impact across multiple fields remains vast, and he will be profoundly missed by many. I am also grateful to my students and collaborators whom I have tried to cite as often as possible and whose work over the years forms the basis for much of what has been covered here. Funded in part, by NIDCR R01 DE016402 and R01 DE025668 to R.A.S.

REFERENCES

Abzhanov, A., Kuo, W. P., Hartmann, C., Grant, B. R., Grant, P. R., and Tabin, C. J. 2006. The calmodulin pathway and evolution of elongated beak morphology in Darwin's finches. *Nature* 442: 563–567.

Abzhanov, A., Protas, M., Grant, B. R., Grant, P. R., and Tabin, C. J. 2004. Bmp4 and morphological variation of beaks in Darwin's finches. *Science* 305: 1462–465.

Abzhanov, A. and Tabin, C. J. 2004. Shh and Fgf8 act synergistically to drive cartilage out-growth during cranial development. *Dev Biol* 273: 134–148.

Akil, O., Hall-Glenn, F., Chang, J., Li, A., Chang, W., Lustig, L. R., Alliston, T., and Hsiao, E. C. 2014. Disrupted bone remodeling leads to cochlear overgrowth and hearing loss in a mouse model of fibrous dysplasia. *PLoS One* 9: e94989.

Alberch, P. 1980a. Adaptation and heterochrony in the evolution of Bolitoglossa (Amphibia: Plethodontidae). PhD thesis, University of California, Berkeley, CA.

Alberch, P. 1980b. Ontogenesis and morphological diversification. *Amer Zool* 20: 653–667.

Alberch, P. 1982a. Developmental constraints in evolutionary processes. In *Evolution and Development* (Ed.) Bonner, J. T., pp. 313–332. Berlin, Germany: Springer-Verlag.

Alberch, P. 1982b. The generative and regulatory roles of development in evolution. In *Environmental Adaptation and Evolution: A Theoretical and Empirical Approach* (Eds.) Mossakowski, D. and Roth, G., pp. 19–36. Stuttgart, Germany: G. Fischer-Verlag.

Alberch, P. 1985. Problems with the interpretation of developmental sequences. *Syst Zool* 34: 46–58.

Alberch, P. 1989. The logic of monsters: Evidence for internal constraint in development and evolution. *Geobios* 22 (Suppl 2): 21–57.

Alberch, P. 1991. From genes to phenotype: Dynamical systems and evolvability. *Genetica* 84: 5–11.

Alberch, P. and Alberch, J. 1981. Heterochronic mechanisms of morphological diversifica-tion and evolutionary change in the neotropical salamander, *Bolitoglossa occidentalis* (Amphibia: Plethodontidae). *J Morphol* 167: 249–264.

Alberch, P. and Gale, E. 1983. Size dependence during the development of the amphibian foot. Colchicine induced digital loss and reduction. *J Embryol Exp Morphol* 76: 177–197.

Alberch, P. and Gale, E. 1985. A developmental analysis of an evolutionary trend: Digital reduction in Amphibians. *Evolution* 39: 8–23.

Alberch, P., Gould, S. J., Oster, G. F., and Wake, D. B. 1979. Size and shape in ontogeny and phylogeny. *Paleobiology* 5: 296–317.

Allmon, W. D. 1994. Patterns and processes of heterochrony in lower tertiary turritelline gastropods U.S. Gulf and Atlantic Coastal Plains. *J Paleont* 68: 80–95.

Anderson, B. G. and Busch, H. L. 1941. Allometry in normal and regenerating antennal seg-ments in daphnia. *Biol Bull* 81: 119–126.

Andres, G. 1949. Untersuchungen an Chimären von *Triton* und *Bombinator*. *Genetica* 24: 387–534.

Arthur, W. 2006. D'Arcy Thompson and the theory of transformations. *Nat Rev Genet* 7: 401–406.

Ashique, A. M., Fu, K., and Richman, J. M. 2002a. Endogenous bone morphogenetic proteins regulate outgrowth and epithelial survival during avian lip fusion. *Development* 129: 4647–4660.

Ashique, A. M., Fu, K., and Richman, J. M. 2002b. Signalling via type IA and type IB bone morphogenetic protein receptors (BMPR) regulates intramembranous bone for-mation, chondrogenesis and feather formation in the chicken embryo. *Int J Dev Biol* 46: 243–253.

Atchley, W. R. 1981. Genetic components of size and shape. II. Multivariate covariance pat-terns in the rat and mouse skull. *Evolution* 35: 1037–1055.

Atchley, W. R. and Hall, B. K. 1991. A model for development and evolution of complex mor-phological structures. *Biol Rev Camb Philos Soc* 66: 101–157.

Atchley, W. R., Riska, B., Kohn, L. A. P., Plummer, A. A., and Rutledge, J. J. 1984. A quan-titative genetic analysis of brain and body size associations, their origin and ontogeny: Data from mice. *Evolution* 38: 1165–1179.

Baker, C. V. and Bronner-Fraser, M. 1997. The origins of the neural crest. Part II: An evolu-tionary perspective. *Mech Dev* 69: 13–29.

Balling, R., Mutter, G., Gruss, P., and Kessel, M. 1989. Craniofacial abnormalities induced by ectopic expression of the homeobox gene Hox-1.1 in transgenic mice. *Cell* 58: 337–347.

Balon, E. K. 1981. Saltatory processes and altricial to precocial forms in the ontogeny of fishes. *Amer Zool* 21: 573–596.

Barlow, A. J., Bogardi, J. P., Ladher, R., and Francis-West, P. H. 1999. Expression of chick *Barx-1* and its differential regulation by FGF-8 and BMP signaling in the maxillary primordia. *Dev Dyn* 214: 291–302.

Barlow, A. J. and Francis-West, P. H. 1997. Ectopic application of recombinant BMP-2 and BMP-4 can change patterning of developing chick facial primordia. *Development* 124: 391–398.

Bateson, W. 1894. *Materials for the Study of Variation, Treated With Especial Regard to the Discontinuity in the Origin of Species*. New York: Macmillan Publishers.

Bateson, W. and Mendel, G. 1902. *Mendel's Principles of Heredity; A Defence*. Cambridge, UK: Cambridge University Press.

Bee, J. and Thorogood, P. 1980. The role of tissue interactions in the skeletogenic differentiation of avian neural crest cells. *Dev Biol* 78: 47–66.

Behonick, D. J., Xing, Z., Lieu, S., Buckley, J. M., Lotz, J. C., Marcucio, R. S., Werb, Z., Miclau, T., and Colnot, C. 2007. Role of matrix metalloproteinase 13 in both endochondral and intramembranous ossification during skeletal regeneration. *PLoS One* 2 (11): e1150.

Belanger, L. F. 1969. Osteocytic osteolysis. *Calcif Tissue Res* 4: 1–12.

Bell, D. M., Leung, K. K., Wheatley, S. C., Ng, L. J., Zhou, S., Ling, K. W., Sham, M. H., Koopman, P., Tam, P. P., and Cheah, K. S. 1997. SOX9 directly regulates the type-II collagen gene. *Nat Genet* 16: 174–178.

Bemis, W. E. 1984. Paedomorphosis and the evolution of the Dipnoi. *Paleobiology* 10: 293–307.

Benson, R. H., Chapman, R. E., and Siegel, A. F. 1982. On the measurement of morphology and its change. *Paleobiology* 8: 328–339.

Bertalanffy, L. and Pirozynski, W. J. 1952. Ontogenetic and evolutionary allometry. *Evolution* 6: 387–392.

Betancur, P., Bronner-Fraser, M., and Sauka-Spengler, T. 2010. Assembling neural crest regulatory circuits into a gene regulatory network. *Ann Rev Cell Devel Biol* 26: 581–603.

Bhullar, B. A., Morris, Z. S., Sefton, E. M., Tok, A., Tokita, M., Namkoong, B., Camacho, J., Burnham, D. A., and Abzhanov, A. 2015. A molecular mechanism for the origin of a key evolutionary innovation, the bird beak and palate, revealed by an integrative approach to major transitions in vertebrate history. *Evolution* 69: 1665–1677.

Blackstone, N. W. and Buss, L. W. 1993. Experimental heterochrony in Hydractiniid Hydroids: Why mechanisms matter. *J Evol Biol* 6: 307–327.

Blanco, M. J. and Alberch, P. 1992. Caenogenesis, developmental variability, and evolution in the carpus and tarsus of the marbled newt *Triturus marmoratus*. *Evolution* 46: 677–687.

Bobick, B. E., Thornhill, T. M., and Kulyk, W. M. 2007. Fibroblast growth factors 2, 4, and 8 exert both negative and positive effects on limb, frontonasal, and mandibular chondrogenesis via MEK-ERK activation. *J Cell Physiol* 211: 233–243.

Bolk, L. 1926. *Das problem der Menschwerdung*. Jena, Germany: Gustav Fischer.

Bolker, J. A. 2000. Modularity in development and why it matters to Evo-Devo. *Amer Zool* 40: 770–776.

Bonner, J. T. (Ed.). 1982. *Evolution and Development*. Berlin, Germany: Springer-Verlag.

Bookstein, F. L. 1978. *The Measurement of Biological Shape and Shape Change*. New York: Springer-Verlag.

Bookstein, F. L. 1990. Multivariate methods. In *Proceedings of the Michgan Morphometrics Workshop* (Eds.). Rohlf, F. J. and Bookstein, F. L., pp. 75–76. Ann Arbor, MI: University of Michigan Museum of Zoology.

Boyle, W. J., Simonet, W. S., and Lacey, D. L. 2003. Osteoclast differentiation and activation. *Nature* 423: 337–342.

Bradamante, Z. and Hall, B. K. 1980. The role of epithelial collagen and proteoglycan in the initiation of osteogenesis by avian neural crest cells. *Anat Rec* 197: 305–315.

Brigandt, I. 2006. Homology and heterochrony: The evolutionary embryologist Gavin Rylands de Beer (1899–1972). *J Exp Zool B Mol Dev Evol* 306: 317–328.

Bronner-Fraser, M. 2008. On the trail of the 'new head' in Les Treilles. *Development* 135: 2995–2999.

Brugmann, S. A., Goodnough, L. H., Gregorieff, A., Leucht, P., ten Berge, D., Fuerer, C., Clevers, H., Nusse, R., and Helms, J. A. 2007. Wnt signaling mediates regional specification in the vertebrate face. *Development* 134: 3283–3295.

Brugmann, S. A., Powder, K. E., Young, N. M., Goodnough, L. H., Hahn, S. M., James, A. W., Helms, J. A., and Lovett, M. 2010. Comparative gene expression analysis of avian embryonic facial structures reveals new candidates for human craniofacial disorders. *Hum Mol Genet* 19: 920–930.

Buckwalter, J. A., Glimcher, M. J., Cooper, R. R., and Recker, R. 1996. Bone biology. II: Formation, form, modeling, remodeling, and regulation of cell function. *Instr Course Lect* 45: 387–399.

Chapman, R. E. 1990. Conventional procrustes approaches. *Proceedings of the Michigan Morphometrics Workshop*, Ann Arbor, MI.

Cheverud, J. M. 1982. Phenotypic, genetic, and environmental morphological integration in the cranium. *Evolution* 36: 499–516.

Choi, Y. H., Ann, E. J., Yoon, J. H., Mo, J. S., Kim, M. Y., and Park, H. S. 2013a. Calcium/calmodulin-dependent protein kinase IV (CaMKIV) enhances osteoclast differentiation via the up-regulation of Notch1 protein stability. *Biochim Biophys Acta* 1833: 69–79.

Choi, Y. H., Choi, J. H., Oh, J. W., and Lee, K. Y. 2013b. Calmodulin-dependent kinase II regulates osteoblast differentiation through regulation of Osterix. *Biochem Biophys Res Commun* 432: 248–255.

Chong, H. J., Young, N. M., Hu, D., Jeong, J., McMahon, A. P., Hallgrimsson, B., and Marcucio, R. S. 2012. Signaling by SHH rescues facial defects following blockade in the brain. *Dev Dyn* 241: 247–256.

Clark, W. E. L. G. and Medawar, P. B. (Eds.). 1945. *Essays on Growth and Form Presented to D'Arcy Wentworth Thompson*. Oxford, UK: Clarendon Press.

Coats, S., Flanagan, W. M., Nourse, J., and Roberts, J. M. 1996. Requirement of p27Kip1 for restriction point control of the fibroblast cell cycle. *Science* 272: 877–880.

Coppinger, L. and Coppinger, R. P. 1982. Livestock–guarding dogs that wear sheep's clothing. *Smithsonian* 13: 64–73.

Coppinger, R. P. and Feinstein, M. 1991. Why dogs bark. *Smithsonian* January: 119–129.

Coppinger, R. P., Glendinning, J., Torop, E., Matthay, C., Sutherland, M., and Smith, C. K. 1987. Degree of behavioral neoteny differentiates canid polymorphs. *Ethology* 75: 89–108.

Coppinger, R. P. and Schneider, R. A. 1995. Evolution of working dogs. In *The Domestic Dog* (Ed.) Serpell, J., pp. 21–47. Cambridge, UK: Cambridge University Press.

Coppinger, R. P. and Smith, C. K. 1989. A model for understanding the evolution of mammalian behavior. In *Current Mammalogy* (Ed.) Genoways, H., pp. 335–374. New York: Plenum Press.

Cordero, D. R., Schneider, R. A., and Helms, J. A. 2002. Morphogenesis of the face. In *Craniofacial Surgery: Science & Surgical Technique* (Eds.) Lin, K. Y., Ogle, R. C., and Jane, J. A., pp. 75–83. Philadelphia, PA: W. B. Saunders Company.

Couly, G., Grapin-Botton, A., Coltey, P., and Le Douarin, N. M. 1996. The regeneration of the cephalic neural crest, a problem revisited: the regenerating cells originate from the contralateral or from the anterior and posterior neural fold. *Development* 122: 3393–3407.

Couly, G., Creuzet, S., Bennaceur, S., Vincent, C., and Le Douarin, N. M. 2002. Interactions between Hox-negative cephalic neural crest cells and the foregut endoderm in patterning the facial skeleton in the vertebrate head. *Development* 129: 1061–1073.

Couly, G. F., Coltey, P. M., and Le Douarin, N. M. 1993. The triple origin of skull in higher vertebrates: A study in quail–chick chimeras. *Development* 117: 409–429.

Creuzet, S., Couly, G., Vincent, C., and Le Douarin, N. M. 2002. Negative effect of Hox gene expression on the development of the neural crest-derived facial skeleton. *Development* 129(18): 4301–4313.

Crumly, C. R. and Sanchez-Villagra, M. R. 2004. Patterns of variation in the phalangeal formulae of land tortoises (Testudinidae): Developmental constraint, size, and phylogenetic history. *J Exp Zool Part B (Mol Dev Evol)* 302B: 134–146.

Crump, J. G., Maves, L., Lawson, N. D., Weinstein, B. M., and Kimmel, C. B. 2004. An essential role for Fgfs in endodermal pouch formation influences later craniofacial skeletal patterning. *Development* 131: 5703–5716.

Darwin, C. 1859. *The Origin of Species*. 1962 ed. New York: The Crowell-Collier Publishing.

David, N. B., Saint-Etienne, L., Tsang, M., Schilling, T. F., and Rosa, F. M. 2002. Requirement for endoderm and FGF3 in ventral head skeleton formation. *Development* 129: 4457–4468.

de Beer, G. R. 1930. *Embryology and Evolution*. Oxford, UK: Clarendon Press.

de Beer, G. R. 1937. *The Development of the Vertebrate Skull*. Chicago, IL: University of Chicago Press.

de Beer, G. R. 1940. *Embryos and ancestors*. Oxford, UK: Clarendon Press.

de Beer, G. R. 1954. *Embryos and Ancestors*. Revised ed. *Monographs on Animal Biology*. Oxford, UK: Clarendon Press.

de Beer, G. R. 1958. *Embryos and Ancestors*. 3rd ed. Oxford, UK: Oxford University Press.

de Crombrugghe, B., Lefebvre, V., Behringer, R. R., Bi, W., Murakami, S., and Huang, W. 2000. Transcriptional mechanisms of chondrocyte differentiation. *Matrix Biol* 19: 389–394.

De Renzi, M. 2009. Developmental and historical patterns at the crossroads in the work of Pere Alberch. In *Pere Alberch: The Creative Trajectory of an Evo-Devo Biologist* (Eds.) Rasskin-Gutman, D. and De Renzi, M., pp. 45–66. Valencia, Spain: Universidad de València.

De Renzi, M., Moya, A., and Peretó, J. 1999. Evolution, development and complexity in Pere Alberch (1954–1998). *J Evol Biol* 12: 624–626.

Deacon, T. W. 1990. Problems of ontogeny and phylogeny in brain-size evolution. *Int J Primatol* 11: 237–282.

Dechambre, E. 1949. La théorie de foetalization et la formation des races de chiens et de porc. *Mammalia* 13: 129–137.

Depew, M. J., Lufkin, T., and Rubenstein, J. L. 2002. Specification of jaw subdivisions by Dlx genes. *Science* 298: 381–385.

Depew, M. J. and Olsson, L. 2008. Symposium on the evolution and development of the vertebrate head. *J Exp Zool Part B (Mol Dev Evol)* 310: 287–293.

Dietrich, M. R. 2000. From hopeful monsters to homeotic effects: Richard Goldschmidt's integration of development, evolution, and genetics. *Amer Zool* 40: 738–747.

Dobzhansky, T. 1937. *Genetics and the Origin of Species, Columbia Biological Series*. New York: Columbia University Press.

Dobzhansky, T. 1951. *Genetics and the Origin of Species*, 3d ed., *Columbia Biological Series*. New York: Columbia University Press.

Doufexi, A. E. and Mina, M. 2008. Signaling pathways regulating the expression of Prx1 and Prx2 in the chick mandibular mesenchyme. *Dev Dyn* 237: 3115–3127.

Drake, A. G. 2011. Dispelling dog dogma: An investigation of heterochrony in dogs using 3D geometric morphometric analysis of skull shape. *Evol Dev* 13: 204–213.

Drissi, H., Hushka, D., Aslam, F., Nguyen, Q., Buffone, E., Koff, A., van Wijnen, A., Lian, J. B., Stein, J. L., and Steinm, G. S. 1999. The cell cycle regulator p27kip1 contributes to growth and differentiation of osteoblasts. *Cancer Res* 59: 3705–3711.

Duboule, D. 1994. Temporal colinearity and the phylotypic progression: A basis for the stability of a vertebrate Bauplan and the evolution of morphologies through heterochrony. *Development* 135–142.

Ducy, P., Starbuck, M., Priemel, M., Shen, J., Pinero, G., Geoffroy, V., Amling, M., and Karsenty, G. 1999. A Cbfa1-dependent genetic pathway controls bone formation beyond embryonic development. *Genes Dev* 13: 1025–1036.

Ducy, P., Zhang, R., Geoffroy, V., Ridall, A. L., and Karsenty, G. 1997. Osf2/Cbfa1: A transcriptional activator of osteoblast differentiation. *Cell* 89: 747–754.

Dunlop, L. L. and Hall, B. K. 1995. Relationships between cellular condensation, preosteoblast formation and epithelial-mesenchymal interactions in initiation of osteogenesis. *Int J Devel Biol* 39: 357–371.

Ealba, E. L., Jheon, A. H., Hall, J., Curantz, C., Butcher, K. D., and Schneider, R. A. 2015. Neural crest-mediated bone resorption is a determinant of species-specific jaw length. *Dev Biol* 408: 151–163.

Ealba, E. L. and Schneider, R. A. 2013. A simple PCR-based strategy for estimating species-specific contributions in chimeras and xenografts. *Development* 140: 3062–3068.

Eames, B. F., de la Fuente, L., and Helms, J. A. 2003. Molecular ontogeny of the skeleton. *Birth Defects Res Part C Embryo Today* 69: 93–101.

Eames, B. F. and Schneider, R. A. 2005. Quail-duck chimeras reveal spatiotemporal plasticity in molecular and histogenic programs of cranial feather development. *Development* 132: 1499–1509.

Eames, B. F. and Schneider, R. A. 2008. The genesis of cartilage size and shape during development and evolution. *Development* 135: 3947–3958.

Eames, B. F., Sharpe, P. T., and Helms, J. A. 2004. Hierarchy revealed in the specification of three skeletal fates by Sox9 and Runx2. *Dev Biol* 274: 188–200.

Egeblad, M., Shen, H. C., Behonick, D. J., Wilmes, L., Eichten, A., Korets, L. V., Kheradmand, F., Werb, Z., and Coussens, L. M. 2007. Type I collagen is a genetic modifier of matrix metalloproteinase 2 in murine skeletal development. *Dev Dyn* 236: 1683–1693.

Ekanayake, S. and Hall, B. K. 1994. Hypertrophy is not a prerequisite for type X collagen expression or mineralization of chondrocytes derived from cultured chick mandibular ectomesenchyme. *Int J Dev Biol* 38: 683–694.

Ekanayake, S. and Hall, B. K. 1997. The in vivo and in vitro effects of bone morphogenetic protein-2 on the development of the chick mandible. *Int J Dev Biol* 41: 67–81.

Eldredge, N. and Gould, S. J. 1972. Punctuated equilibria: An alternative to phyletic gradualism. In *Models in Paleobiology* (Ed.) Schopf, T. J. M., pp. 82–115. San Francisco, CA: Freeman Cooper.

Engsig, M. T., Chen, Q. J., Vu, T. H., Pedersen, A. C., Therkidsen, B., Lund, L. R., Henriksen, K. et al. 2000. Matrix metalloproteinase 9 and vascular endothelial growth factor are essential for osteoclast recruitment into developing long bones. *J Cell Biol* 151: 879–889.

Enlow, D. H., Moyers, R. E., and Merow, W. W. 1975. *Handbook of Facial Growth.* Philadelphia, PA: Saunders.

Ettinger, L. and Doljanski, F. 1992. On the generation of form by the continuous interactions between cells and their extracellular matrix. *Biol Rev Camb Philos Soc* 67: 459–489.

Etxeberria, A. and De la Rosa, L. N. 2009. A world of opportunity within constraint: Pere Alberch's early evo-devo. In *Pere Alberch: The Creative Trajectory of an Evo-Devo Biologist* (Eds.) Rasskin-Gutman, D. and De Renzi, M., pp. 21–44. Valencia, Spain: Universidad de València.

Fassler, P. E. 1996. Hans Spemann (1869–1941) and the Freiburg School of Embryology. *Int J Dev Biol* 40: 49–57.

Ferguson, C. A., Tucker, A. S., and Sharpe, P. T. 2000. Temporospatial cell interactions regulating mandibular and maxillary arch patterning. *Development* 127: 403–412.

Filvaroff, E. and Derynck, R. 1998. Bone remodelling: A signalling system for osteoclast regulation. *Curr Biol* 8:R679–R682.

Fish, J. L. and Schneider, R. A. 2014a. On the origins of species-specific size. *The Node.* http://thenode.biologists.com/on-the-origins-of-species-specific-size/research/.

Fish, J. L. and Schneider, R. A. 2014b. Assessing species-specific contributions to craniofacial development using quail-duck chimeras. *J Vis Exp* 87(e51534): 1–6.

Fish, J. L. and Schneider, R. A. 2014c. Neural crest-mediated tissue interactions during craniofacial development: The origins of species-specific pattern. In *Neural Crest Cells* (Ed.) Trainor, P. A., pp. 101–124. Boston, MA: Academic Press.

Fish, J. L., Sklar, R. S., Woronowicz, K. C., and Schneider, R. A. 2014. Multiple developmental mechanisms regulate species-specific jaw size. *Development* 141: 674–684.

Fisher, R. A. 1930. *The Genetical Theory of Natural Selection*. Oxford, UK: The Clarendon Press.

Fondon, J. W., and Garner, H. R. 2004. Molecular origins of rapid and continuous morphological evolution. *Proc Natl Acad Sci USA* 101: 18058–18063.

Foppiano, S., Hu, D., and Marcucio, R. S. 2007. Signaling by bone morphogenetic proteins directs formation of an ectodermal signaling center that regulates craniofacial development. *Dev Biol* 312: 103–114.

Foster, D. W. and Kaesler, R. L. 1988. Shape analysis: Ideas from the Ostracoda. In *Heterochrony in Evolution: A Multidisciplinary Approach* (Ed.) McKinney, M. L., pp. 53–69. New York: Plenum Press.

Fowler, T. W., Acevedo, C., Mazur, C. M., Hall-Glenn, F., Fields, A. J., Bale, H. A., Ritchie, R. O., Lotz, J. C., Vail, T. P., and Alliston, T. 2017. Glucocorticoid suppression of osteocyte perilacunar remodeling is associated with subchondral bone degeneration in osteonecrosis. *Sci Rep* 7: 44618.

Francis-West, P. H., Robson, L., and Evans, D. J. 2003. Craniofacial development: The tissue and molecular interactions that control development of the head. *Adv Anat Embryol Cell Biol* 169: 1–138.

Francis-West, P. H., Tatla, T., and Brickell, P. M. 1994. Expression patterns of the bone morphogenetic protein genes Bmp-4 and Bmp-2 in the developing chick face suggest a role in outgrowth of the primordia. *Dev Dyn* 201: 168–178.

Francis-West, P., Ladher, R., Barlow, A., and Graveson, A. 1998. Signalling interactions during facial development. *Mech Dev* 75: 3–28.

Galilei, G. 1914. *Dialogues Concerning Two New Sciences*. Translated from the Italian and Latin into English by Crew, H. and De Salvio, A. New York: Macmillan.

Galindo, M., Pratap, J., Young, D. W., Hovhannisyan, H., Im, H. J., Choi, J. Y., Lian, J. B., Stein, J. L., Stein, G. S., and van Wijnen, A. J. 2005. The bone-specific expression of Runx2 oscillates during the cell cycle to support a G1-related antiproliferative function in osteoblasts. *J Biol Chem* 280: 20274–20285.

Gans, C. and Northcutt, R. G. 1983. Neural crest and the origin of vertebrates: A new head. *Science* 220: 268–274.

Garrod, D. R. 1986. Specific inductive flypaper. *Bioessays* 5: 172–173.

Garstang, W. 1928. The morphology of the Tunicata, and its bearings on the phylogeny of the Chordata. *Quart J Microsc Sci* 75: 51–187.

Garstang, W. 1922. The theory of recapitulation: A critical restatement of the biogenic law. *Je Linn Soc Lond, Zool* 35: 81–101.

Gayon, J. 2000. History of the concept of allometry. *Amer Zool* 40: 748–758.

Geist, V. 1987. On speciation in ice age mammals, with special reference to cervids and caprids. *Can J Zool* 65: 1067–1084.

Gendron-Maguire, M., Mallo, M., Zhang, M., and Gridley, T. 1993. Hoxa-2 mutant mice exhibit homeotic transformation of skeletal elements derived from cranial neural crest. *Cell* 75: 1317–1331.

Gilbert, S. F., Opitz, J. M., and Raff, R. A. 1996. Resynthesizing evolutionary and developmental biology. *Dev Biol* 173: 357–372.

Godfrey, L. R. and Sutherland, M. R. 1995a. What's growth got to do with it? Process and product in the evolution of ontogeny. *J Human Evol* 29: 405–431.

Godfrey, L. R. and Sutherland, M. R. 1995b. Flawed Inference: Why size-based tests of heterochronic processes do not work. *J Theor Biol* 172: 43–61.

Goldschmidt, R. 1938. *Physiological Genetics*. New York: McGraw-Hill.

Goldschmidt, R. 1940. *The Material Basis of Evolution*. 1982 ed. New Haven, CT: Yale University Press.

Goldschmidt, R. 1953. Homeotic mutants and evolution. *Acta Biotheoretica* 10: 87–104.

Gorlin, R.J., Cohen, M. M., and Levin, L. S. 1990. *Syndromes of the Head and Neck*. 3rd edition, 1 vols. Vol. 1, *Oxford Monographs on Medical Genetics*. New York: Oxford University Press.

Gould, S. J. 1966. Allometry and size in ontogeny and phylogeny. *Biol Rev Camb Philos Soc* 41: 587–640.

Gould, S. J. 1971. Geometric similarity in allometric growth: A contribution to the problems of scaling in the evolution of size. *Amer Nat* 105: 113–136.

Gould, S. J. 1977. *Ontogeny and Phylogeny*. Cambridge, MA: Harvard University Press.

Gould, S. J. 1981. *The Mismeasure of Man*. New York: W. W. Norton & Company.

Gould, S. J. 1982a. Change in developmental timing as a mechanism of macroevolution. In *Evolution and Development* (Ed.) Bonner, J., pp. 333–346. Berlin, Germany: Springer-Verlag.

Gould, S. J. 1982b. The uses of heresy: An introduction to Richard Goldschmidt's *The Material Basis of Evolution*. In *The Material Basis of Evolution*, pp. 13–42. New Haven, CT: Yale University.

Gould, S. J. 2002. *The Structure of Evolutionary Theory*. Cambridge, MA: Belknap Press of Harvard University Press.

Gould, S. J. and Lewontin, R. C. 1979. The spandrels of San Marco and the Panglossian paradigm: a critique of the adaptionist programme. *Proc R Soc Lond* B205: 581–598.

Gould, S. J. and Vrba, E. S. 1982. Exaptation-a missing term in the science of form. *Paleobiology*: 4–15.

Govindarajan, V. and Overbeek, P. A. 2006. FGF9 can induce endochondral ossification in cranial mesenchyme. *BMC Dev Biol* 6: 7.

Graham, A. 2003. The neural crest. *Curr Biol* 13: R381–R384.

Graham, A., Begbie, J., and McGonnell, I. 2004. Significance of the cranial neural crest. *Dev Dyn* 229: 5–13.

Grammatopoulos, G. A., Bell, E., Toole, L., Lumsden, A., and Tucker, A. S. 2000. Homeotic transformation of branchial arch identity after Hoxa2 overexpression. *Development* 127: 5355–5365.

Grant, P. R., Grant, B. R., and Abzhanov, A. 2006. A developing paradigm for the development of bird beaks. *Biol J Linn Soc Lond* 88: 17–22.

Gregory, W. K. 1934. Polyisomerism and anisomerism in cranial and dental evolution among vertebrates. *Proc Natl Acad Sci USA* 20: 1–9.

Gunter, H. M., Koppermann, C., and Meyer, A. 2014. Revisiting de Beer's textbook example of heterochrony and jaw elongation in fish: Calmodulin expression reflects heterochronic growth, and underlies morphological innovation in the jaws of belonoid fishes. *Evodevo* 5: 8.

Hafner, J. C. and Hafner, M. S. 1988. Heterochrony in rodents. In *Heterochrony in Evolution: A Multidisciplinary Approach* (Ed.) McKinney, M. L., pp. 217–235. New York: Plenum Press.

Haldane, J. B. S. 1926. On being the right size. *Harper's Magazine* 152: 424–427.

Haldane, J. B. S. 1932. *The Causes of Evolution*, 1990 Reprint ed. Princeton, NJ: Princeton University Press.

Hall, B. K. 1978. Initiation of osteogenesis by mandibular mesenchyme of the embryonic chick in response to mandibular and non-mandibular epithelia. *Arch Oral Biol* 23: 1157–1161.

Hall, B. K. 1980. Tissue interactions and the initiation of osteogenesis and chondrogenesis in the neural crest-derived mandibular skeleton of the embryonic mouse as seen in isolated murine tissues and in recombinations of murine and avian tissues. *J Embryol Exp Morphol* 58: 251–264.

Hall, B. K. 1981. The induction of neural crest-derived cartilage and bone by embryonic epithelia: An analysis of the mode of action of an epithelial-mesenchymal interaction. *J Embryol Exp Morphol* 64: 305–320.

Hall, B. K. 1982a. Distribution of osteo- and chondrogenic neural crest-derived cells and of osteogenically inductive epithelia in mandibular arches of embryonic chicks. *J Embryol Exp Morphol* 68: 127–136.

Hall, B. K. 1982b. The role of tissue interactions in the growth of bone. In *Factors and Mechanisms Influencing Bone Growth* (Eds.) Dixon, A. D. and Sarnat, B. G., pp. 205–215. New York: Alan R. Liss.

Hall, B. K. 1984. Developmental processes underlying heterochrony as an evolutionary mechanism. *Can J Zool* 62: 1–7.

Hall, B. K. 1987. Tissue interactions in the development and evolution of the vertebrate head. In *Developmental and Evolutionary Aspects of the Neural Crest* (Ed.) Maderson, P. F. A., pp. 215–260. New York: John Wiley & Sons.

Hall, B. K. 1999. *The Neural Crest in Development and Evolution.* New York: Springer.

Hall, B. K. 2000a. Balfour, Garstang and de Beer: The first century of evolutionary embryology. *Amer Zool* 40: 718–728.

Hall, B. K. 2000b. The neural crest as a fourth germ layer and vertebrates as quadroblastic not triploblastic. *Evol Dev* 2: 3–5.

Hall, B. K. 2005. *Bones and Cartilage: Developmental and Evolutionary Skeletal Biology.* San Diego, CA: Elsevier Academic Press.

Hall, B. K. and Hörstadius, S. 1988. *The Neural Crest.* London, UK: Oxford University Press.

Hall, B. K. and T. Miyake. 1992. The membranous skeleton: the role of cell condensations in vertebrate skeletogenesis. *Anat Embryol (Berl)* 186: 107–124.

Hall, B. K. and Miyake, T. 1995. Divide, accumulate, differentiate: Cell condensation in skeletal development revisited. *Int J Dev Biol* 39: 881–893.

Hall, B. K. and Miyake, T. 2000. All for one and one for all: Condensations and the initiation of skeletal development. *Bioessays* 22: 138–147.

Hall, B. K. and Van Exan, R. J. 1982. Induction of bone by epithelial cell products. *J Embryol Exp Morphol* 69: 37–46.

Hall, B. K., Van Exan, R. J., and Brunt, S. L. 1983. Retention of epithelial basal lamina allows isolated mandibular mesenchyme to form bone. *J Craniofac Genet Dev Biol* 3: 253–267.

Hall, J., Jheon, A. H., Ealba, E. L., Eames, B. F., Butcher, K. D., Mak, S. S., Ladher, R., Alliston, T., and Schneider, R. A. 2014. Evolution of a developmental mechanism: Species-specific regulation of the cell cycle and the timing of events during craniofacial osteogenesis. *Dev Biol* 385: 380–395.

Hallgrimsson, B., Percival, C. J., Green, R., Young, N. M., Mio, W., and Marcucio, R. 2015. Morphometrics, 3D imaging, and craniofacial development. *Curr Top Dev Biol* 115: 561–597.

Haluska, F. and Alberch, P. 1983. The cranial development of Elaphe obsoleta (Ophidia, Colubridae). *J Morphol* 178: 37–55.

Hancox, N. M. 1949. The osteoclast. *Biol Rev Camb Philos Soc* 24: 448–471.

Hanken, J. 1989. Development and evolution of the neural crest. *Evolution* 43: 1337–1338.

Hanken, J. and Hall, B. K. 1984. Variation and timing of the cranial ossification sequence of the Oriental fire-bellied toad, *Bombina orientalis* (Amphibia, Discoglossidae). *J Morphol* 182: 245–255.

Hanken, J. and Hall, B. K. (Eds.) 1993. *The Skull*. Chicago, IL: University of Chicago Press.

Harrison, R. G. 1933. Some difficulties of the determination problem. *Amer Nat* 67: 306–321.

Havens, B. A., Rodgers, B., and Mina, M. 2006. Tissue-specific expression of Fgfr2b and Fgfr2c isoforms, Fgf10 and Fgf9 in the developing chick mandible. *Arch Oral Biol* 51: 134–145.

Havens, B. A., Velonis, D., Kronenberg, M. S., Lichtler, A. C., Oliver, B., and Mina, M. 2008. Roles of FGFR3 during morphogenesis of Meckel's cartilage and mandibular bones. *Dev Biol* 316: 336–349.

Haworth, K. E., Wilson, J. M., Grevellec, A., Cobourne, M. T., Healy, C., Helms, J. A., Sharpe, P. T., and Tucker, A. S. 2007. Sonic hedgehog in the pharyngeal endoderm controls arch pattern via regulation of Fgf8 in head ectoderm. *Dev Biol* 303: 244–258.

Healy, C., Uwanogho, D., and Sharpe, P. T. 1996. Expression of the chicken Sox9 gene marks the onset of cartilage differentiation. *Ann N Y Acad Sci* 785: 261–262.

Healy, C., Uwanogho, D., and Sharpe, P. T. 1999. Regulation and role of Sox9 in cartilage formation. *Devel Dyn* 215: 69–78.

Helms, J. A. and Schneider, R. A. 2003. Cranial skeletal biology. *Nature* 423: 326–331.

Hersh, A. H. 1934. Evolutionary allometric growth in the titanotheres. *Amer Nat* 68: 537–561.

Hirano, K., Hirano, M., Zeng, Y., Nishimura, J., Hara, K., Muta, K., Nawata, H., and Kanaide, H. 2001. Cloning and functional expression of a degradation-resistant novel isoform of p27Kip1. *Biochem J* 353: 51–57.

Hoberg, E. P. 1987. Recognition of larvae of the Tetrabothriidae Eucestoda: Implications for the origin of tapeworms in marine homeotherms. *Can J Zool* 65: 997–1000.

Hu, D. and Marcucio, R. S. 2009. Unique organization of the frontonasal ectodermal zone in birds and mammals. *Dev Biol* 325: 200–210.

Hu, D. and Marcucio, R. S. 2012. Neural crest cells pattern the surface cephalic ectoderm during FEZ formation. *Dev Dyn* 241: 732–740.

Hu, D., Marcucio, R. S., and Helms, J. A. 2003. A zone of frontonasal ectoderm regulates patterning and growth in the face. *Development* 130: 1749–1758.

Hu, D., Young, N. M., Li, X., Xu, Y., Hallgrimsson, B., and Marcucio, R. S. 2015a. A dynamic Shh expression pattern, regulated by SHH and BMP signaling, coordinates fusion of primordia in the amniote face. *Development* 142: 567–574.

Hu, D., Young, N. M., Xu, Q., Jamniczky, H., Green, R. M., Mio, W., Marcucio, R. S., and Hallgrimsson, B. 2015b. Signals from the brain induce variation in avian facial shape. *Dev Dyn* 244: 1133–1143.

Hunt, P., Clarke, J. D., Buxton, P., Ferretti, P., and Thorogood, P. 1998. Stability and plasticity of neural crest patterning and branchial arch Hox code after extensive cephalic crest rotation. *Devel Biol* 198: 82–104.

Hunt, P., Ferretti, P., Krumlauf, R., and Thorogood, P. 1995. Restoration of normal Hox code and branchial arch morphogenesis after extensive deletion of hindbrain neural crest. *Dev Biol* 168: 584–597.

Huxley, J. S. 1924. Constant differential growth-ratios and their significance. *Nature* 114: 895–896.

Huxley, J. S. 1932. *Problems of Relative Growth*. London, UK: Methuen.

Huxley, J. S. 1942. *Evolution, the Modern Synthesis*. London, UK: G. Allen & Unwin.

Huxley, J. S. 1950. Relative growth and form transformation. *Proc R Soc Lond (B)* 137: 465–469.

Huxley, J. S. and de Beer, G. R. 1934. *The Elements of Experimental Embryology*. Cambridge, UK: The University Press.

Huxley, J. S., and Teissier, G. 1936. Terminology of relative growth. *Nature* 137: 780–781.

Hwang, S. Y. and Putney, J. W. Jr. 2011. Calcium signaling in osteoclasts. *Biochim Biophys Acta* 1813: 979–983.

Jauregui, E. J., Akil, O., Acevedo, C., Hall-Glenn, F., Tsai, B. S., Bale, H. A., Liebenberg, E. et al. 2016. Parallel mechanisms suppress cochlear bone remodeling to protect hearing. *Bone* 89: 7–15.

Jheon, A. H. and Schneider, R. A. 2009. The cells that fill the bill: Neural crest and the evolution of craniofacial development. *J Dent Res* 88: 12–21.

Johansson, N., Saarialho-Kere, U., Airola, K., Herva, R., Nissinen, L., Westermarck, J., Vuorio, E., Heino, J., and Kähäri, V. M. 1997. Collagenase-3 (MMP-13) is expressed by hypertrophic chondrocytes, periosteal cells, and osteoblasts during human fetal bone development. *Dev Dyn* 208: 387–397.

Jotereau, F. V. and Le Douarin, N. M. 1978. The development relationship between osteocytes and osteoclasts: A study using the quail-chick nuclear marker in endochondral ossification. *Devel Biol* 63: 253–265.

Kahn, J., Shwartz, Y., Blitz, E., Krief, S., Sharir, A., Breitel, D. A., Rattenbach, R. et al. 2009. Muscle contraction is necessary to maintain joint progenitor cell fate. *Dev Cell* 16: 734–743.

Kajiya, H. 2012. Calcium signaling in osteoclast differentiation and bone resorption. *Adv Exp Med Biol* 740: 917–932.

Kavumpurath, S. and Hall, B. K. 1990. Lack of either chondrocyte hypertrophy or osteogenesis in Meckel's cartilage of the embryonic chick exposed to epithelia and to thyroxine in vitro. *J Craniofac Genet Devel Biol* 10: 263–275.

Kermack, K. A. and Haldane, J. B. S. 1950. Organic correlation and allometry. *Biometrika* 37: 30–41.

Keyte, A. L. and Smith, K. K. 2014. Heterochrony and developmental timing mechanisms: Changing ontogenies in evolution. *Semin Cell Dev Biol* 34: 99–107.

Kikuchi, Y., Agathon, A., Alexander, J., Thisse, C., Waldron, S., Yelon, D., Thisse, B., and Stainier, D. Y. 2001. Casanova encodes a novel Sox-related protein necessary and sufficient for early endoderm formation in zebrafish. *Genes Dev* 15: 1493–1505.

Kimmel, C. B., Miller, C. T., Kruze, G., Ullmann, B., BreMiller, R. A., Larison, K. D., and Snyder, H. C. 1998. The shaping of pharyngeal cartilages during early development of the zebrafish. *Dev Biol* 203: 245–263.

Kimmel, C. B., Ullmann, B., Walker, C., Wilson, C., Currey, M., Phillips, P. C., Bell, M. A., Postlethwait, J. H., and Cresko, W. A. 2005. Evolution and development of facial bone morphology in threespine sticklebacks. *Proc Natl Acad Sci USA* 102: 5791–5796.

Kirschner, M. and Gerhart, J. 1998. Evolvability. *Proc Natl Acad Sci USA* 95: 8420–8427.

Klingenberg, C. P. and Spence, J. R. 1993. Heterochrony and allometry: Lessons from the water strider genus *Limnoporus*. *Evolution* 47: 1834–1853.

Klingenberg, C. P. 1998. Heterochrony and allometry: The analysis of evolutionary change in ontogeny. *Biol Rev Camb Philos Soc* 73: 79–123.

Kollmann, J. 1885. Das Ueberwintern von europäischen Frosch- und Triton-larven und die Umwandlung des mexikanischen Axolotl. *Verhandlungen der Naturforschenden Gesellschaft in Basel* 7: 387–398.

Komori, T., Yagi, H., Nomura, S., Yamaguchi, A., Sasaki, K., Deguchi, K., Shimizu, Y. et al. 1997. Targeted disruption of Cbfa1 results in a complete lack of bone formation owing to maturational arrest of osteoblasts. *Cell* 89: 755–764.

Köntges, G. and Lumsden, A. 1996. Rhombencephalic neural crest segmentation is preserved throughout craniofacial ontogeny. *Development* 122: 3229–3242.

Lande, R. 1979. Quantitative genetic analysis of multivariate evolution, applied to brain: Body size allometry. *Evolution* 33: 402–416.

Langille, R. M. and Hall, B. K. 1993. Pattern formation and the neural crest. In *The Skull* (Eds.) Hanken, J. and Hall, B. K., pp. 77–111. Chicago, IL: University of Chicago Press.

Le Douarin, N. M., Creuzet, S., Couly, G., and Dupin, E. 2004. Neural crest cell plasticity and its limits. *Development* 131: 4637–4650.

Le Lièvre, C. S. 1978. Participation of neural crest-derived cells in the genesis of the skull in birds. *J Embryol Exp Morphol* 47: 17–37.

Le Lièvre, C. S. and Le Douarin, N. M. 1975. Mesenchymal derivatives of the neural crest: Analysis of chimaeric quail and chick embryos. *J Embryol Exp Morphol* 34: 125–154.

Leevers, S. J. and McNeill, H. 2005. Controlling the size of organs and organisms. *Curr Opin Cell Biol* 17: 604–609.

Lezot, F., Chesneau, J., Battaglia, S., Brion, R., Castaneda, B., Farges, J. C., Heymann, D., and Redini, F. 2014. Preclinical evidence of potential craniofacial adverse effect of zoledronic acid in pediatric patients with bone malignancies. *Bone* 68: 146–152.

Liu, W., Selever, J., Murali, D., Sun, X., Brugger, S. M., Ma, L., Schwartz, R. J., Maxson, R., Furuta, Y., and Martin, J. F. 2005. Threshold-specific requirements for Bmp4 in mandibular development. *Dev Biol* 283: 282–293.

Lord, K., Schneider, R. A., and Coppinger, R. P. 2016. Evolution of working dogs. In *The Domestic Dog* (Ed.) Serpell, J., pp. 42–68. Cambridge, UK: Cambridge University Press.

Lufkin, T., Mark, M., Hart, C., Dollé, P., Lemeur, M., and Chambon, P. 1992. Homeotic transformation of the occipital bones of the skull by ectopic expression of a homeobox gene. *Nature* 359: 835–841.

Lumer, H. 1940. Evolutionary allometry in the skeleton of the domesticated dog. *Amer Nat* 76: 439–467.

Lumer, H., Anderson, B. G., and Hersh, A. H. 1942. On the Significance of the Constant b in the Law of Allometry y = bx^a. *Amer Nat* 76: 364–375.

Lumer, H. and Schultz, A. H. 1941. Relative growth of the limb segments and tail in macaques. *Human Biol* 13: 283–305.

Lumsden, A. G. 1988. Spatial organization of the epithelium and the role of neural crest cells in the initiation of the mammalian tooth germ. *Development* 103: 155–169.

Lwigale, P. Y. and Schneider, R. A. 2008. Other chimeras: Quail-duck and mouse-chick. *Methods Cell Biol* 87: 59–74.

MacDonald, M. E., Abbott, U. K., and Richman, J. M. 2004. Upper beak truncation in chicken embryos with the cleft primary palate mutation is due to an epithelial defect in the frontonasal mass. *Dev Dyn* 230: 335–349.

MacDonald, M. E., and Hall, B. K. 2001. Altered timing of the extracellular-matrix-mediated epithelial-mesenchymal interaction that initiates mandibular skeletogenesis in three inbred strains of mice: Development, heterochrony, and evolutionary change in morphology. *J Exp Zool* 291: 258–273.

Maderson, P. F. A. 1987. *Developmental and Evolutionary Aspects of the Neural Crest.* New York: John Wiley & Sons.

Maderson, P. F. A., Alberch, P., Goodwin, B. C., Gould, S. J., Hoffman, A., Murray, J. D., Raup, D. M. et al. 1982. The role of development in macroevolutionary change. Evolution and development. In *Evolution and Development* (Ed.) Bonner, J. T., pp. 279–312. Berlin, Germany: Springer-Verlag.

Maeno, T., Moriishi, T., Yoshida, C. A., Komori, H., Kanatani, N., Izumi, S., Takaoka, K., and Komori, T. 2011. Early onset of Runx2 expression caused craniosynostosis, ectopic bone formation, and limb defects. *Bone* 49: 673–682.

Marcucio, R. S., Cordero, D. R., Hu, D., and Helms, J. A. 2005. Molecular interactions coordinating the development of the forebrain and face. *Dev Biol* 284: 48–61.

Marcucio, R. S., Young, N. M., Hu, D., and Hallgrimsson, B. 2011. Mechanisms that underlie co-variation of the brain and face. *Genesis* 49: 177–189.

Marcus, L. F. 1996. *Advances in Morphometrics, NATO ASI Series. Series A, Life Sciences; v. 284.* New York: Plenum Press.

Martin, T. J. and Ng, K. W. 1994. Mechanisms by which cells of the osteoblast lineage control osteoclast formation and activity. *J Cell Biochem* 56: 357–366.

Maunz, M. and German, R. Z. 1996. Craniofacial heterochrony and sexual dimorphism in the short-tailed opossum (*Monodelphis domestica*). *J Mammal* 77: 992–1005.

Mayr, E. 1963. *Animal Species and Evolution.* Cambridge, MA: Harvard University Press.

Mayr, E. 1983. How to carry out the adaptionist program? *Amer Nat* 121: 24–34.

McKinney, M. L. 1988a. Classifying heterochrony: Allometry, size, and time. In *Heterochrony in Evolution: A Multidisciplinary Approach* (Ed.) McKinney, M. L., pp. 17–34. New York: Plenum Press.

McKinney, M. L. 1988b. *Heterochrony in Evolution: A Multidisciplinary Approach.* New York: Plenum Press.

Merrill, A. E., Eames, B. F., Weston, S. J., Heath, T., and Schneider, R. A. 2008. Mesenchyme-dependent BMP signaling directs the timing of mandibular osteogenesis. *Development* 135: 1223–1234.

Miller, C. T., Schilling, T. F., Lee, K., Parker, J., and Kimmel, C. B. 2000. sucker encodes a zebrafish Endothelin-1 required for ventral pharyngeal arch development. *Development* 127: 3815–3828.

Mina, M., Gluhak, J., Upholt, W. B., Kollar, E. J., and Rogers, B. 1995. Experimental analysis of Msx-1 and Msx-2 gene expression during chick mandibular morphogenesis. *Dev Dyn* 202: 195–214.

Mina, M., Upholt, W. B., and Kollar, E. J. 1994. Enhancement of avian mandibular chondro-genesis in vitro in the absence of epithelium. *Arch Oral Biol* 39: 551–562.

Mina, M., Wang, Y. H., Ivanisevic, A. M., Upholt, W. B., and Rodgers, B. 2002. Region- and stage-specific effects of FGFs and BMPs in chick mandibular morphogenesis. *Dev Dyn* 223: 333–352.

Minkin, C. 1982. Bone acid phosphatase: Tartrate-resistant acid phosphatase as a marker of osteoclast function. *Calcif Tissue Int* 34: 285–290.

Minot, C. S. 1908. *The Problem of Age, Growth, and Death: A Study of Cytomorphosis, Based on Lectures at the Lowell Institute, March, 1907, The progressive science series.* London, UK: John Murray.

Mitgutsch, C., Wimmer, C., Sanchez-Villagra, M. R., Hahnloser, R., and Schneider, R. A. 2011. Timing of ossification in duck, quail, and zebra finch: Intraspecific variation, heterochronies, and life history evolution. *Zool Sci* 28: 491–500.

Mitsiadis, T. A., Caton, J., and Cobourne, M. 2006. Waking-up the sleeping beauty: Recovery of the ancestral bird odontogenic program. *J Exp Zoolog B Mol Dev Evol* 306: 227–233.

Mitsiadis, T. A., Cheraud, Y., Sharpe, P., and Fontaine-Perus, J. 2003. Development of teeth in chick embryos after mouse neural crest transplantations. *Proc Natl Acad Sci USA* 100: 65416545.

Moore, W. J. 1981. *The Mammalian Skull* (Eds.) Harrison, R. J. and McMinn, R. M., Vol. 8, *Biological Structure and Function.* Cambridge, UK: Cambridge University Press.

Morgan, T. H. 1919. *The Physical Basis of Heredity, Monographs on Experimental Biology.* Philadelphia, PA: Lippincott.

Morgan, T. H., Bridges, C. V., Muller, H. J., and Sturtevant, A. H. 1915. *The Mechanism of Mendelian Heredity.* London, UK: Constable & Robinson.

Murakami, S., Kan, M., McKeehan, W. L., and de Crombrugghe, B. 2000. Up-regulation of the chondrogenic Sox9 gene by fibroblast growth factors is mediated by the mitogen-activated protein kinase pathway. *Proc Natl Acad Sci USA* 97: 1113–1118.

Nagai, H., Mak, S. S., Weng, W., Nakaya, Y., Ladher, R., and Sheng, G. 2011. Embryonic development of the emu, Dromaius novaehollandiae. *Dev Dyn* 240: 162–175.

Needham, J. and Lerner, M. I. 1940. Terminology of relative growth-rates. *Nature* 146: 618.

Nguyen, J., Tang, S. Y., Nguyen, D., and Alliston, T. 2013. Load regulates bone formation and Sclerostin expression through a TGFbeta-dependent mechanism. *PLoS One* 8: e53813.

Nikitina, N., Sauka-Spengler, T., and Bronner-Fraser, M. 2008. Dissecting early regulatory relationships in the lamprey neural crest gene network. *Proc Natl Acad Sci USA* 105: 20083–20088.

Noden, D. M. 1978. The control of avian cephalic neural crest cytodifferentiation. I. Skeletal and connective tissues. *Dev Biol* 67: 296–312.

Noden, D. M. 1982. Patterns and organization of craniofacial skeletogenic mesenchyme: A perspective. In *Factors and Mechanisms Influencing Bone Growth* (Eds.) Dixon A. D., and Sarnat, B. G., pp. 168–203. New York: Alan R. Liss.

Noden, D. M. 1983. The role of the neural crest in patterning of avian cranial skeletal, connective, and muscle tissues. *Devel Biol* 96: 144–165.

Noden, D. M. and Schneider, R. A. 2006. Neural crest cells and the community of plan for craniofacial development: Historical debates and current perspectives. *Adv Exp Med Biol* 589: 1–23.

Noden, D. M. and Trainor, P. A. 2005. Relations and interactions between cranial mesoderm and neural crest populations. *J Anat* 207: 575–601.

Northcutt, R. G. 2005. The new head hypothesis revisited. *J Exp Zool Part B (Mol Devel Evol)* 304B: 274–297.

Nunn, C. L. and Smith, K. K. 1998. Statistical analyses of developmental sequences: The craniofacial region in marsupial and placental mammals. *Amer Nat* 152: 82–101.

O'Brien, C. A., Plotkin, L. I., Galli, C., Goellner, J. J., Gortazar, A. R., Allen, M. R., Robling, A. G. et al. 2008. Control of bone mass and remodeling by PTH receptor signaling in osteocytes. *PLoS One* 3: e2942.

Oster, G. and Alberch, P. 1982. Evolution and bifurcation of developmental programs. *Evolution* 36: 444–459.

Oster, G. F., Shubin, N., Murray, J. D., and Alberch, P. 1988. Evolution and morphogenetic rules–The shape of the vertebrate limb in ontogeny and phylogeny. *Evolution* 42: 862–884.

Otto, F., Thornell, A. P., Crompton, T., Denzel, A., Gilmour, K. C., Rosewell, I. R., Stamp, G. W. et al. 1997. Cbfa1, a candidate gene for cleidocranial dysplasia syndrome, is essential for osteoblast differentiation and bone development. *Cell* 89: 765–771.

Parsons, K. J. and Albertson, R. C. 2009. Roles for Bmp4 and CaM1 in shaping the jaw: Evo-devo and beyond. *Annu Rev Genet* 43: 369–388.

Pasqualetti, M., Ori, M., Nardi, I., and Rijli, F. M. 2000. Ectopic Hoxa2 induction after neural crest migration results in homeosis of jaw elements in *Xenopus. Development* 127: 5367–5378.

Petiot, A., Ferretti, P., Copp, A. J., and Chan, C. T. 2002. Induction of chondrogenesis in neural crest cells by mutant fibroblast growth factor receptors. *Dev Dyn* 224: 210–221.

Piotrowski, T. and Nusslein-Volhard, C. 2000. The endoderm plays an important role in patterning the segmented pharyngeal region in zebrafish (*Danio rerio*). *Dev Biol* 225: 339–356.

Pointer, M. A., Kamilar, J. M., Warmuth, V., Chester, S. G., Delsuc, F., Mundy, N. I., Asher, R. J., and Bradley, B. J. 2012. RUNX2 tandem repeats and the evolution of facial length in placental mammals. *BMC Evol Biol* 12: 103.

Powder, K. E., Ku, Y. C., Brugmann, S. A., Veile, R. A., Renaud, N. A., Helms, J. A., and Lovett, M. 2012. A cross-species analysis of microRNAs in the developing avian face. *PLoS One* 7:e35111.

Pratap, J., Galindo, M., Zaidi, S. K., Vradii, D., Bhat, B. M., Robinson, J. A., Choi, J. Y. et al. 2003. Cell growth regulatory role of Runx2 during proliferative expansion of preosteoblasts. *Cancer Res* 63: 5357–5362.

Qing, H., Ardeshirpour, L., Pajevic, P. D., Dusevich, V., Jahn, K., Kato, S., Wysolmerski, J., and Bonewald, L. F. 2012. Demonstration of osteocytic perilacunar/canalicular remodeling in mice during lactation. *J Bone Miner Res* 27: 1018–1029.

Qiu, M., Bulfone, A., Ghattas, I., Meneses, J. J., Christensen, L., Sharpe, P. T., Presley, R., Pedersen, R. A., and Rubenstein, J. L. 1997. Role of the Dlx homeobox genes in proximodistal patterning of the branchial arches: mutations of Dlx-1, Dlx-2, and Dlx-1 and -2 alter morphogenesis of proximal skeletal and soft tissue structures derived from the first and second arches. *Dev Biol* 185: 165–184.

Radlanski, R. J. and Klarkowski, M. C. 2001. Bone remodeling of the human mandible during prenatal development. *J Orofac Orthop* 62: 191–201.

Radlanski, R. J., Renz, H., Lajvardi, S., and Schneider, R. A. 2004. Bone remodeling during prenatal morphogenesis of the human mental foramen. *Eur J Oral Sci* 112: 301–310.

Raff, R. A. 1996. *The Shape of Life: Genes, Development, and the Evolution of Animal Form*. Chicago, IL: University of Chicago Press.

Raff, R. A. and Kaufman, T. C. 1983. *Embryos, Genes, and Evolution*. New York: Macmillan Publishing.

Reeve, E. C. R. 1950. Genetical aspects of size allometry. *Proc Roy Soc Lond Series B Biol Sci* 137: 515–518.

Reiss, J. O., Burke, A. C., Archer, C. W., De Renzi, M., Dopazo, H., Etxeberría, A., Gale, E. A. et al. 2008. Pere alberch: Originator of evodevo. *Biol Theory* 3: 351–356.

Rensch, B. 1948. Histological changes correlated with evolutionary changes of body size. *Evolution* 2: 218–230.

Reponen, P., Sahlberg, C., Munaut, C., Thesleff, I., and Tryggvason, K. 1994. High expression of 92-kD type IV collagenase (gelatinase B) in the osteoclast lineage during mouse development. *J Cell Biol* 124: 1091–1102.

Richardson, M. K. 1995. Heterochrony and the phylotypic period. *Dev Biol* 172: 412–421.

Richardson, M. K., Hanken, J., Gooneratne, M. L., Pieau, C., Raynaud, A., Selwood, L., and Wright, G. M. 1997. There is no highly conserved embryonic stage in the vertebrates: Implications for current theories of evolution and development. *Anat Embryol* 196: 91–106.

Richman, J. M., Herbert, M., Matovinovic, E., and Walin, J. 1997. Effect of fibroblast growth factors on outgrowth of facial mesenchyme. *Dev Biol* 189: 135–147.

Richman, J. M. and Tickle, C. 1992. Epithelial-mesenchymal interactions in the outgrowth of limb buds and facial primordia in chick embryos. *Dev Biol* 154: 299–308.

Richman, J. M. and Tickle, C. 1989. Epithelia are interchangeable between facial primordia of chick embryos and morphogenesis is controlled by the mesenchyme. *Dev Biol* 136: 201–210.

Ricklefs, R. E. and Starck, J. M. 1998. Embryonic growth and development. In *Avian Growth and Development: Evolution within the Altricial-Precocial Spectrum* (Eds.) Matthias Starck, J. and Ricklefs, R. E., pp. 31–58. New York: Oxford University Press.

Ridley, M. 1985. Embryology and classical zoology in Great Britain. In *A History of Embryology: The Eighth Symposium of the British Society for Developmental Biology* (Eds.) Horder, T. J., Witkowski, J., and Wylie, C. C., pp. 35–67. Cambridge, UK: Cambridge University Press.

Rijli, F. M., Mark, M., Lakkaraju, S., Dierich, A., Dolle, P., and Chambon, P. 1993. A homeotic transformation is generated in the rostral branchial region of the head by disruption of *Hoxa-2*, which acts as a selector gene. *Cell* 75: 1333–1349.

Riska, B. 1986. Some models for development, growth. and morphometric correlation. *Evolution* 40: 1303–1311.

Riska, B. and Atchley, W. R. 1985. Genetics of growth predict patterns of brain-size evolution. *Science* 229: 668–671.

Robb, R. C. 1935. A study of mutations in evolution. *J Genet* 31: 39–46.

Rohlf, F. J. and Bookstein, F. L. Eds. 1990. *Proceedings of the Michigan Morphometrics Workshop*. Vol. Special Publication Number 2. Ann Arbor, MI: University of Michgan Museum of Zoology.

Roth, G. and Wake, D. B. 1989. Conservatism and Innovation in the evolution of feeding in vertebrates. In *Complex Organismal Functions: Integration and Evolution in Vertebrates* (Eds.) Wake, D. B. and Roth, G., pp. 7–21. New York: John Wiley & Sons.

Roth, V. L. 1984. How elephants grow: Heterochrony and the calibration of developmental stages in some living and fossil species. *J Vert Paleont* 4: 126–145.

Roth, V. L. and Mercer, J. M. 2000. Morphometrics in development and evolution. *Amer Zool* 40: 801–810.

Rowe, A., Richman, J. M., and Brickell, P. M. 1992. Development of the spatial pattern of retinoic acid receptor-beta transcripts in embryonic chick facial primordia. *Development* 114: 805–813.

Russell, E. S. 1916. *Form and Function: A Contribution to the History of Animal Morphology*. London, UK: John Murray Publishers.

Sanchez-Villagra, M. R., Geiger, M., and Schneider, R. A. 2016. The taming of the neural crest: A developmental perspective on the origins of morphological covariation in domesticated mammals. *R Soc Open Sci* 3: 160107.

Sanford, G. M., Lutterschmidt, W. I., and Hutchison, V. H. 2002. The comparative method revisited. *Bioscience* 52: 830–836.

Santagati, F. and Rijli, F. M. 2003. Cranial neural crest and the building of the vertebrate head. *Nat Rev Neurosci* 4: 806–818.

Sasano, Y., Zhu, J. X., Tsubota, M., Takahashi, I., Onodera, K., Mizoguchi, I., and Kagayama, M. 2002. Gene expression of MMP8 and MMP13 during embryonic development of bone and cartilage in the rat mandible and hind limb. *J Histochem Cytochem* 50: 325–332.

Scherson, T., Serbedzija, G., Fraser, S., and Bronner-Fraser, M. 1993. Regulative capacity of the cranial neural tube to form neural crest. *Development* 118: 1049–1062.

Schilling, T. F. 1997. Genetic analysis of craniofacial development in the vertebrate embryo. *Bioessays* 19: 459–468.

Schlosser, G. and Wagner, G. P. 2004. *Modularity in Development and Evolution*. Chicago, IL: University of Chicago Press.

Schmalhausen, I. I. 1949. *Factors of Evolution: The Theory of Stabilizing Selection*. Translated by I. Dordick. (Ed.) Dobzhansky, T. Chicago, IL: University of Chicago Press.

Schneider, R. A. 1999. Neural crest can form cartilages normally derived from mesoderm during development of the avian head skeleton. *Devel Biol* 208: 441–455.

Schneider, R. A. 2005. Developmental mechanisms facilitating the evolution of bills and quills. *J Anat* 207: 563–573.

Schneider, R. A. 2007. How to tweak a beak: Molecular techniques for studying the evolution of size and shape in Darwin's finches and other birds. *Bioessays* 29: 1–6.

Schneider, R. A. 2015. Regulation of jaw length during development, disease, and evolution. *Curr Top Dev Biol* 115: 271–298.

Schneider, R. A. and Helms, J. A. 2003. The cellular and molecular origins of beak morphology. *Science* 299: 565–568.

Schneider, R. A., Hu, D., and Helms, J. A. 1999. From head to toe: Conservation of molecular signals regulating limb and craniofacial morphogenesis. *Cell Tissue Res* 296: 103–109.

Schneider, R. A., Hu, D., Rubenstein, J. L., Maden, M., and Helms, J. A. 2001. Local retinoid signaling coordinates forebrain and facial morphogenesis by maintaining FGF8 and SHH. *Development* 128: 2755–2767.

Schowing, J. 1968. Influence inductrice de l'encéphale embryonnaire sur le développement du crâne chez le Poulet. *J Embryol Exp Morphol* 19: 9–32.

Seales, E. C., Micoli, K. M., and McDonald, J. M. 2006. Calmodulin is a critical regulator of osteoclastic differentiation, function, and survival. *J Cell Biochem* 97: 45–55.

Sears, K. E., Goswami, A., Flynn, J. J., and Niswander, L. A. 2007. The correlated evolution of Runx2 tandem repeats, transcriptional activity, and facial length in carnivora. *Evol Dev* 9: 555–565.

Sechrist, J., Nieto, M. A., Zamanian, R. T., and Bronner-Fraser, M. 1995. Regulative response of the cranial neural tube after neural fold ablation: spatiotemporal nature of neural crest regeneration and up-regulation of Slug. *Development* 121: 4103–4115.

Shea, B. T. 1983. Paedomorphosis and neoteny in the pygmy chimpanzee. *Science* 222: 521–522.

Shea, B. T. 1985. Ontogenetic allometry and scaling: A discussion based on the growth and form of the skull in African Apes In *Size and Scaling in Primate Biology* (Ed.) Jungers, W. L., pp. 175–205. New York: Springer.

Shea, B. T. 1989. Heterochrony in human evolution: The case for neoteny reconsidered. *Yearbook Physical Anthropol* 32: 69–101.

Shigetani, Y., Nobusada, Y., and Kuratani, S. 2000. Ectodermally derived FGF8 defines the maxillomandibular region in the early chick embryo: Epithelial-mesenchymal interactions in the specification of the craniofacial ectomesenchyme. *Dev Biol* 228: 73–85.

Shubin, N. H. and Alberch, P. 1986. A morphogenetic approach to the origin and basic organization of the tetrapod limb. In *Evolutionary Biology* (Eds.) Hecht, M. K. and Wallace, B., pp. 319–387. New York: Plenum Press.

Siegel, A. F. and Benson, R. H. 1982. A robust comparison of biological shapes. *Biometrics* 38: 341–350.

Sinnott, E. W. and Dunn, L. C. 1932. *Principles of Genetics: A Textbook with Problems*. London, UK: McGraw-Hill.

Slatkin, M. 1987. Quantitative genetics of heterochrony. *Evolution* 41: 799–811.

Smith, F. J., Percival, C. J., Young, N. M., Hu, D., Schneider, R. A., Marcucio, R. S., and Hallgrimsson, B. 2015. Divergence of craniofacial developmental trajectories among avian embryos. *Dev Dyn* 415: 188–197.

Smith, K. K. 1997. Comparative patterns of craniofacial development in Eutherian and Metatherian Mammals. *Evolution* 51: 1663–1678.

Smith, K. K. 2001. Heterochrony revisited: The evolution of developmental sequences. *Biol J Linn Soc Lond* 73: 169–186.

Smith, K. K. 2002. Sequence heterochrony and the evolution of development. *J Morphol* 252: 82–97.

Smith, K. K. 2003. Time's arrow: Heterochrony and the evolution of development. *Int J Dev Biol* 47: 613–621.

Smith, K. K. and Schneider, R. A. 1998. Have gene knockouts caused evolutionary reversals in the Mammalian first arch? *BioEssays* 20: 245–255.

Smith, M. M. and Hall, B. K. 1990. Development and evolutionary origins of vertebrate skeletogenic and odontogenic tissues. *Biol Rev Camb Philos Soc* 65: 277–373.

Sohal, G. S. 1976. Effects of reciprocal forebrain transplantation on motility and hatching in chick and duck embryos. *Brain Res* 113: 35–43.

Solem, R. C., Eames, B. F., Tokita, M., and Schneider, R. A. 2011. Mesenchymal and mechanical mechanisms of secondary cartilage induction. *Dev Biol* 356: 28–39.

Spemann, H. 1938. *Embryonic Development and Induction*. New Haven, CT: Yale University Press.

Spemann, H. and Mangold, H. 1924. *Induction of Embryonic Primordia by Implantation of Organizers from a Different Species* (Eds.) Willier, B. H. and Oppenheimer, J. M., *Foundations of Experimental Embyology*. New York: Hafner Press.

Spemann, H. and Schotté, O. 1932. Über xenoplastische Transplantation als Mittel zur Analyse der embryonvalen Induktion. *Naturwissenschaften* 20: 463–467.

Starck, J. M. 1989. Zeitmuster der Ontogenesen bei nestflüchtenden und nesthockenden Vögeln. *Cour Forsch-Inst Senckenberg* 114: 1–319.

Starck, J. M. and Ricklefs, R. E. 1998. *Avian Growth and Development: Evolution within the Altricial-Precocial Spectrum, Oxford Ornithology Series; 8.* New York: Oxford University Press.

Stern, D. L. and Emlen, D. J. 1999. The developmental basis for allometry in insects. *Development* 126: 1091–1101.

Szabo-Rogers, H. L., Geetha-Loganathan, P., Nimmagadda, S., Fu, K. K., and Richman, J. M. 2008. FGF signals from the nasal pit are necessary for normal facial morphogenesis. *Dev Biol* 318: 289–302.

Tang, S. Y., Herber, R. P., Ho, S. P., and Alliston, T. 2012. Matrix metalloproteinase-13 is required for osteocytic perilacunar remodeling and maintains bone fracture resistance. *J Bone Min Res* 27: 1936–1950.

Teitelbaum, S. L. 2000. Bone resorption by osteoclasts. *Science* 289: 1504–1508.

Teitelbaum, S. L., Tondravi, M. M., and Ross, F. P. 1997. Osteoclasts, macrophages, and the molecular mechanisms of bone resorption. *J Leukocyte Biol* 61: 381–388.

Thomas, D. M., Johnson, S. A., Sims, N. A., Trivett, M. K., Slavin, J. L., Rubin, B. P., Waring, P. et al. 2004. Terminal osteoblast differentiation, mediated by runx2 and p27KIP1, is disrupted in osteosarcoma. *J Cell Biol* 167: 925–934.

Thompson, D. W. 1917. *On Growth and Form.* Cambridge, UK: Cambridge University Press.

Thompson, D. W. 1952. *On Growth and Form.* 2nd edition. Cambridge, UK: Cambridge University Press.

Thorogood, P. 1987. Mechanisms of morphogenetic specification in skull development. In *Mesenchymal-Epithelial Interactions in Neural Development* (Eds.) Sievers, J. and Berry, M., Wolff, J. R., pp. 141–152. Berlin, Germany: Springer-Verlag.

Thorogood, P. 1988. The developmental specification of the vertebrate skull. *Development* 103: 141–153.

Thorogood, P. 1993. Differentiation and morphogenesis of cranial skeletal tissues. In *The Skull* (Eds.) Hanken, J. and Hall, B. K., pp. 112–152. Chicago, IL: University of Chicago Press.

Thorogood, P., Bee, J., and von der Mark, K. 1986. Transient expression of collagen type II at epitheliomesenchymal interfaces during morphogenesis of the cartilaginous neurocranium. *Dev Biol* 116: 497–509.

Tokita, M., Kiyoshi, T., and Armstrong, K. N. 2007. Evolution of craniofacial novelty in parrots through developmental modularity and heterochrony. *Evol Dev* 9: 590–601.

Tokita, M. and Schneider, R. A. 2009. Developmental origins of species-specific muscle pattern. *Dev Biol* 331: 311–325.

Trainor, P. A., Melton, K. R., and Manzanares, M. 2003. Origins and plasticity of neural crest cells and their roles in jaw and craniofacial evolution. *Int J Dev Biol* 47: 541–553.

Tucker, A. S. and Lumsden, A. 2004. Neural crest cells provide species-specific patterning information in the developing branchial skeleton. *Evol Dev* 6: 32–40.

Tucker, A. S., Yamada, G., Grigoriou, M., Pachnis, V., and Sharpe, P. T. 1999. Fgf-8 determines rostral-caudal polarity in the first branchial arch. *Development* 126: 51–61.

Tyler, M. S. 1978. Epithelial influences on membrane bone formation in the maxilla of the embryonic chick. *Anat Rec* 192: 225–233.

Tyler, M. S. 1983. Development of the frontal bone and cranial meninges in the embryonic chick: An experimental study of tissue interactions. *Anat Rec* 206: 61–70.

Tyler, M. S. and Hall, B. K. 1977. Epithelial influences on skeletogenesis in the mandible of the embryonic chick. *Anat Rec* 188: 229–240.

Vaglia, J. L. and Smith, K. K. 2003. Early differentiation and migration of cranial neural crest in the opossum, *Monodelphis domestica. Evol Dev* 5: 121–135.

Van Exan, R. J. and Hall, B. K. 1984. Epithelial induction of osteogenesis in embryonic chick mandibular mesenchyme studied by transfilter tissue recombinations. *J Embryol Exp Morphol* 79: 225–242.

Veitch, E., Begbie, J., Schilling, T. F., Smith, M. M., and Graham, A. 1999. Pharyngeal arch patterning in the absence of neural crest. *Curr Biol* 9: 1481–1484.

von Bonin, G. 1937. Brain-weight and body-weight of mammals. *J General Psychol* 16: 379–389.

Waddington, C. H. 1939. *An Introduction to Modern Genetics*. London: G. Allen & Unwin.

Waddington, C. H. 1940. *Organisers & Genes*. Cambridge, UK: Cambridge University Press.

Waddington, C. H. 1950. The biological foundations of measurements of growth and form. *Proc RoySoc Lond. Ser B, Biol Sci* 137: 509–515.

Waddington, C. H. 1957a. *The Strategy of the Genes*. London, UK: George Allan Unwin.

Waddington, C. H. 1957b. *The Strategy of the Genes; A Discussion of some Aspects of Theoretical Biology*. London, UK: Allen & Unwin.

Waddington, C. H. 1962. *New Patterns in Genetics and Development, Columbia Biological Series*. New York: Columbia University Press.

Waddington, C. H. 1975. *The Evolution of an Evolutionist*. Ithaca, NY: Cornell University Press.

Wagner, G. 1959. Untersuchungen an *Bombinator-Triton*-Chimaeren. *Roux' Archiv Entwicklungsmech der Organismen* 151: 136–158.

Wake, D. B. 1998. Pere Alberch (1954–98). *Nature* 393: 632.

Wake, D. B. and Roth, G. (Eds.) 1989. *Complex Organismal Functions: Integration and Evolution in Vertebrates, Complex Organismal Functions: Integration and Evolution in Vertebrates*. New York: John Wiley & Sons.

Wall, N. A. and Hogan, B. L. 1995. Expression of bone morphogenetic protein-4 (BMP-4), bone morphogenetic protein-7 (BMP-7), fibroblast growth factor-8 (FGF-8) and sonic hedgehog (SHH) during branchial arch development in the chick. *Mech Dev* 53: 383–392.

Wang, Y. H., Rutherford, B., Upholt, W. B., and Mina, M. 1999. Effects of BMP-7 on mouse tooth mesenchyme and chick mandibular mesenchyme. *Dev Dyn* 216: 320–335.

West-Eberhard, M. J. 2003. *Developmental Plasticity and Evolution*. New York: Oxford University Press.

Wilson, J. and Tucker, A. S. 2004. Fgf and Bmp signals repress the expression of Bapx1 in the mandibular mesenchyme and control the position of the developing jaw joint. *Dev Biol* 266: 138–150.

Wright, S. 1931. Evolution in Mendelian populations. *Genetics* 16: 97–159.

Wu, P., Jiang, T. X., Shen, J. Y., Widelitz, R. B., and Chuong, C. M. 2006. Morphoregulation of avian beaks: Comparative mapping of growth zone activities and morphological evolution. *Dev Dyn* 235: 1400–1412.

Wu, P., Jiang, T. X., Suksaweang, S., Widelitz, R. B., and Chuong, C. M. 2004. Molecular shaping of the beak. *Science* 305: 1465–1466.

Xia, S. L. and Ferrier, J. 1996. Localized calcium signaling in multinucleated osteoclasts. *J Cell Physiol* 167: 148–155.

Xiong, J. and O'Brien, C. A. 2012. Osteocyte RANKL: new insights into the control of bone remodeling. *J Bone Miner Res* 27: 499–505.

Xiong, J., Piemontese, M., Thostenson, J. D., Weinstein, R. S., Manolagas, S. C., and O'Brien, C. A. 2014. Osteocyte-derived RANKL is a critical mediator of the increased bone resorption caused by dietary calcium deficiency. *Bone* 66: 146–154.

Yamashita, T. and Sohal, G. S. 1986. Development of smooth and skeletal muscle cells in the iris of the domestic duck, chick and quail. *Cell Tissue Res* 244: 121–131.

Young, D. W., Hassan, M. Q., Pratap, J., Galindo, M., Zaidi, S. K., Lee, S. H., Yang, X. et al. 2007. Mitotic occupancy and lineage-specific transcriptional control of rRNA genes by Runx2. *Nature* 445: 442–446.

Young, D. L., Schneider, R. A., Hu, D., and Helms, J. A. 2000. Genetic and teratogenic approaches to craniofacial development. *Crit Rev Oral Biol Med* 11: 304–317.

Young, N. M., Chong, H. J., Hu, D., Hallgrimsson, B., and Marcucio, R. S. 2010. Quantitative analyses link modulation of sonic hedgehog signaling to continuous variation in facial growth and shape. *Development* 137: 3405–3409.

Young, N. M., Hu, D., Lainoff, A. J., Smith, F. J., Diaz, R., Tucker, A. S., Trainor, P. A., Schneider, R. A., Hallgrimsson, B., and Marcucio, R. S. 2014. Embryonic bauplans and the developmental origins of facial diversity and constraint. *Development* 141: 1059–1063.

Zavitz, K. H. and Zipursky, S. L. 1997. Controlling cell proliferation in differentiating tissues: Genetic analysis of negative regulators of G_1-->S-phase progression. *Curr Opin Cell Biol* 9: 773–781.

Zayzafoon, M. 2006. Calcium/calmodulin signaling controls osteoblast growth and differentiation. *J Cell Biochem* 97: 56–70.

Zelditch, M. 2004. *Geometric Morphometrics for Biologists: A Primer*. Boston, MA: Elsevier Academic Press.

Zelditch, M. L., Bookstein, F. L., and Lundrigan, B. A. 1992. Ontogeny of integrated skull growth in the cotton rat *Sigmodon fulviventer*. *Evolution* 46: 1164–1180.

Zhang, W., Bergamaschi, D., Jin, B., and Lu, X. 2005. Posttranslational modifications of p27kip1 determine its binding specificity to different cyclins and cyclin-dependent kinases in vivo. *Blood* 105: 3691–3698.

Zhao, Q., Eberspaecher, H., Lefebvre, V., and De Crombrugghe, B. 1997. Parallel expression of Sox9 and Col2a1 in cells undergoing chondrogenesis. *Deel Dyn* 209: 377–386.

8 Cellular Basis of Evolution in Animals
An Evo-Devo Perspective

R. Craig Albertson

CONTENTS

8.1 SUMMARY

With respect to multicellular evolution, cells occupy the gray area between genotype and phenotype. As a result, this level of biological organization has remained understudied. However, as we move beyond the "gene" to characterize the emergent properties that contribute to adaptive phenotypic variation, cells should figure more prominently. This chapter examines the roles of cells in morphological evolution. More accurately, it re-examines the recent evo-devo literature, with an eye not for genes and signal transduction pathways, but rather for the cellular behaviors these signals mediate. Most examples, taken from key events in animal evolution, illustrate the potential for cells and cell biology to expand the current evolutionary paradigm.

The overall conclusion of this chapter is consistent with the general theme of this book, which is that cells mediate the genotype–phenotype connection, and therefore deserve to be brought more explicitly into the fold of modern evolutionary research and theory.

8.2 WHO'S DRIVING THE BUS? CELLS AS DRIVERS OR PASSENGERS OF EVOLUTION?

In the prevailing evolutionary paradigms of the past century, cells have existed in a virtual no-man's land, especially with respect to multicellular evolution. During the neo-Darwinian synthesis of the early to mid-1900s the focus was on alleles and populations (Mayr 1993). Recent efforts in evo-devo have also focused almost exclusively on making connections between genotype and phenotype (Raff 2000; Gilbert 2003; Hall 2003). Since cells are neither genotype nor phenotype, they have been largely ignored by evolutionary biologists for the past 100+ years.

The rise of evo-devo has been especially damaging to the cell. This research area has advanced our understanding of the molecular basis of evolutionary change in new and exciting ways, but it has also galvanized a gene-centric view of evolution (Parsons and Albertson 2013). The genesis of this field, in its modern incarnation, was in many ways an artifact of research into the genetic "tool kit" of animal development (e.g., Carroll et al. 2005). As researchers began to reveal the logic of development in worms, flies and mice, they noted a remarkable overlap in the genes that regulate these processes. This accumulated knowledge led naturally to the main question that nucleated the field and still drives it today: If the developmental tool kit is conserved across animals, then how do differences in form and function arise (Carroll et al. 2005)? Variation in gene expression was a logical place to begin the search, and proved fruitful. The field is now dominated by studies into the various ways genes can be differentially expressed during development to produce variation in developmental outcomes. With technological advances, the resolution and breadth with which we can ask this question continues to increase—from RNA-seq to Chip-seq and beyond.

Two unforeseen consequences have emerged from this paradigm. The first is a *focus on the gene*. The second is a *focus on the embryo*. Both have established a bias in the literature, which, in turn, has led to what is likely an oversimplified view of evolutionary processes. Evidence for this comes from large-scale genome-wide association studies (GWAS) in humans, which have only been able to account for a minority of variation in most complex traits (Manolio et al. 2009; Eichler et al. 2010; Gibson 2010). Moreover, the loci that are detected are often distinct from toolkit genes implicated in "building" traits (Hallgrimsson et al. 2014). Finally, organisms can "recover" from molecular perturbation early in development (Powder et al. 2015), underscoring the concept of robustness in developmental systems (Mestek Boukhibar and Barkoulas 2016). Thus, to fully appreciate the mechanisms that underlie evolutionary change we must (1) move past transcription factors and embryos, and (2) recognize that larger developmental windows and other levels of biological organization are relevant to the generation of phenotypic variation, which is the primary resource for evolution.

8.3 BACK TO THE FUTURE: A RETURN TO THE IDEA OF STRUCTURAL GENES

In an important early book, Raff and Kaufman (1983) stated that structural genes may "provide little in the way of regulatory information," but quickly explained that these types of genes should not be ignored, as their products "provide the actual machinery for cell shape-change and cell movements' directly underlying morpho-genesis." Although this is still a gene-centric argument, it implicates structural genes that build and sensitize the cell, rather than toolkit genes that build the organism. More recently, Hallgrimsson and colleagues (2014) pointed out that toolkit genes implicated in early developmental processes are underrepresented in GWAS data aimed at identifying loci that underlie phenotypic variation. They suggest that these genes are under such intense stabilizing selection during early embryogenesis that variation is kept to a minimum. In terms of variation in skeletal shape, they suggest that an important source may come from structural genes that help to sensitize bone progenitor cells such that they can better sense and respond to local mechanical load (e.g., chewing, running, and throwing; more on this below).

In this chapter, I do not seek to dismiss genes or the developmental toolkit, as these factors can be a primary and important source of variation. Rather I consider them to be one biological level in many, and ultimately argue that where mutations occur in the genome may be less relevant than the cellular outputs that they mediate (e.g., genes influence signal transduction pathways, which, in turn, induce a cellular response). In this way, cells can be considered as the drivers of the bus, as they represent a critical gateway of phenotypic variation, acting at a level above the genome, but below the phenotype.

This discussion takes an explicitly evo-devo perspective with an emphasis on the development and evolution of animal form, and thus does not cover many salient aspects of cell biology related to (for example) the physiological or neuro-logical basis of adaptation. These are no less important, but lie outside the scope of this chapter (and the boundary of my expertise!). Nevertheless, the concepts and principals explored here may easily be applied to these other realms of cell biology.

8.4 CELLS AND CELLULAR BEHAVIORS IN EVOLUTIONARY CHANGE

In the following sections, I utilize key events in the evolutionary history of animals (metazoans) to illustrate the importance of cell behavior in facilitating these transi-tions. Indeed, all of these events, from major evolutionary innovations to more subtle degrees of phenotypic variation, can be traced to differential cell behavior. In addi-tion to these more detailed examples, Table 8.1 provides a longer list of cell behaviors in morphological evolution.

TABLE 8.1

A List of Cell Behaviors That Underlie the Evolution of Various Traits/Phenotypes, Including the Lineage(s) in Which the Event Has Occurred, the Class of Evolutionary Change (Novelty vs. Variation), and Example References. The Class of Evolution Is Meant to Be a General Designation of Whether It Is Associated with a Macro- (i.e., Novelty) or Micro- (i.e., Variation) Evolutionary Event

Cell Behavior	Trait/Tissue/Cells	Lineage	Process	References
Epithelial–mesenchyme transition (EMT)	Neural crest cells	Vertebrates	Evolutionary novelty	reviewed in Muñoz and Trainor 2015; Hall 2009
	Mesoderm	Metazoa	Evolutionary novelty	reviewed in Baum et al. 2008
	Sclerotome/Dermomyotome	Chordates	Evolutionary novelty	reviewed in Thiery et al. 2009
Proliferation	Multicellularity	Metazoa	Evolutionary novelty	King et al. 2008
	Secondary palate	Mammals	Evolutionary novelty	reviewed in Bush and Jiang 2012
	Limb	Marsupials	Evolutionary novelty	Sears et al. 2012; Beiriger and Sears 2014
Migration	Aggregative multicellularity	Dictyostelium	Evolutionary novelty	Du et al. 2015
	Neural crest cells	Vertebrates	Evolutionary novelty	reviewed in Muñoz and Trainor 2015; Hall 2009
	Secondary palate	Mammals	Evolutionary novelty	reviewed by Bush and Jiang 2012
	Jaw skeleton	Cichlids	Phenotypic variation	Powder et al. 2014
Differentiation	Multicellularity	Metazoa	Evolutionary novelty	King et al. 2008
	Pigmentation/chromatophores	Cichlids	Phenotypic variation	Albertson et al. 2014
	Head skeleton	Cichlids	Phenotypic variation	Parsons et al. 2014
	Leaf adaxial–abaxial specification	Plants	Phenotypic variation	reviewed in Yamaguchi et al. 2012
Polarity	Multicellularity	Metazoa	Evolutionary novelty	King et al. 2008
	Early embryogenesis/axis specification	Nematodes	Phenotypic variation	Brauchle et al. 2009
	Plasticity in growth and morphology	Plants	Phenotypic variation	reviewed in Korbei and Luschnig 2011

(Continued)

TABLE 8.1 (Continued)

A List of Cell Behaviors That Underlie the Evolution of Various Traits/Phenotypes, Including the Lineage(s) in Which the Event Has Occurred, the Class of Evolutionary Change (Novelty vs. Variation), and Example References. The Class of Evolution Is Meant to Be a General Designation of Whether It Is Associated with a Macro- (i.e., Novelty) or Micro- (i.e., Variation) Evolutionary Event

Cell Behavior	Trait/Tissue/Cells	Lineage	Process	References
Communication[*]	Multicellularity	Metazoa	Evolutionary novelty	King et al. 2008; Pincus et al. 2008; Du et al. 2015
	Secondary palate	Mammals	Evolutionary novelty	reviewed in Bush and Jiang 2012
	Tooth shape	Vertebrates	Phenotypic variation	reviewed in Jernvall and Thesleff 2012
	Leaf adaxial–abaxial specification	Plants	Phenotypic variation	reviewed in Yamaguchi et al. 2012
Adhesion	Multicellularity	Metazoa, Algae, Dictyostelium	Evolutionary novelty	King et al. 2008; Abedin and King 2010
	Tissue/organ complexity	Metazoa	Evolutionary novelty	reviewed in Magie and Martindale 2008
	Secondary palate	Mammals	Evolutionary novelty	reviewed in Bush and Jiang 2012
Apoptosis	Multicellularity	Metazoa	Evolutionary novelty	King et al. 2008
	Secondary palate	Mammals	Evolutionary novelty	reviewed in Bush and Jiang 2012
	Wing	Bats	Evolutionary novelty	Weatherbee et al. 2006
	Tail reduction	Tunicates	Evolutionary novelty	Chambon et al. 2002
	Metamorphosis	Amphibians	Evolutionary novelty	Ishizuya-Oka et al. 2010
	Digit reduction	Mammals	Phenotypic variation	Cooper et al. 2014
	Eye loss	Astyanax	Phenotypic variation	reviewed in Jeffery 2009

*Signal transduction is a key mechanism through which cells communicate with one another. It is also a major focus of modern evo-devo research (e.g., Carroll et al. 2005). Therefore, many (perhaps the majority) of evo-devo studies involve cell–cell communication.

A   THE ORIGINS OF METAZOANS AND THE EVOLUTION OF MULTICELLULARITY

Multicellularity can be achieved via two basic means: coming together or failing to separate (King 2004). The former, as illustrated by the slime model *Dictyostellium* (Williams et al. 2005), is not the norm with respect to multicellular eukaryotes. The latter represents a much more generalized event across eukaryotic kingdoms, including the lineage leading to animals. This initial step in animal evolution highlights several important transitions in cell behavior (King 2005; King et al. 2008).

It is important to note that the eukaryotic branch of life is rife with multicellular forms, with up to 16 independent origins of multicellularity (Barton et al. 2007, and see Chapter 4 by Nanjundiah, Ruiz-Trillo, and Kirk in this volume. This suggests that multicellular evolution from a unicellular ancestor is not a prohibitively difficult transition; even in its most basic form, multicellularity offers several advantages in terms of projection from the physical and biotic (e.g., unicellular predators) environment. For instance, many large unicellular predators engulf their prey via phagocytosis. In response to this pressure, many protozoans form transient colonies which make it difficult for such "gape-limited" predators to prey upon them. The evolution of relatively simple multicellular life offered a permanent solution to this challenge. In contrast to small and relatively simple multicellular life, the evolution of large, complex forms is limited to relatively few lineages (e.g., plants, animals, and brown algae), which suggest that this is a more complicated evolutionary transition (King 2004).

Many cellular behaviors are fundamental to multicellular development. For instance:

- Cell *division* is required to amass enough cellular material to build complex organisms.
- Stable cell *adhesion* allows daughter cells to stay together.
- Cell *migration* is required for various morphogenic processes as the embryo takes "shape."
- Cell *polarity* allows cells to orient themselves in the context of tissues and organs.
- Cell *differentiation* is necessary to define regions of the embryo with distinct functions.
- Finally, cell–cell *communication* allows individual cells to "know" where they are in the context of the multicellular embryo.

Importantly, none of these cell behaviors needed to evolve *de novo* in multicellular lineages. Protozoans exhibit all of these behaviors, albeit in different contexts and to different degrees.

For instance, cell *division* is how unicellular organisms reproduce. Cell *adhesion* allows protozoans to form transient colonies and attach substrates. Actin-based *migration* through complex three dimensional substrates is common among prokaryotes. Cell *polarity* enables unicellular organisms to orient themselves with respect to their physical environment. Protozoans also *differentiate* over time (e.g., shifts in mating-type or locomotive-type) rather than spatially. Finally, *cell–cell* communication between unicellular organisms allows them to recognize conspecifics (e.g., to exchange genetic material), or heterospecifics (e.g., potential predators), or to coordinate movements (e.g., during colony formation). Thus, the basic cellular machinery necessary for multicellular development was likely already present in protozoan ancestors, and only needed to be elaborated upon to achieve multicellularity.

Evidence for this prediction came from genomic comparisons between metazoans and their unicellular sister group, choanoflagellates (King et al. 2008). These analyses showed, for example, that the choanoflagellate genome contains genes with both cell-adhesion receptor domains and extracellular matrix domains. Again, the product of these genes likely enables choanoflagellate to adhere to the substrate and to each other during colony formation. Choanoflagellates also possess components of some but not all metazoan-specific signal transduction pathways, which are an important means by which cell–cell communication occurs, as they transduce an intercellular signal into the nucleus to induce a transcriptional response.

Tyrosine phosphorylation (pTyr) signaling provides a key example of the importance of cell–cell communication in the evolution of multicellularity (Pincus et al. 2008). This molecular pathway involves three basic components: (1) tyrosine kinases (TyrK) that transfer phosphate groups onto proteins, (2) Src Homology 2 (SH2) domain proteins that "read" this signal, and (3) tyrosine phosphatases (PTP), that remove the phosphate group thereby terminating the signal. The balance between TyrK and PTP activity within cells fine tunes the signal, and enables the regulation of a diversity of biological functions. Notably, closely related, but unicellular eukaryotes have a small number of SH2 and PTP domain proteins but lack TyrK proteins. The latter appears to have evolved in the lineage that includes both metazoans and choanoflagellates, and its appearance is coincident with a marked expansion in all three pTyr components. A greater diversity of pTyr signaling components should, in theory, enable a greater diversity of cellular outputs. Thus, an expansion of this basic cell behavior poised this lineage to evolve complex cell–cell communication mechanisms, a necessary step in the transition to complex multicellular life.

B SECONDARY PALATE FORMATION IN MAMMALS AS AN EXAMPLE OF PROLIFERATION, SURVIVAL, AND ADHESION IN EVOLUTION

The evolution of the secondary palate in mammals marks a key event in their evolution, as it enabled the evolution of two mammal-specific innovations. The first is the manipulation and chewing of food before swallowing. The second is the suckling of neonates. The evolution of the secondary palate is critical to both as it enables air to bypass the mouth, such that mammals may breathe while chewing or suckling (Liem et al. 2001).

The development of the secondary palate is a dynamic but well-studied process (Figure 8.1), because cleft palate is one of the most common human birth defects (Lan et al. 2015). This literature provides insights into the cell behaviors that were necessary for the evolution of this structure (Bush and Jiang 2012). An *outgrowth* of the palatal shelf is the first stage in this process and involves the secretion of Sonic hedgehog (SHH) from the oral epithelium to induce proliferation of the underlying mesenchyme. A feedback loop between SHH and Fibroblast growth factor (Fgf) signaling maintains proliferation. Fgf signaling is also necessary for cell survival in this (and other) tissues. *Fusion* is another critical step in palatogenesis.

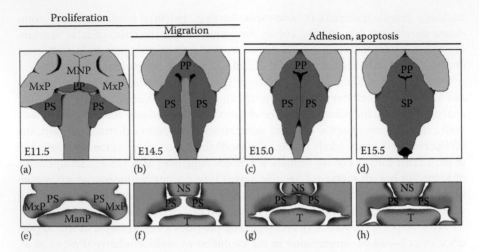

FIGURE 8.1 Cell behaviors during secondary palate formation in mice. After Bush and Jiang (2012). (a–d) Ventral view of the developing palate with representative stages in the lower left. (e–h) Schematic cross-sections of the same developmental stages. Examples of distinct cell behaviors involved in each stage are shown at the top. For example, initiation of outgrowth of the primary and secondary palatal shelves requires proliferation of the palatal mesenchyme (a, e). Dynamic patterns of palatal mesenchyme migration are also associated with shelf outgrowth (b, f). Finally, cell adhesion and apoptosis of the midline epithelial seam are necessary for shelf fusion (c, g) and ultimate formation of the intact secondary palate (d, h). Abbreviations: ManP, mandibular process; MNP, medial nasal process; MxP, maxillary process, NS, nasal septum; PP, primary palate; PS, palatal shelf; SP, secondary palate; T, tongue.

There should be strong selection for palatal fusion before birth as neonates with unfused palates cannot suckle, and cell adhesion molecules figure prominently in this process. Support for this conclusion comes from two lines of evidence. When early stages of palatogenesis are disrupted (e.g., an outgrowth of the palatal shelf), clefting occurs and is accompanied by aberrant fusion of tissues (e.g., between the palate and tongue). Thus, adhesion molecules are still active, but fusion occurs between the wrong tissues. Secondly, in TGF-beta- or Bmp-deficient mice, the developing palatal shelves make contact but fail to fuse. Both signaling molecules lead to the synthesis of extracellular matrix proteins, which are necessary for cell adhesion (d'Amaro et al. 2012).

This basic knowledge of palate development hints at the cellular behaviors that were recruited during the evolution of the palate. Moreover, palate evolution in tetrapods has been well studied at a structural level (Kimmel et al. 2009). However, more focused analyses into how this structure evolved from a reptile-like palate at the cellular-level are lacking. For instance, did the stem mammal lineage experience an expansion of cell-adhesion and/or extracellular matrix domains within its genome, or was there simply recruitment of these proteins to the palatal shelf during development? Either way, an interesting and integrative evo-devo story likely waits to be told here; one that necessarily requires an examination at genomic, genetic, and cellular levels.

C LIMB DEVELOPMENT AND EVOLUTION TO ILLUSTRATE PROLIFERATION AND APOPTOSIS IN ADAPTATION

Limb development is another well studied and remarkably conserved developmental process across tetrapods. Development of limbs even shares many processes with fin development in teleosts because of the "deep" homology of fins and limbs (Nakamura et al. 2016).

Despite a common developmental program, tetrapod limbs have diverged extensively across vertebrates, with lineages often being defined by specific limb morphologies. For instances, owing to their unique early life-history, marsupials exhibit accelerated forelimb development relative to hindlimb development and relative to placental mammals. Functional forelimbs are necessary for newborn neonates to craw from the birth canal to the nipple within the pouch where they complete development while suckling. Recent work has demonstrated that *differential cell proliferation*, driven by Insulin-like growth factor 1 (Igf1), underlies differences in fore- and hindlimb growth at different stages during development (Sears et al. 2012; Beiriger and Sears 2014).

Cell death is another important regulator of limb development and adaptive variation in limb form and function. For most stages of development, the distal portion of the limb (the autopod) is paddle-shaped, with extensive webbing between the developing digits. In most tetrapods, digits are sculpted via targeting programmed cell death (apoptosis) in this webbing. Animals that retain interdigital webbing as adults do so by blocking this cell behavior. Bone morphogenetic proteins (Bmps) signaling is an important signal for apoptosis. In ducks, hindlimb webbing is retained via expression between digits of the Bmp-antagonist, Gremlin, thus blocking Bmp-signaling and cell death. In bats, where webbing is retained in the forelimbs, two signals appear necessary for the retention of interdigital webbing. As in ducks, Gremlin is expressed between the digits in the developing bat forelimbs (but not hindlimbs). However, this signal does not appear to be sufficient to fully repress the signal (because downstream transcriptional targets of Bmp are still expressed in the webbing). This might be because Bmps are also necessary for digit growth and elongation, which is especially pronounced in bat forelimbs (Sears et al. 2006). Thus higher than normal levels of Bmp signaling is needed for the development of elongated digits in bats. The second signal recruited for the retention of interdigital webbing is Fgf, which promotes cell survival. Thus, the current model for retention of interdigital webbing in bats involves both downregulation of Bmp signaling and upregulation of Fgf in the autopod (Weatherbee et al. 2006).

Programmed cell death has also been implicated in other aspects of limb adaptation. Extant tetrapods possess five digits (pentadactyl) and in spite of the fact that some basal tetrapods had up to 10 digits, the lineage leading to modern tetrapods has never evolved more than five digits. As an illustration for how strong this constraint is, consider that extra digits in moles and the panda "thumb" are not true digits, but rather ossified ligaments along the posterior margin of the hand.

Alternatively, many lineages have evolved digit loss as species adapted to distinct modes of locomotion. Cooper and colleagues recently showed that digit loss in some mammals occurs at least in part via apoptosis (Cooper et al. 2014).

Specifically, in the authopod of the three-toed jerboa, two-toed camel, and one-toed horse apoptosis not only occurs between digits but also within the putative cartilage condensations of lost digits. Increased apoptosis is correlated with expanded Bmp signaling in the autopod of mammals with reduced digit numbers, which suggests that the evolution of digit loss in these lineages involved expanding the apoptotic signal during limb development from between the digits to within cartilage condensations of these digits.

D NEURAL CREST CELLS AS AN EXAMPLE OF DIFFERENTIATION
AND MIGRATION IN NOVELTY AND ADAPTATION

Perhaps the most celebrated cell type in vertebrate biology is the neural crest cell. From a developmental perspective, this cell type holds such potency potential that it has been referred to as the 4th germ layer. Specifically:

- Neural crest cells (NCCs) contribute to the development of a vast number of tissues, exceeding those derived from mesoderm (Hall 2000).
- From the perspective of cell biology, NCCs exhibit several notable behaviors (Hall 2009).
- They are one of a relatively few cell-types that undergo *epithelial–mesenchymal transition* (EMT) during normal development.
- NCCs also exhibit extensive, dynamic and coordinated patterns of migration, which involves cell-autonomous, nonautonomous, and even semiautonomous behaviors.
- NCCs are also exquisitely sensitive to the environment, as the developmental fate of each population of NCCs depends on inductive cues from surrounding tissues.
- From an evolutionary perspective, the NCC is considered an innovation that underlies the origin and evolutionary success of vertebrate.

In much the same way that the evolution of mesoderm in bilaterians is credited for elevated levels of disparity in this group, the evolution of NCCs is thought to underlie increased complexity in vertebrate body plans. Many vertebrate-specific innovations, including the craniofacial skeleton and teeth, develop from NCCs. Further, many features associated with speciation, while not restricted to vertebrates, are also NCC derived such as pigmentation.

NCC and the New Head Hypothesis

The evolution of the vertebrate skull is a paramount innovation in the history of animals, and the origin and consequences of this evolutionary novelty have been the focus of research for over 100 years.

In 1983, Gans and Northcutt offered the "new head hypothesis" (NHH) to help focus this line of research (Gans and Northcutt 1983). Among other propositions, the NHH suggested that NCCs are a vertebrate-specific innovation and that structures developing from these cells directly enabled a shift from passive filter feeding to active predation. Many claims of the NHH have not stood the test of time, especially that NCCs are a vertebrate novelty (Northcutt 2005). For instance, there is increasing evidence to suggest that nonvertebrate chordates possess neural crest-like cells, which suggests that they did not evolve *de novo* in vertebrates. Tunicate larvae, for example, possess cells that meet some, but not all, of the defining criteria of NCCs. They originate from around the central nervous system, are migratory, become pigment, and express the NCC- specific antigens HNK1 and Zic1 (Jeffery et al. 2004).

In addition, many NCC-specific genes are expressed in the neural-non neural boarder of the dorsal ectoderm in amphioxus. However, these cells do not undergo epithelial–mesenchymal transition (EMT) or migrate away from the neural tube (Holland and Holland 2001). There's also good evidence to suggest that skeletogenetic potential of proto-NCCs involved the evolution of new transcription factor binding sites in *SoxE* (Jandzik et al. 2015). This gene is expressed in both mesodermally derived oral cartilages in amphioxus and vertebrate NCC-derived pharyngeal cartilages. Notably, when expressed in a zebrafish embryo, amphioxus *SoxE* is restricted to the neural tube, but is reduced by NCC migration stages.

In all, these data argue for a preadaptation model for NCC evolution, wherein the specific cellular attributes of NCCs evolved independently in nonvertebrate deuterostomes, but became integrated within the lineage leading to vertebrates. Regardless of the specific details regarding their evolutionary origins, the core claim of the NHH that the evolution of NCCs enabled an evolutionary arms race between predator and prey remains roundly supported (Northcutt 2005). Whether NCC-derived tissues are inherently more flexible from an evolutionary perspective than other tissues is a difficult question to assess. What is clear however is that NCC-derived tissues appear to be large targets for natural selection as species continue to adapt and evolve. Some examples of this are provided below.

Pigmentation in Fishes

Vertebrate pigmentation is largely derived from NCCs and is an important trait for the origins and maintenance of biodiversity. Color helps to mediate both ecological and behavioral interactions. For instance, it is critical for predator avoidance via both crypsis and enhanced coloration to announce toxicity. It also mediates reproductive success via intra- and intersexual selection.

The Pax3/7 gene family plays an ancient and conserved role in NCC formation (Basch et al. 2006; Maczkowiak et al. 2010; Murdoch et al. 2012). Members of this gene family also are necessary for pigment formation (Lacosta et al. 2005, 2007). Fish possess three pigment types—melanocytes (brown–black), xanthophores (yellow–orange), and iridophore (blue). All three are thought to arise from a common NCC-derived chromatoblast precursor population. Work in zebrafish has demonstrated roles for these genes in fish NCC and pigment development (Minchin and Hughes 2008). In particular, loss of *Pax3a* function results in a reduction in some, but not all, NCC-derivatives, which supports the idea that the fate of these cells is determined by a distinct set of genetic signals.

With respect to pigmentation, loss of *Pax3a* leads to an reduction in xanthophore number and increased melanocyte number. These data support a model wherein *Pax3a* mediates a fate switch in a common chromatoblast population, with higher levels leading to a greater number of xanthophores and fewer melanophores, and lower levels leading to more melanophores and fewer xanthophores. Genetic variation at the *Pax3a* locus in cichlid fish supports this model (Albertson et al. 2014). First, genetic mapping implicated a genome region that contains *Pax3a* in mediating levels of xanthophore-based pigmentation across the body. Next, a genetic polymorphism in the regulatory region (i.e., 5' UTR) of this gene was associated with variation in pigmentation, such that animals with more yellow and less black pigmentation possessed one allele, while species with more black and less yellow pigment possessed the opposite allele. This mode of action is consistent with the putative role of this gene in mediating a xanthophore-melanocyte fate switch in zebrafish. Finally, allele-specific expression demonstrated that F_1 hybrid animals with a greater amount of yellow pigmentation expressed a higher level of the yellow *Pax3a* allele. These data implicate a switch in NCCs fate decision in evolved differences in vertebrate pigmentation.

In related work, *Pax7a* was implicated in the evolution of a discrete pigmentation polymorphism in cichlids (Roberts et al. 2009). In several species, females develop an orange blotched (OB) phenotype wherein the normal pigmentation pattern is completely abolished and melanocytes that normally organize in discrete and regular vertical bars are instead aggregated in randomly spaced and sized blotches. This pattern matches the pocked background of the shallow rocky habitat in Lake Malawi, and it is thought that the polymorphism is maintained by selection for crypsis. This phenotype is due to a *cis*-regulatory mutation in *Pax7a*, with higher levels of expression being associated with increases in melanophore cell death and the development of the OB phenotype. Notably, different *Pax7a* alleles are associated with different levels of expression and the development of discrete OB morphs, with higher transcript levels associated with the development of few small blotches, and relatively lower levels associated with a greater number of large black blotches (Roberts et al. 2017). Taken together, these investigations point to the Pax3/7 gene family in mediating phenotypic evolution via the regulation of distinct NCC behaviors (e.g., fate-switch and survival).

NCCs in Skull Shape Evolution

While pigmentation is a common feature to all metazoans, the bones of the anterior skull and face are specific to vertebrates and derived exclusively from NCCs. An important and ongoing question is whether differences in skull shape among species can be traced to alternate NCC behaviors. In other words, how much "evolutionary potential" is contained within this cell population? Both experimental embryology and genetic/genomic mapping approaches have proven especially useful in addressing this question.

Most of the facial skeleton originates from cranial NCCs, and vertebrates have evolved marked diversity in the facial skeleton as lineages adapted to different modes of foraging. It stands to reason therefore that NCCs themselves possess the genetic memory to develop species-specific morphologies. Alternatively, it is possible that variation in shape arises not from the NCCs themselves but from inductive signals from surrounding tissue. That is, perhaps NCCs are just especially sensitive to such paracrine signals. Distinguishing between these alternate hypotheses was a major focus at the turn of the century.

One of the first studies to empirically demonstrate the evolutionary potential of the NC involved chimeras between duck and quail tissues (Schneider and Helms 2003).

Specifically, pre-migratory NCCs from duck embryos were transplanted into quail hosts and vice versa. Given the marked differences in beak shape between these two species, the experiment provided insights into how much of this difference could be attributed to the NC. The morphological effects of the transformations were widespread. Not only were NCC-derived tissues transformed in host embryos to match donor morphologies, but non-NCC tissues were also transformed. Thus, NCCs possess autonomous molecular cues to regulate themselves, as well as the surrounding tissues to generate species-specific shape variation. This study demonstrated the power of the chimera system in addressing questions about the roles of NCCs in shaping the face. Since then it has been used to address more specific NCC behaviors during species-specific craniofacial development, including cell cycle regulation (Hall et al. 2014), NCC allocation (Fish et al. 2014), osteogenesis (Merrill et al. 2008), and bone resorption (Ealba et al. 2015). While this body of work has provided amazing insights into how the facial skeleton is shaped, duck and quail are distantly related (i.e., last common ancestor >50 mya), and thus this work cannot speak to the process of evolution *per se*. For this a comparison between more closely related species is necessary.

Cichlid fish represent a paramount evolutionary model, as they exhibit tremendous diversity, which has arisen in a very brief amount of evolutionary time. Lake Malawi, for example, contains an approximated 1,000 cichlid species that have evolved in 1–2 million years. Importantly this classic adaptive radiation involves diversification in a number of NCC-derived traits including pigment (see above), dentition (Streelman and Albertson 2006; Fraser et al. 2009, 2013), and craniofacial shape (Albertson and Kocher 2001; Albertson et al. 2003, 2005; Cooper et al. 2010; Hu and Albertson 2014; Parsons et al. 2014).

In a recent study, Powder et al. (2014) sought to investigate the developmental mechanisms by which species adapt their jaws to accommodate different modes of feeding. A major axes of diversification in the lake involves recurrent shifts between long jaw and short jaw eco-types (Cooper et al. 2010). All things being equal, long jaws enable greater speed during jaw rotation, which is optimal for foraging on fast and/or elusive prey. Short jaws, on the other hand, enable the production of greater force, which is optimal for foraging on hard and/or tough prey. This line of research began 15 years ago with an unbiased pedigree-based genetic mapping approach to identify genetic intervals associated with alternate (e.g., short vs long) jaw shapes (Albertson et al. 2003, 2005). Population genome scans were used more recently to narrow intervals and identify candidate loci (Powder et al. 2014). These analyses identified a nonsynonymous single nucleotide polymorphism (SNP) in the gene Limb bud and heart homolog (*Lbh*), which results in a substitution at a highly conserved amino acid residue. This protein is not well studied, but had been reported to be expressed in mouse migratory NCCs (Briegel and Joyner 2001).

Subsequent analyses showed that *Lbh* is also expressed in fish and frog NCCs, suggesting conserved roles for this gene in NCC development. Knockdown of this protein in fish and frog resulted in marked craniofacial defects that were traced to disrupt NCC migration, especially to those NCC streams that contribute to the upper and lower jaw. These data are consistent with differential patterns of NCC migration in cichlid embryos, which show more NCC being allocated to the mandibular arch in the long-jaw species, and less NCC allocated to the anterior stream in the short jaw species.

Remarkably, when the long-jaw form of *Lbh* was overexpressed in frog NCCs, more cells were shunted to the mandibular stream. When the short-jaw form was overexpressed, NCCs were shunted away from the mandibular stream. Collectively, these data point to evolution via a discrete cellular behavior—that is, the differential allocation of progenitor cells to the organ under selection (Figure 8.2). That *Lbh* is the gene that

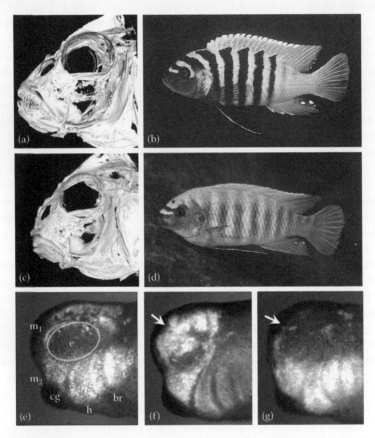

FIGURE 8.2 Differential neural crest cell migration is associated with variation in craniofacial form. (a–b) *Maylandia zebra* exhibits relatively long oral jaws. (c–d) Comparatively, *Labeotropheus fuelleborni* exhibits much shorter jaws. Genetic/genomic mapping implicated a coding change in the amino acid sequence of the gene *Limb bud and heart homolog (Lbh)* in mediating this difference. *Lbh* is expressed in migrating cranial neural crest (CNC) cells, and when knocked-down migration is inhibited (Powder et al. 2014). The effect is especially pronounced in anteriorly migrating cells, which ultimately contribute to the oral jaws. Over expressing the different forms of *Lbh* (i.e., long-jaw versus short-jaw variants) in Xenopus embryos produced markedly different effects in terms of cell behavior. (e) GFP-labeled CNC in a control Xenopus embryo. Abbreviations: cg, cement gland; br, branchial streams; e, eye; h, hyoid stream; m_1, mandibular stream 1 (over the eye); m_2, mandibular stream 2 (under the eye). (f) When the long-jaw form of Lbh is expressed, more cells migrate anteriorly. (g) When the short-jaw variant is expressed, fewer cells migrate anteriorly. As with the knock-down phenotype, the effect is more pronounced in the anterior-most CNC stream (white arrow), due at least in part, to differential migration of neural crest progenitor cells. (After Powder, K.E. et al., *Mol. Biol. Evol.*, 31, 3113–3124, 2014.)

mediates this response is interesting, but it is the cellular response that ultimately leads to adaptive variation within and between species.

Notably, results in fish are consistent with those in birds, where more NCCs are allocated to the mandibular stream in long-jawed ducks compared to short-jawed quail (Fish et al. 2014). That a similar cellular mechanism may underlie patterns of divergence at both the macro- and micro-evolutionary levels, and across widely divergent species, is notable and may hint at a common mechanism through which species adapt to distinct foraging environments.

8.5 FEEDBACK ON THE SYSTEM: CELLULAR BEHAVIORS IN MEDIATING PHENOTYPIC PLASTICITY

Phenotypic plasticity is the capacity of an organism's phenotype to vary in distinct environments. The ability of an organism to change its phenotype in different environments may increase its fitness in fluctuating environments. If true, plasticity should be adaptive and therefore subject to selection itself.

In addition, phenotypic plasticity is thought to play important roles in several evolutionary phenomena, including the origins of novel traits (Moczek 2008), speciation (Price et al. 2003; West-Eberhard 2005; Pfennig et al. 2010), and adaptive radiations (West-Eberhard 2003; Wund et al. 2008). This has led to an increased interest in plasticity as a driver of phenotypic evolution in general. While sufficient levels of genetic variation have been documented for plasticity to respond to selection, a strict genetic basis for this process/mechanism has remained elusive (Pigliucci 2005). This uncertainty about the molecular nature of plasticity has hindered progress into understanding the mechanisms through which plasticity may evolve or influence the evolutionary process (Via et al. 1995). Thus, phenotypic plasticity is recognized as an important process in evolution, but we still lack a general understanding of many fundamental aspects of its biology.

One way that plasticity is thought to influence phenotypic evolution is through the process of *genetic assimilation* (Pigliucci et al. 2006). Under this scenario, a population is exposed to a novel environment, and individuals respond by changing their phenotype via plasticity. If the environment becomes stable, then genetic mutations that "fix" the phenotypic response will be favored by selection. Thus, plasticity has the potential to bias the direction of an evolutionary response by determining the types of variation exposed to selection. Alternatively, if the environment becomes predictably variable (e.g., seasons), then mutations that "fix" a plastic response will be favored. This is referred to as *genetic accommodation*. Theories that involve genetic assimilation and accommodation provide a compelling framework to study phenotypic evolution. However, they have not brought us closer to understanding the process at the molecular level. They also, again, represent a gene-centric view of phenotypic evolution, which may in fact hinder progress by not focusing on more salient levels of biological organization.

Are genes sensitive to the environment? Certainly epigenetic modifications to the genome can occur due to environmental stimuli, and these can dramatically influence patterns and levels of gene expression (Huang et al. 2014). Given the focus of the field on this level of biological organization, epigenetics has been brought squarely into the fold of evolutionary biology (Mendizabal et al. 2014; Skinner 2015). However, there is a long history of studies into mechanosensing in cells. Several organelles and subcellular structures help cells to sense and response to their environment, including the cytoskeleton, cell–cell junctions, ion channels, and the primary cilia. These structures and processes are well studied by cell and developmental biologists, but have not yet been brought into the fold of evolutionary biologists (at least not in the context of evo-devo). I submit that, while toolkit genes help to build the organism and have been the focus of evo-devo research, structural genes that build mechanosensing organelles will be a fruitful place to focus for analyses into developmental plasticity.

The primary cilia are sensory organelles that are tuned to several different environmental variables, including both kinetic (e.g., fluid flow and pressure) and chemical (e.g., ions and hormones) stimuli (Zaghloul and Brugmann 2011; Yuan et al. 2015). They transduce these signals into the cell to elicit various responses, including changes in proliferation, differentiation, migrations, polarity, tissue morphology and transcription. Thus, cilia allow cells and tissue to both interpret and respond to their local environment.

For example, bone is a dynamic tissue that is exquisitely sensitive to its mechanical environment. The primary cilia on bone cells are thought to play critical roles in mediating this process (Xiao et al. 2006; Papachroni et al. 2009; Nguyen and Jacobs 2013). Mice lacking functional cilia in bone precursor cells exhibit normal larval skeletal patterning, but impaired growth and remodeling at later stages. Specifically, mice in which the cilia motility factor Kif3a has been knocked down in osteoblasts, exhibit early onset osteopenia (Qiu et al. 2012), as well as a reduced ability to form bone in response to mechanical loading (Temiyasathit et al. 2012). Both phenotypes may be tied to expanded proliferation and impaired osteoblast differentiation in conditional knockout mice. These and other laboratory studies (Nguyen and Jacobs 2013; Yuan et al. 2015), which are consistent with the prevalence of skeletal abnormalities in human ciliopathies (Zaghloul and Brugmann 2011), collectively suggest that the primary cilia are critical for proper development, growth and mechanical load-induced remodeling of the skeleton.

Within the context of plasticity and genetic assimilation/accommodation, such mechanisms ought to be considered. Here again bone biology may help illuminate such mechanisms. Vertebrates exposed to new environments are often exposed to new kinematic challenges. Whether novel prey items or new modes of locomotion, these shifts will feed back onto the skeleton, which should, in turn, remodel itself to accommodation these shifts. If organelles, such as the primary cilia, underlie these plastic responses, then tuning of the system may occur via mutations in structural genes. The search for such loci will not be trivial, as cilia are one of the most complex organelles yet described. Nevertheless, mutations that enhance the sensitivity of the primary cilia to environmental change may be the source of genetic accommodation, whereas those that "lock-in" sensitivities may underlie the assimilation of plastic phenotypes. If true, then the search for such genes should expand into the structural realm.

8.6 LINKING PLASTICITY AND EVOLUTION: SYNTHESIS AND CONCLUSIONS

In consideration of the current trajectory of modern evolutionary theory, I envision the cell playing a more prominent role. Natural selection acts on phenotypic variation, and the genotype is a fundamental source of that variation. While the cell is neither genotype nor phenotype within this paradigm, it represents a critical link between the two levels.

In many ways, cells are less constrained by the parameters of the central dogma, as they operate above the level of the genotype, and thus have many more degrees of freedom through which to generate variation. For example, there is no gene in the body that encodes for the speed or direction of NCC migration *per se*. Rather such cellular processes are dictated by higher-order, emergent properties of development; properties of the extracellular environment, kinetics of actin-filament assembly/disassembly. In this way, genomic variation can be thought of as simply tuning the system. It does so by regulating the timing, efficiency, and switching of cellular behaviors. Evidence for this comes from the observation that these behaviors can be markedly different in different environments. Thus, the same genotype will result in a differentially "tuned" system, depending on environmental context. Moreover, different genotypes result in developmental systems that are differentially sensitive to the environment, and this variation in phenotypic plasticity can be tied directly to cell-biology (see above).

A key theory that may help to link genes, cells and phenotypic variation is the flexible stem hypothesis (West-Eberhard 2003), which posits that plasticity in ancestral populations will bias the direction of adaptive response by exposing specific variants (generated via plasticity) to natural selection. An explicit prediction of this theory is that the mechanisms involved in regulating a plasticity response will also be involved in adaptation and ultimately speciation. Specifically, mutations that "fix" a plastic trait should occur in genes that help to tune the system in the first place. Given the multitude of ways in which cells can sense and respond to environmental stimuli, this level of biological organization will be of paramount importance as we seek mechanistic support for the flexible stem hypothesis. An important question along these lines is what type of genes will be "assimilated."

To circle back to the beginning of this chapter and a discussion of regulatory versus structural genes, there is little doubt that toolkit genes—genes that regulate signal transduction pathways and developmental outcomes—are vital sources of phenotypic variation. But what about structural genes which build the cell and help to regulate cell function? I submit that there is strong and accumulating evidence to suggest that this category of genes holds great potential for evolutionary biology. Unlike "genetic/genomic evolution," which permeates the field, the area of "cellular evolution" is relatively untapped with respect to multicellular evolution. As we move forward in our search for the mechanisms that promote and maintain adaptive phenotypic variation, cells and cell-biology will be a critical source of information. Indeed, cells represent a critical link between genotype and phenotype, as well as the gateway through which environmental stimuli may induce a transcriptional response.

REFERENCES

Abedin, M. and King, N. 2010. Diverse evolutionary paths to cell adhesion. *Trends Cell Biol* 20: 734–742.

Albertson, R. C. and Kocher, T. D. 2001. Assessing morphological differences in an adaptive trait: A landmark-based morphometric approach. *J Exp Zool* 289: 385–403.

Albertson, R. C., Powder, K. E., Hu, Y. A., Coyle, K. A., Roberts, R. B., and Parsons, K. J. 2014. Genetic basis of continuous variation in the levels and modular inheritance of pigmentation in cichlid fishes. *Mol Ecol* 23: 5135–5150.

Albertson, R. C., Streelman, J. T., and Kocher, T. D. 2003. Directional selection has shaped the oral jaws of Lake Malawi cichlid fishes. *Proc Natl Acad Sci USA* 100: 5252–5257.

Albertson, R. C., Streelman, J. T., Kocher, T. D., and Yelick, P. C. 2005. Integration and evolution of the cichlid mandible: The molecular basis of alternate feeding strategies. *Proc Natl Acad Sci USA* 102: 16287–16292.

Barton, N. H., Briggs, D. E. G., Eisen, J. A., Goldstein, D. B., and Patel, N. H. 2007. *Evolution*. Woodbury, NY: Cold Spring Harbor Laboratory Press.

Basch, M. L., Bronner-Fraser, M., and Garcia-Castro, M. I. 2006. Specification of the neural crest occurs during gastrulation and requires Pax7. *Nature* 441: 218–222.

Baum, B., Settleman, J., and Quinlan, M. P. 2008. Transitions between epithelial and mesenchymal states in development and disease. *Semin Cell Dev Biol* 19: 294–308.

Beiriger, A. and Sears, K. E. 2014. Cellular basis of differential limb growth in postnatal gray short-tailed opossums (*Monodelphis domestica*). *J Exp Zool B Mol Dev Evol* 322: 221–229.

Brauchle, M., Kiontke, K., MacMenamin, P., Fitch, D. H., and Piano, F. 2009. Evolution of early embryogenesis in rhabditid nematodes. *Dev Biol* 335: 253–262.

Briegel, K. J. and Joyner, A. L. 2001. Identification and characterization of Lbh, a novel conserved nuclear protein expressed during early limb and heart development. *Dev Biol* 233: 291–304.

Bush, J. O. and Jiang, R. 2012. Palatogenesis: Morphogenetic and molecular mechanisms of secondary palate development. *Development* 139: 231–243.

Carroll, S. B., Grenier, J. K., and Weatherbee, S. D. 2005. *From DNA to Diversity: Molecular Genetics and the Evolution of Animal Design*. Malden, MA: Blackwell.

Chambon, J. P., Soule, J., Pomies, P., Fort, P., Sahuquet, A., Alexandre, D., Mangeat, P. H., and Baghdiguian, S. 2002. Tail regression in *Ciona intestinalis* (Prochordate) involves a Caspase-dependent apoptosis event associated with ERK activation. *Development* 129: 3105–3114.

Cooper, K. L., Sears, K. E., Uygur, A., Maier, J., Baczkowski, K. S., Brosnahan, M., Antczak, D., Skidmore, J. A., and Tabin, C. J. 2014. Patterning and post-patterning modes of evolutionary digit loss in mammals. *Nature* 511: 41–45.

Cooper, W. J., Parsons, K., McIntyre, A., Kern, B., McGee-Moore, B., and Albertson, R. C. 2010. Bentho-pelagic divergence of cichlid feeding architecture was prodigious and consistent during multiple adaptive radiations within African rift-lakes. *PLoS One* 5: e9551.

d'Amaro, R., Scheidegger, R., Blumer, S., Pazera, P., Katsaros, C., Graf, D., and Chiquet, M. 2012. Putative functions of extracellular matrix glycoproteins in secondary palate morphogenesis. *Front Physiol* 3: e377.

Du, Q., Kawabe, Y., Schilde, C., Chen, Z. H., and Schaap, P. 2015. The evolution of aggregative multicellularity and cell-cell communication in the Dictyostelia. *J Mol Biol* 427: 3722–3733.

Ealba, E. L., Jheon, A. H., Hall, J., Curantz, C., Butcher, K. D., and Schneider, R. A. 2015. Neural crest-mediated bone resorption is a determinant of species-specific jaw length. *Dev Biol* 408: 151–163.

Eichler, E. E., Flint, J., Gibson, G., Kong, A., Leal, S. M., Moore, J. H., and Nadeau, J. H. 2010. Missing heritability and strategies for finding the underlying causes of complex disease. *Nat Rev Genet* 11: 446–450.

Fish, J. L., Sklar, R. S., Woronowicz, K. C., and Schneider, R. A. 2014. Multiple developmental mechanisms regulate species-specific jaw size. *Development* 141: 674–684.

Fraser, G. J., Bloomquist, R. F., and Streelman, J. T. 2013. Common developmental pathways link tooth shape to regeneration. *Dev Biol* 377: 399–414.

Fraser, G. J., Hulsey, C. D., Bloomquist, R. F., Uyesugi, K., Manley, N. R., and Streelman, J. T. 2009. An ancient gene network is co-opted for teeth on old and new jaws. *PLoS Biol* 7: e31.

Gans, C. and Northcutt, R. G. 1983. Neural crest and the origin of vertebrates: A new head. *Science* 220: 268–273.

Gibson, G. 2010. Hints of hidden heritability in GWAS. *Nat Genet* 42: 558–560.

Gilbert, S. F. 2003. The morphogenesis of evolutionary developmental biology. *Int J Dev Biol* 47: 467–477.

Hall, B. K. 2000. The neural crest as a fourth germ layer and vertebrates as quadroblastic not triploblastic. *Evol Dev* 2: 3–5.

Hall, B. K. 2003. Evo-Devo: Evolutionary developmental mechanisms. *Int J Dev Biol* 47: 491–495.

Hall, B. K. 2009. *The Neural Crest and Neural Crest Cells in Vertebrate Development and Evolution*. New York: Springer.

Hall, J., Jheon, A. H., Ealba, E. L., Eames, B. F., Butcher, K. D., Mak, S. S., Ladher, R., Alliston, T., and Schneider, R. A. 2014. Evolution of a developmental mechanism: Species-specific regulation of the cell cycle and the timing of events during craniofacial osteogenesis. *Dev Biol* 385: 380–395.

Hallgrimsson, B., Mio, W., Marcucio, R. S., and Spritz, R. 2014. Let's face it-complex traits are just not that simple. *PLoS Genet* 10: e1004724.

Holland, L. Z. and Holland, N. D. 2001. Evolution of neural crest and placodes: Amphioxus as a model for the ancestral vertebrate? *J Anat* 199: 85–98.

Hu, Y. and Albertson, R. C. 2014. Hedgehog signaling mediates adaptive variation in a complex functional system in the cichlid skull. *Integ Comp Biol* 54: E95.

Huang, B., Jiang, C., and Zhang, R. 2014. Epigenetics: The language of the cell? *Epigenomics* 6: 73–88.

Ishizuya-Oka, A., Hasebe, T., and Shi, Y. B. 2010. Apoptosis in amphibian organs during metamorphosis. *Apoptosis* 15: 350–364.

Jandzik, D., Garnett, A. T., Square, T. A., Cattell, M. V., Yu, J. K., and Medeiros, D. M. 2015. Evolution of the new vertebrate head by co-option of an ancient chordate skeletal tissue. *Nature* 518: 534–537.

Jeffery, W. R. 2009. Evolution and development in the cavefish Astyanax. *Curr Top Dev Biol* 86: 191–221.

Jeffery, W. R., Strickler, A. G., and Yamamoto, Y. 2004. Migratory neural crest-like cells form body pigmentation in a urochordate embryo. *Nature* 431: 696–699.

Jernvall, J. and Thesleff, I. 2012. Tooth shape formation and tooth renewal: Evolving with the same signals. *Development* 139: 3487–3497.

Kimmel, C. B., Sidlauskas, B., and Clack, J. A. 2009. Linked morphological changes during palate evolution in early tetrapods. *J Anat* 215: 91–109.

King, N. 2004. The unicellular ancestry of animal development. *Dev Cell* 7: 313–325.

King, N. 2005. Choanoflagellates. *Curr Biol* 15: R113–R114.

King, N., Westbrook, M. J., Young, S. L., Kuo, A., Abedin, M., Chapman, J., Fairclough, S. et al. 2008. The genome of the choanoflagellate *Monosiga brevicollis* and the origin of metazoans. *Nature* 451: 783–788.

Korbei, B. and Luschnig, C. 2011. Cell polarity: PIN it down! *Curr Biol* 21: R197–R199.

Lacosta, A. M., Canudas, J., Gonzalez, C., Muniesa, P., Sarasa, M., and Dominguez, L. 2007. Pax7 identifies neural crest, chromatophore lineages and pigment stem cells during zebrafish development. *Int J Dev Biol* 51: 327–331.

Lacosta, A. M., Muniesa, P., Ruberte, J., Sarasa, M., and Dominguez, L. 2005. Novel expression patterns of Pax3/Pax7 in early trunk neural crest and its melanocyte and non-melanocyte lineages in amniote embryos. *Pigment Cell Res* 18: 243–251.

Lan, Y., Xu, J., and Jiang, J. 2015. Cellular and molecular mechanisms of palatogenesis. *Curr Top Dev Biol* 115: 59–84.

Liem, K. F., Bemis, W. E., Walker, W. F., and Grande, L. 2001. *Functional Anatomy of the Vertebrates: An Evolutionary Perspective*. Belmont, CA: Brooks/Cole-Thomas Learning.

Maczkowiak, F., Mateos, S., Wang, E., Roche, D., Harland, R., and Monsoro-Burq, A. H. 2010. The Pax3 and Pax7 paralogs cooperate in neural and neural crest patterning using distinct molecular mechanisms, in *Xenopus laevis* embryos. *Dev Biol* 340: 381–396.

Magie, C. R. and Martindale, M. Q. 2008. Cell-cell adhesion in the cnidaria: Insights into the evolution of tissue morphogenesis. *Biol Bull* 214: 218–232.

Manolio, T. A., Collins, F. S., Cox, N. J., Goldstein, D. B., Hindorff, L. A., Hunter, D. J., McCarthy, M. I. et al. 2009. Finding the missing heritability of complex diseases. *Nature* 461: 747–753.

Mayr, E. 1993. What was the evolutionary synthesis? *Trends Ecol Evol* 8: 31–33.

Mendizabal, I., Keller, T. E., Zeng, J., and Yi, S. V. 2014. Epigenetics and evolution. *Integr Comp Biol* 54: 31–42.

Merrill, A. E., Eames, B. F., Weston, S. J., Heath, T., and Schneider, R. A. 2008. Mesenchyme-dependent BMP signaling directs the timing of mandibular osteogenesis. *Development* 135: 1223–1234.

Mestek Boukhibar, L. and Barkoulas, M. 2016. The developmental genetics of biological robustness. *Ann Bot* 117: 699–707.

Minchin, J. E. and Hughes, S. M. 2008. Sequential actions of Pax3 and Pax7 drive xanthophore development in zebrafish neural crest. *Dev Biol* 317: 508–522.

Moczek, A. P. 2008. On the origins of novelty in development and evolution. *Bioessays* 30: 432–447.

Murdoch, B., DelConte, C., and Garcia-Castro, M. I. 2012. Pax7 lineage contributions to the mammalian neural crest. *PLoS One* 7: e41089.

Muñoz, W. A. and Trainor, P. A. 2015. Neural crest cell evolution: How and when did a neural crest cell become a neural crest cell. *Curr Top Dev Biol* 111: 3–26.

Nakamura, T., Gehrke, A. R., Lemberg, J., Szymaszek, J., and Shubin, N. H. 2016. Digits and fin rays share common developmental histories. *Nature* 537: 225–228.

Nguyen, A. M. and Jacobs, C. R. 2013. Emerging role of primary cilia as mechanosensors in osteocytes. *Bone* 54: 196–204.

Northcutt, G. R. 2005. The new head hypothesis revisited. *J Exp Zool B Mol Dev Evol* 304: 274–297.

Papachroni, K. K., Karatzas, D. N., Papavassiliou, K. A., Basdra, E. K., and Papavassiliou, A. G. 2009. Mechanotransduction in osteoblast regulation and bone disease. *Trends Mol Med* 15: 208–216.

Parsons, K. J. and Albertson, R.C. 2013. From black and white to shades of gray: Unifying evo-devo through the integration of molecular and quantitative approaches. *Advances in Evolution and Development* (Ed.) Streelman, J. T., pp. 81–109. Hoboken, NJ: John Wiley & Sons.

Parsons, K. J., Trent Taylor, A., Powder, K. E., and Albertson, R. C. 2014. Wnt signalling underlies the evolution of new phenotypes and craniofacial variability in Lake Malawi cichlids. *Nat Commun* 5: 3629.

Pfennig, D. W., Wund, M. A., Snell-Rood, E. C., Cruickshank, T., Schlichting, C. D., and Moczek, A. P. 2010. Phenotypic plasticity's impacts on diversification and speciation. *Trends Ecol Evol* 25: 459–467.

Pigliucci, M. 2005. Evolution of phenotypic plasticity: Where are we going now? *Trends Ecol Evol* 210: 481–486.

Pigliucci, M., Murren, C. J., and Schlichting, C. D. 2006. Phenotypic plasticity and evolution by genetic assimilation. *J Exp Biol* 209: 2362–2367.

Pincus, D., Letunic, I., Bork, P., and Lim, W. A. 2008. Evolution of the phospho-tyrosine signaling machinery in premetazoan lineages. *Proc Natl Acad Sci USA* 105: 9680–9684.

Powder, K. E., Cousin, H., McLinden, G. P., and Albertson, R. C. 2014. A nonsynonymous mutation in the transcriptional regulator *lbh* is associated with cichlid craniofacial adaptation and neural crest cell development. *Mol Bio Evol* 31: 3113–3124.

Powder, K. E., Milch, K., Asselin, G., and Albertson, R. C. 2015. Constraint and diversification of developmental trajectories in cichlid facial morphologies. *EvoDevo* 6: 25.

Price, T. D., Qvarnstrom, A., and Irwin, D. E. 2003. The role of phenotypic plasticity in driving genetic evolution. *Proc R Soc Lond B Biol Sci* 270: 1433–1440.

Qiu, N., Xiao, Z., Cao, L., Buechel, M. M., David, V., Roan, E., and Quarles, L. D. 2012. Disruption of Kif3a in osteoblasts results in defective bone formation and osteopenia. *J Cell Sci* 125: 1945–1957.

Raff, R. A. 2000. Evo-devo: The evolution of a new discipline. *Nat Rev Genet* 1: 74–79.

Raff, R. A. and Kaufman, T. C. 1983. *Embryos, Genes, and Evolution. The Developmental-Genetic Basis of Evolutionary Change.* New York: Collier Macmillan.

Roberts, R. B., Moore, E. C., and Kocher, T. D. 2017. An allelic series at *pax7a* is associated with colour polymorphism diversity in Lake Malawi cichlid fish. *Mol Ecol* 26: 2625–2639.

Roberts, R. B., Ser, J. R., and Kocher, T. D. 2009. Sexual conflict resolved by invasion of a novel sex determiner in Lake Malawi cichlid fishes. *Science* 326: 998–1001.

Schneider, R. A. and Helms, J. A. 2003. The cellular and molecular origins of beak morphology. *Science* 299: 565–568.

Sears, K. E., Behringer, R. R., Rasweiler, J. J. T., and Niswander, L. A. 2006. Development of bat flight: Morphologic and molecular evolution of bat wing digits. *Proc Natl Acad Sci USA* 103: 6581–6586.

Sears, K. E., Patel, A., Hubler, M., Cao, X., Vandeberg, J. L., and Zhong, S. 2012. Disparate Igf1 expression and growth in the fore- and hind limbs of a marsupial mammal (Monodelphis domestica). *J Exp Zool B Mol Dev Evol* 318: 279–293.

Skinner, M. K. 2015. Environmental epigenetics and a unified theory of the molecular aspects of evolution: A neo-Lamarckian concept that facilitates neo-Darwinian evolution. *Genome Biol & Evol* 7: 1296–1302.

Streelman, J. T. and Albertson, R. C. 2006. Evolution of novelty in the cichlid dentition. *J Exp Zool B Mol Dev Evol* 306: 216–226.

Temiyasathit, S., Tang, W. J., Leucht, P., Anderson, C. T., Monica, S. D., Castillo, A. B., Helms, J. A., Stearns, T., and Jacobs, C. R. 2012. Mechanosensing by the primary cilium: Deletion of Kif3A reduces bone formation due to loading. *PLoS One* 7: e33368.

Thiery, J. P., Acloque, H., Huang, R. Y., and Nieto, M. A. 2009. Epithelial-mesenchymal transitions in development and disease. *Cell* 139: 871–890.

Via, S., Gomulkiewicz, R., Dejong, G., Scheiner, S. M., Schlichting, C. D., and Vantienderen, P. H. 1995. Adaptive phenotypic plasticity–consensus and controversy. *Trends Ecol Evol* 10: 212–217.

Weatherbee, S. D., Behringer, R. R., Rasweiler, J. J., and Niswander, L. A. 2006. Interdigital webbing retention in bat wings illustrates genetic changes underlying amniote limb diversification. *Proc Natl Acad Sci USA* 103: 15103–15107.

West-Eberhard, M. J. 2003. *Developmental Plasticity and Evolution.* New York: Oxford University Press.

West-Eberhard, M. J. 2005. Developmental plasticity and the origin of species differences. *Proc Natl Acad Sci USA* 102: 6543–6549.

Williams, J. G., Noegel, A. A., and Eichinger, L. 2005. Manifestations of multicellularity: *Dictyostelium* reports in. *Trends Genet* 21: 392–398.

Wund, M. A., Baker, J. A., Clancy, B., Golub, J. L., and Fosterk, S. A. 2008. A test of the Flexible stem model of evolution: Ancestral plasticity, genetic accommodation, and morphological divergence in the threespine stickleback radiation. *Amer Nat* 172: 449–462.

Xiao, Z., Zhang, S., Mahlios, J., Zhou, G., Magenheimer, B. S., Guo, D., Dallas, S. L. et al. 2006. Cilia-like structures and polycystin-1 in osteoblasts/osteocytes and associated abnormalities in skeletogenesis and Runx2 expression. *J Biol Chem* 281: 30884–30895.

Yamaguchi, T., Nukazuka, A., and Tsukaya, H. 2012. Leaf adaxial-abaxial polarity specification and lamina outgrowth: Evolution and development. *Plant Cell Physiol* 53: 1180–1194.

Yuan, X., Serra, R. A., and Yang, S. 2015. Function and regulation of primary cilia and intraflagellar transport proteins in the skeleton. *Ann N Y Acad Sci* 1335: 78–99.

Zaghloul, N. A. and Brugmann, S. A. 2011. The emerging face of primary cilia. *Genesis* 49: 231–246.

9 Dynamical Patterning Modules Link Genotypes to Morphological Phenotypes in Multicellular Evolution

Stuart A. Newman and Karl J. Niklas

CONTENTS

9.1 INTRODUCTION

Unicellular life existed for many millions of years before multicellular forms appeared in the biosphere. Depending on the criteria applied (cell–cell attachment, cell communication, division of cell labor) multicellularity evolved on anywhere between 10 and 25 independent occasions in at least 10 major clades (reviewed in Niklas and Newman 2013). Single-celled forms became highly complex by evolution at the biosynthetic and metabolic levels and in some cases by endosymbiosis. Unicellular organisms invariably release small molecules and synthesize proteins, which are either targeted to their enclosing membranes or are secreted. These components provided a means for ancient cell–cell interactions and communication in ancestral forms. This, in turn, provided a basis in present-day single-celled prokaryotes and eukaryotes, for collective activities such as quorum sensing in bacteria, the formation of pseudohyphae in yeast, and ecologically regulated aggregation and morphogenesis in social amoebae.

The activities of the single-celled ancestors of animals, plants, and fungi and the molecules they produced were necessary ingredients for what later became embryogenesis and analogous morphogenetic processes, but they were not sufficient. Multicellular tissues are viscoelastic liquids and gels, or fluid–solid composites, depending on the type of organism, and the shapes and forms they assume are outcomes of the physical processes and forces that organize such materials. It is the goal of this chapter to provide a framework for understanding how cell behaviors and molecular components that first arose in the single-cell world were mobilized by the physics of *mesoscale* (i.e., between the quantum- and gravitation-determined levels) matter to generate the typical structures of multicellular organisms.

Common notions about morphological evolution hold that the phenotypic variants that ultimately change the complexions of populations represent gradual changes of preexisting forms and are autonomously generated with no preferred directions. These ideas correspond to the rejection by the modern evolutionary synthesis of saltation (abrupt changes of phenotype), plasticity (external influences on phenotype), and orthogenesis (Mayr 1982). As we discuss below, each of these factors has in fact contributed to the generation of novel forms during the history of multicellular life. Their absence from the standard model had its origins in a lack of scientific understanding during the period in which Charles Darwin and Alfred Russel Wallace devised the theory of natural selection of the forces that mold and organize the middle- or meso-scale materials (including living tissues). While some earlier thinkers (e.g., Johann Wolfgang von Goethe and Étienne Geoffroy Saint-Hilaire) speculated about *laws of form* (Webster and Goodwin 1996), such ideas were excluded from the dominant evolutionary model, as was any attempt to connect morphological development with evolution (Newman and Linde-Medina 2013).

When the focus is on biochemical phenotypes, evolutionary change is uncontroversially saltational, directional, and dependent on externalities for the realization of phenotypic outcomes. Clinical phenomena such as inborn errors of metabolism entered the corpus of evolutionary theory early in the twentieth century when enzymology and other chemical sciences were already established (Rosenberg 2008). Whatever the claims of the Darwin–Wallace theory concerning morphological evolution, it was indisputable that phylogenetic change at the cellular and physiological levels was constrained by the laws of chemistry and macromolecular structure, the making or breaking of chemical bonds, and the formation of novel protein folds or assemblages. The effects of the mutations that lead to sickle cell disease or phenylketonuria, for example, are chemically predicable, and their phenotypic effects are, depending on contextual factors, potentially all-or-none.

However, unlike the reconfiguration of metabolic pathways by chemical processes familiar to biologists, morphogenesis and morphological evolution involve the reshaping of living matter by less familiar physical effects, specifically, by the physics of the mesoscale. Mature animal tissues are semisolid materials, with cells anchored in place by intercellular junctions or stiff extracellular matrices. During embryogenesis or wound repair and regeneration, however, the individual mobility and collective cohesiveness of the cells cause developing primordia to behave like viscous or viscoelastic liquids. In contrast, plant and fungal tissues are solid by virtue of the rigid walls that surround their cells. In these systems, morphogenesis is a function of differential growth, expansion, or loss of cells rather than cell rearrangement. In each case, the "generic" (i.e., liquid or solid) properties of the cell collective determine major morphological motifs (multilayering, segmentation, branching, and anastomosis), while depending on, or being modulated by, earlier-evolved biological properties of their living cellular subunits. The materials that constitute multicellular organisms have thus been referred to as *biogeneric* (Newman 2014a, 2016a).

In addition to their identity as viscoelastic or solid materials, animal, plant, and fungal tissues are "excitable media." Such materials are characterized by the ability to store mechanical, chemical or electrical energy and to yield it up under certain conditions (Levine and Ben-Jacob 2004; Sinha and Sridhar 2015). For a material to be excitable, there must be a way for excitations to propagate globally. Thus, even though individual cells are excitable, for a tissue to exhibit integrated global excitation, its cells must relay their states. Nearly without exception, the means for this are different from those found in nonliving excitable media, for example, the spiral pattern-producing Belousov–Zhabotinsky chemical reaction (Winfree 1994). In the case of animal tissues, mechanical strains are relayed via junctionally coupled cells capable of generating amplifying contractile forces (Borghi et al. 2010), while chemical signals are propagated transcellularly by secreted molecules (*morphogens*) sometimes passively, but often accompanied by induced release and uptake of the morphogens (Lander 2007). The developing tissues of animals of most phyla also sustain electrical field gradients that mediate long-range ion transport and the formation of anatomical templates (Levin 2014). In land plants, global excitability is mediated by long-range hormone (e.g., auxin) transport (Dhonukshe 2009; Kutschera and Niklas 2016) and *plasmodesmata* (Brunkard and Zambryski 2016), which are cytoplasmic bridges that facilitate the rapid transfer of developmental signals including transcription factors. The hyphae of growing fungi can be syncytial or contain porous septa (Cole 1996) so that global communication across the tissue mass is intracellular.

Despite these differences in global coordination mechanisms among the three categories of tissue, their behavior as excitable media is governed by laws first described (as were the laws of mesoscale matter) in the mid-twentieth century. The generic patterns predicted by such laws enable the understanding of phenomena like the spacing of integumentary bristles, the placement of leaf buds, and the formation of the gills of a fruiting body. As we describe, approaching morphogenesis in terms of physical forces and processes, mobilized in the multicellular context by specific gene products and pathways, permits the integrated treatment of embryonic development and the origination of organic forms, a program-agenda that has been missing in the standard evolutionary framework (Forgacs and Newman 2005, Niklas and Spatz 2012, Niklas 2016).

This physicogenetic account of the origin of multicellular forms does not conflict with the recognition that the establishment of novel organisms in an ecological setting requires that they be functionally adequate to their environment. However, ecological niches are no longer theorized as preexisting features of the natural world that organisms strive to adapt to and occupy. Instead, organisms play active roles in constructing, defining, and potentially inventing their niches, and more generally, their modes of life (Odling-Smee et al. 2003). According to the modern synthesis, new forms supplant old ones because they are better at meeting some challenge (adaptationism), but they only do so in marginal steps, because any organism with a markedly changed phenotype could not persist in the niche in which it arose (gradualism) (Mayr 1988). But existing niche-adapted forms can thrive in environments entirely different from the ones in which they evolved (Sax et al. 2013), and novel forms can depart from their point of origin and establish themselves in more suitable settings (Rieseberg et al. 1999, 2003). We can, therefore, with theoretical consistency, study mechanisms for the generation of novel forms independently of any obvious competitive advantage that may promote their survival. This applies no less to mechanisms that produce abrupt morphological change when mutated than to those that gradually modify the phenotype.

Our discussion of the role of physical forces and processes in the origination and evolution of animals, plants, and multicellular fungi is framed in terms of *dynamical patterning modules* (DPMs). These are defined as units consisting of specific physical effects along with the specific molecules that mobilize them in the multicellular context so as to promote morphogenesis and pattern formation (Newman and Bhat 2008, 2009, Hernández-Hernández et al. 2012).

The DPM framework incorporates the dual recognition that physical forces that shape tissues are central to multicellular development and its evolution, but that these determinants cannot be considered independently of the actual materials (cell collectivities and their molecules) that they act on. It also avoids the misconception that because the physical laws pertaining to mesoscale materials and excitable media exist independently of living organisms, they, therefore, apply equally to all kinds of matter (Newman 2011b). Not all masses of cells are susceptible to the same array of physical effects: To take two simple examples, cells without attachment molecules will not adhere to one another, and cells that do not influence their neighbors cannot differentiate into reproducible spatial patterns.

In the following sections, we discuss separately the DPMs of animals and of plants, using comparative phylogenomics to infer the transition to the multicellular organization in ancestral forms. Although less is known on the subject, we follow this line of analysis with the multicellular fungi as well. Finally, we conclude with a discussion of how the assemblages of morphological motifs characteristic of animals and plants have become integrated and transformed by further evolution to produce morphologies of present-day complex organisms.

9.2 "LIQUID TISSUES" AND THE ORIGIN OF THE ANIMALS

The animals (Animalia or Metazoa)—are members of a phylogenetic group, the opisthokonts, which also includes the fungi and modern unicellular and transiently colonial organisms (Nielson 2012; Budd and Jensen 2015) The multicellular state in present-day metazoans is mediated by certain members of the cadherin family of homophilic Ca^{2+}-dependent cell adhesion molecules (CAMs) (Halbleib and Nelson 2006). When these proteins bind a group of cells together the resulting mass behaves as if it was a liquid (Foty et al. 1994, Steinberg 2007) (Figure 9.1, a and b).

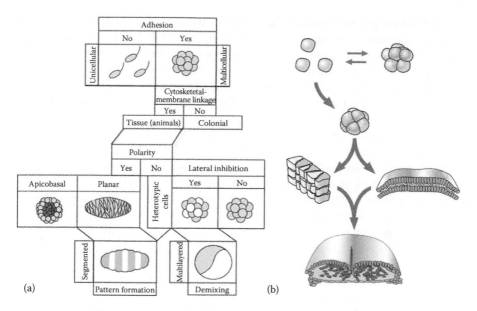

FIGURE 9.1 Morphospace and phylogenetic perspectives on the role of various DPMs in animal evolution and development. (a) Morphological motifs of animal body plans involve choices between modes of multicellular organization specified by the presence or absence of molecular components or pathways mobilizing specific physical forces or effects. Holozoan cells can form clusters if they express adhesive proteins (e.g., protocadherins). The clusters are either simple colonies or, if they contain linkage between adhesive and motile functions (via classical cadherins), form the liquid tissues characteristic of animals. If they undergo polarization (DPMs: POL_a and POL_p; usually induced by Wnt), they can exhibit lumens (via A/B polarization) and/or elongate (via PCP). For cells capable of assuming alternative states of differentiation, a stable or balanced distribution of cell types in the multicellular entity depends on LAT (DPM: *LAT*; usually mediated by the Notch pathway). Stable populations of differentially adhesive cells will sort out, forming immiscible layers (DPM: *DAD*), a driving force of gastrulation and organogenesis. If the adhesive state is regulated in a temporally or spatially periodic fashion (DPM: *OSC*; typically based on Hes1 expression oscillations), the tissue mass can become segmented. (Modified from Niklas, K.J., and Newman, S.A., *Evol. Dev.*, 15, 41–52, 2013. With permission; Newman, S.A. and Bhat, R., *Phys. Biol.*, 5, 15008, 2008. With permission; Newman, S.A. and Bhat, R., *Int. J. Dev. Biol.*, 53, 693–705, 2009. With permission.) (b) Major evolutionary transitions in animal body form (i.e., radiation of superphyla) were generated by the emergence of successively novel "biogeneric" materials. First and second rows: ancestral holozoan colonies became protometazoans by conversion to the liquid tissue state by the appearance of classical cadherins. Third row: basal metazoans were converted to diploblasts by the appearance of PCP and consequent potential for elongation by convergent extension, and stiff, flexible basal laminae (stippling). (Fourth row), triploblasts (a chicken embryo is depicted) emerged from some diploblastic forms by the appearance of EMT-inducing ECMs. (Courtesy of Newman, S.A., *Phil. Trans. R. Soc. Lond. B Sci.*, 371, 20150444, 2016a.)

Liquids have generic properties, regardless of what they are made of (consider the physical similarities among, despite dramatic molecular differences between, mercury, water, and molten paraffin), properties also characteristic of liquid tissues. A substance will only assume the liquid state if its subunits (e.g., atoms, simple molecules, or polymers) simultaneously have a strong affinity for one another and are independently mobile due to the transience of their interactions. Aggregates of cells will not generally have this property, but connections between the cell's attachment capabilities and its motile activities enable cells to act as the subunits of a liquid (Miller et al. 2013). This makes the liquid tissue state a novel form of living matter. (Although some other forms of multicellular life (e.g., social amoebae, see Chapter 4 by Nanjundiah, Ruiz-Trillo, and Kirk in this volume) have liquid-like properties, none achieve it in precisely the same way that animal tissues do.) Unlike nonliving liquids, however, in which constituent subunits change position by Brownian motion (entirely independent of their internal structure), for liquid tissues, active processes of the cell interior are essential, making this category of matter *biogeneric* (Newman 2016a, 2014a).

The outside–inside linkage in liquid tissues is based on a reversible engagement of the cadherin ectodomain with the cytoskeleton. This is mediated by the cytoplasmic portion of transmembrane cadherin proteins with cytoskeletal adaptor proteins such as β-catenin and vinculin (Halbleib and Nelson 2006). The affinity between the cellular subunits is not due solely to physicochemical forces of adhesion mediated by the cadherin ectodomains. The tension the cytoskeleton exerts on the interior surface of the cell membrane influences the extent of their contact with their neighbors, affecting adhesion from the inside (Krieg et al. 2008, Maître et al. 2012). Intracellular tension, moreover, affects the conformation of the surface proteins, enabling cadherins on one cell to bind in a specific fashion to those on an adjoining cell (Halbleib and Nelson 2006). These effects, in conjunction with the quantities of cadherins and other CAMs on the cell surfaces, regulate the strength of intercellular attachment (Amack and Manning 2012; Heisenberg and Bellaiche 2013).

Premetazoan ancestors are inferred—by the presence of protocadherins in non-metazoan holozoans (nonfungal opisthokonts) (Nichols et al. 2012)—to have contained surface proteins of the cadherin family. Protocadherins lack the cytoplasmic domain of classic cadherins (exclusive to, and universal in, the animals), and non-metazoans, in any case, do not have the adaptor proteins that would link it to the cytoskeleton (Abedin and King 2008). Protocadherins do not mediate metazoan-type multicellularity. Nonmetazoan holozoans can be unicellular (*Monosiga brevicollis*; (Abedin and King 2008) or, if multicellular, employ non-cadherin means of cell–cell attachment (*Salpingoeca rosetta*; (Dayel et al. 2011); *Capsaspora owczarzaki*; (Sebé-Pedrós et al. 2013).

The earliest metazoans thus appear to have arisen by repurposing or *neofunctionalization* of cadherins by the acquisition of DNA sequences specifying new proteins and protein modules. This may have occurred in gradual steps or abruptly by horizontal transfer from other unicellular lineages (Tucker 2013), or through other mechanisms of gene innovation (Long et al. 2013). The mobilization by classical cadherins of the force of *adhesion* to bind cells together is the most fundamental metazoan DPM, termed *ADH*; its occurrence in a population of unicellular opisthokont ancestors marked the origin of the animals.

In general, embryonic cells in present-day animals, which are clonally derived from a zygote or fertilized egg, remain attached to one another via one type of cadherin which, as development proceeds, is then replaced by other types in those tissues (termed *epithelioid* or *epithelial*) whose cells remain directly attached to each other. As discussed below, morphogenesis in early-stage metazoan embryos (and in inferred ancestors) leading to characteristic structural motifs such as multiple tissue layers, body cavities, and body and organ primordia elongation, depends on tissue liquidity. When cells disaggregate to form more loosely organized *mesenchymal* and *connective* tissues, they continue to cohere, but this occurs via the binding of a different class of proteins on their surfaces (usually of the integrin family) to secreted extracellular matrix (ECM) molecules (reviewed in Forgacs and Newman 2005). While mesenchymal and connective tissues have their own characteristic physical properties (they can be viscous, elastic, or even solid), these differ from the liquid tissue state. They are essential, however, for organogenesis in animals more complex than the most basal metazoans.

By considering the inherent properties of liquid tissues, we can infer a kind of ground state of morphogenetic capabilities for all animals, extant and ancestral. From comparative phylogenomics, we also have a plausible scenario for the history of the appearance of the genes of the so-called developmental genetic toolkit, and particularly the subset of them that mediate cell–cell interactions (the *interaction toolkit*; Newman et al. 2009). Combining these we can construct a set of metazoan DPMs that generate anatomical elements common to all animals. As will quickly be seen, however, the morphological motifs produced by the earliest-appearing DPMs cannot account for the complex body plans and organs characteristic of most animal phyla. For this, additional DPMs, employing evolutionary novel proteins and previously unrealized physical effects were necessary. We argue that a DPM-based combinatorial hierarchy overlaps with, and can in effect place on a more rigorous footing, the phylum concept for Metazoa.

A MULTILAYERING AND LUMEN FORMATION

Liquids (e.g., oil and water) can undergo phase separation, forming an interface across which they do not mix if their cohesivities (determined by the strength of interaction among their respective subunits) are sufficiently different. The animals with the simplest body plans, the sandwich-like Placozoa (represented by the single extant species *Trichoplax adhaerens*; Miller and Ball 2005) and the labyrinthine demosponges (the largest class of Porifera; Degnan et al. 2015), contain distinct layers, due in part to immiscibility of the two or more liquid tissues each contains. These basal metazoans (which are not literally ancestral to more complex forms, but inferred on the basis of the fossil record and phylogenomics to resemble early-emerging animals) lack the distinct sheet-like epithelia found in the more elaborate eumetazoans that develop from two- and three-layered embryos (Lanna 2015). Their tissues are instead epithelioid, the motile cells being in direct contact with one another, but not in planar arrangements (Figure 9.1, right).

Cells within the same organism can become functionally differentiated from each other, in the expression of cadherins, for instance, by exposure to different levels of an external gene regulatory molecule. Molecules of this sort, Sonic hedgehog

(SHH) and bone morphogenetic protein (Bmp), for example, were carried over to the earliest metazoans from their unicellular holozoan ancestors. In the multicellular context, where they could assume spatially nonuniform concentrations, they came to act as morphogens. From these descriptions we can discern the presumed emergence of two additional universal metazoan DPMs (*ADH* being the first)—*differential adhesion* (*DAD*), enabling multilayering of tissues, and *morphogen activity* (*MOR*), enabling the emergence of regional differentiation. A fourth DPM distinguishes sponges and all other animal groups from the placozoan. This is *lateral inhibition* (*LAT*), mediated by the Notch pathway, which is absent in *T. adhaerens*. This pathway, employing complementary ligand–receptor pairs (e.g., Notch-Delta) on the surfaces of participating cells and intracellular mediators such as Hes1 gives cells the means to enforce a different differentiated state from its own on an otherwise equivalent neighbor. It thus stabilizes cell type differences (e.g., differential expression of a cadherin) that may be reversibly induced by a morphogen (Newman and Bhat 2008, 2009, Newman 2016b).

A class of cell behaviors with ancient evolutionary roots that contributes to all morphogenetic change is *polarization* (Nance 2014) (Figure 9.1, left and right). In each case, rearrangements of the cytoskeleton are involved, mediated by the intracellular PAR proteins (Nance and Zallen 2011), and induced by the ECM or short-range morphogens. Some cells in every placozoan and sponge exhibit apicobasal (A/B) polarization, that is, regulated regional differences in cell surface properties, typically reflected in distinct domains of cadherin distribution (Karner et al. 2006a). This permits them to attach simultaneously to their neighbors laterally and to acellular substrata (a syncytial middle layer in the case of the placozoan, an ECM, the *mesohyl*, in the case of sponges) basally. The DPM designated POL_a, in which A/B polarization occurs in a multicellular context (Newman and Bhat 2008), is regulated by Wnt-family morphogens, structurally novel proteins entirely exclusive to the metazoans (Loh et al. 2016). Wnt employs membrane receptors and mechanisms of cytoskeletal reorganization that predated the metazoans in the unicellular world, but its effect in the multicellular context is to enable the formation of spaces and lumens within a tissue mass. Here the cells, instead of forming a solid aggregate, will orient themselves so that their adhesive portions bind to each other while their less adhesive regions enclose an internal free space or lumen (Tsarfaty et al. 1992, 1994), a phenomenon analogous to the formation of micelles enclosing an interior fluid-filled space in nonliving liquids containing amphipathic molecules. What appear to be small, hollow cell clusters identified in Precambrian fossil beds in China, suggests that lumens were among the earliest innovations of metazoan evolution (Chen et al. 2004, Hagadorn et al. 2006).

In the embryos of present-day animals, early development typically proceeds by employing the DPMs described above: cohesion of the products of cleavage or cell division to form blastulae, morulae or inner cell masses, establishment of morphogen gradients and consolidation of cell state differences by lateral inhibition, formation of immiscible layers and lumens (Gilbert and Barresi 2016). For embryos to move past this point, for example, to gastrulate, elongate, and segment, and thus, to generate the more complex body plans of most metazoans (Figure 9.1), additional DPMs, acquired later in the phylogeny of animals, are required (Figure 9.2).

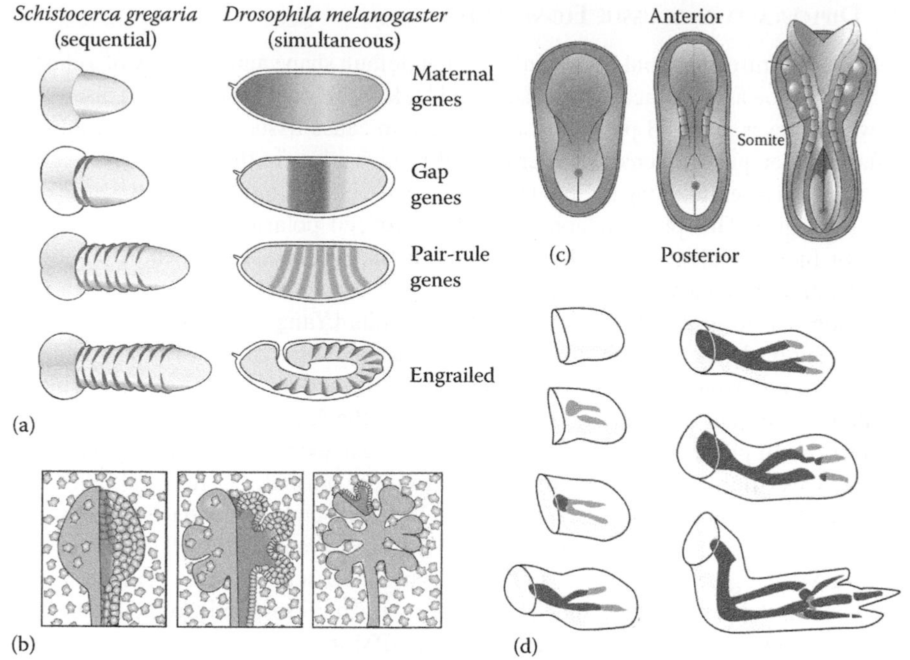

FIGURE 9.2 Segmentation and organogenesis in triploblastic metazoans driven by characterized DPMs. (a) short- and long- germ band modes of segmentation in insects. In short-germ band embryos like that of the grasshopper, temporally periodic expression of some early expressed genes (*OSC*) in conjunction with convergent extension (*POL*$_p$) leads to successively elaborated stripes of the Hox protein engrailed, which mediates boundary formation. In long-germ band embryos like that of the fruit-fly, a hierarchy of spatial patterns of early expressed gene products leads to seven-stripe patterns of the even-skipped and other "pair-rule" gene products. These induce engrailed, leading to a presegmental stage nearly identical to that of the short-germ band forms. It has been suggested that the phylogenetically "derived" long-germ band mechanism was based on an ancestral *OSC*-related Turing-type process (*TUR* DPM) involving pair-rule autoregulation and diffusion (Salazar-Ciudad et al. 2001). (c) somitogenesis in vertebrate embryos (a human embryo is depicted) involves interaction among periodically expressed Hes1 (*OSC*), gradients of the morphogens FGF and Wnt (*MOR* DPM; high point at the tail tip), and tissue elongation (*POL*$_p$). (b) three stages of salivary gland morphogenesis, dependent on various manifestations of the *ECM* DPM. An epithelial tube covered by a well-formed basal lamina is embedded in dispersed mesenchyme. Clefting and branching of the epithelium are induced by the mechanical effects of local breakdown and deposition of new basement membrane. (d) development of the chicken forelimb between days 3 and 7 of development. Early cartilage, including precartilage condensations, is shown in gray; definitive cartilage is shown in black. The cartilage primordia are replaced by bone later in development in most tetrapod species. Proximodistal (body wall to digit tip) progression of skeletal development depends on the elongation of the limb bud (utilizing the *POL*$_p$ DPM), which, in turn, depends on interaction between the apical ectodermal ridge (AER) and the underlying mesenchyme. Skeletal pattern formation depends on one or more Turing-type reaction-diffusion systems (Newman et al. 2018) (*TUR* DPM). (See text for details.) (Adapted from Forgacs, G. and Newman, S.A., *Biological Physics of the Developing Embryo*, Cambridge University Press, Cambridge, UK, 2005.)

B   DIPLOBLASTY AND TISSUE ELONGATION

The liquid nature of animal tissues implies that default shape and topology of a cluster of cells will be a topologically solid spheroid lacking an internal cavity. As discussed above, however, the A/B polarization of cells can cause tissues to diverge from this default and acquire lumens. Another type of polarization, in this case of cell shape, can cause tissues to elongate, deviating from the other liquid default (Figure 9.1, left and right). This phenomenon, termed planar cell polarity (PCP; Karner et al. 2006b), has an analogy in the physics of liquids: *liquid crystals* formed by polymers or anisotropic nanoparticles have resting droplet shapes that are elongated in one direction due to oriented packing of their subunits (Yang et al. 2005; Croll et al. 2006). Similarly, cells oriented by PCP can align and intercalate with one another, leading to a narrowing of the tissue mass in the direction of intercalation. This is termed *convergent extension* and it is intrinsic to the embryogenesis of all eumeta-zoans (i.e., all phyla but the placozoans and sponges); examples are the formation of appendages (limbs, wings) and other organs in those animals (Keller et al. 2008).

Like A/B polarity, PCP is induced by Wnt but employs a partially distinct (*noncanonical*) signaling pathway (reviewed in (Newman 2016b). The membrane protein Van Gogh/Strabismus (Vang/Stbm), which is essential to this pathway, is not specified in the placozoan or sponge genomes, which corresponds to the absence of convergent extension and the respective POL_p DPM in these groups.

The POL_p DPM is not the only morphogenetic functionality that distinguishes basal metazoans from eumetazoans. As mentioned, neither the placozoan nor (with certain exceptions) sponges have basal laminae, but all diploblasts and triploblast do. This planar ECM underlies all true epithelia and the associated DPM (termed *ECM*), in conjunction with convergent extension, enables the formation of elongated bodies and appendages (Figure 9.1). There is no inherent mechanistic connection between the convergent extension and basal lamina formation, but each animal type has both or neither, suggestive of an evolutionary bottleneck in the distant past. Formation of a basal lamina depends on the crosslinking of subunits of type IV collagen by the enzyme peroxidasin, which is exclusive to metazoans. Although *T. adhaerens* and some categories of sponges express portions of the collagen, they either lack the enzyme or their type IV collagen moieties lack the cross-linkable residues (Fidler et al. 2014).

Cells that produce basal laminae invariably undergo A/B polarization. The basal surfaces of these cells array themselves along the basal lamina forming true epithe-lia, in contrast to the protean epithelioid masses of sponges. The morphologically simplest organisms with epithelia are *diploblasts*, which contain two such layers. Their facing basal laminae are associated with an intervening gel-like matrix similar to the mesohyl of sponges, forming the *mesoglea*. Among extant forms, only cteno-phores (comb jellies) and cnidarians (e.g., hydra, jellyfish, and corals) are diploblastic as mature organisms, but embryos of most eumetazoans pass through a two-layered stage (reviewed in Newman 2016b).

Finally, the diploblasts and the more complex triploblasts (except for echino-derms: sea urchins, starfish, etc.), couple their metabolic activities (at least among the cells of some tissues) via direct passage of ions and other small molecules through structural pores (gap junctions) in their adjoining membranes. These junctions are formed by a family of multisubunit proteins, the innexins, whose function has been replaced by an unrelated family, the connexins, in chordates (Abascal and Zardoya 2013). The global coordination of physiological and morphological state afforded by this electrical coupling (which we term the *ELE* DPM) is utilized in tissue regeneration and repair in some groups, such as planarians, where the gap junction-dependent generation of electrical gradients provides templates for the restoration of lost structures (Levin 2014).

C   Triploblasty and Organogenesis

Diploblasts preceded triploblasts evolutionarily, the latter probably arising from a subgroup of the former (reviewed in Newman 2016b). One of the two germ layers outpockets, separates or disaggregates into a mesenchymal third germ layer. The evolutionary innovation of epithelial–mesenchymal transformation (EMT; Savagner 2015) occurred via the addition of molecules to the mesoglea which promoted dispersal of one of the epithelial layers and ingression of some of its cells into the middle zone. Gastrulation, the tissue movements that establish the body plan, involves epithelial folding and elongation in diploblasts (Figure 9.1 right, third row). In triploblasts, the intrusion of a middle tissue layer is also part of this process (Figure 9.1 right, fourth row).

Some proteins that promote the formation of the third, mesodermal layer are phylum-specific innovations (e.g., fibronectin and tenascin in chordates; Adams, Chiquet-Ehrismann, and Tucker 2015), whereas others have deeper evolutionary roots. The thrombospondin type 1 repeat (TSR) superfamily of proteins, for example, which variously stimulate cell adhesion, migration and breakdown of ECM has members throughout the animal phyla and in unicellular holozoans (Tucker 2004). Galectins, carbohydrate-binding proteins present in most metazoan phyla and some multicellular fungi, but in no unicellular opisthokonts (Kaltner and Gabius 2012), appear to be involved in mesoderm formation or patterning during gastrulation in a wide range of triploblastic organisms, including echinoderms and vertebrates (reviewed in Newman 2016b).

The liquid-tissue state is abrogated by EMT, since the coupling of cohesiveness of the cell mass with cell motility fostered by the cadherin-cytoskeletal coaction no longer exists in mesenchymes. Although disaggregated cells residing in ECMs may retain some motility during development, what dominates the material properties of the resulting tissues are the compositional and organizational properties of the cells' microenvironments. Exoskeletons (as in molluskan shells and arthropod integuments) and endoskeletons (as in the ossicles of sea urchins and the ribs and digits of vertebrates) are examples of solidified ECMs from within a range of rheological states including liquids (blood plasma) and gels (the center of the intervertebral disk). Physical models for each of these materials can be used to understand their behaviors, but the respective tissues generally do not exhibit the classic liquid-like behaviors (surface tension, phase separation) of epithelioid tissues. In particular, the separation of cells in mesenchymal tissues and the capacity of regional variations in ECM composition to draw them together enables the formation of *mesenchymal condensations*

(Figure 9.2, lower right quadrant), discrete clusters of interacting cells that act as initiators and primordia of epithelial appendages and endoskeletal elements (Widelitz and Chuong 1999; Hall and Miyake 2000; Yang 2009). The molecular components of these matrices of triploblasts and the physical processes they mobilize in the multicellular context are exemplars, together with the mesohyl and mesoglea of the sponges and diploblasts, of the previously mentioned *ECM* DPM.

Convergent extension promotes tissue reshaping and elongation in all eumetazoans, but in triploblasts the mechanical consequence of interpolation of a middle layer is to flatten the elongating body, resulting in a bilaterally symmetric form, at least at early developmental stages. The triploblasts are thus, coincident with the *bilaterians*.

Triploblasts have more complex body plans than diploblasts and are the only animals to have true organs. Acoelomate triploblasts such as planaria and other flatworms produce a small number of distinct organs—ovaries and testes containing, respectively, female and male germ cells, and ganglionic clusters of neurons. Organ complexity increased dramatically with the emergence of coelomic spaces in some lineages, leading to topological separation of an internal tubular primordium from the body wall. The surface (ectodermal) epithelium, with its underlying mesenchymal tissue then became the locus of appendages such as bristles, hairs, feathers, teeth and limbs, while the lining (endodermal) epithelium and its overlying mesoderm become the intrinsic (villi, crypts) and extrinsic (liver, pancreas) elaborations of the digestive tube. Interactions between the epithelium and mesenchyme in these regions drive the formation of the cardiovascular, pulmonary, urogenital and other organs, as well as various glands (Newman 2016b) (Figure 9.2, bottom quadrants).

In physical terms, the interactions between flexible epithelial sheets with mesenchymal masses of varying ECM-dependent consistency enabled a range of branched, clefted, coiled and alveolar structures not possible with either type of tissue in isolation. Interactions between different epithelial layers—particularly between epithelia and mesenchymes—was facilitated by members of the Fgf family of diffusible signals and their cognate Fgf receptors, none of which are found in the sponges (Rebscher et al. 2009; Bertrand et al. 2014). Unlike earlier emerging morphogen families (Shh, Bmp), which act on the same tissues that produce them, the Fgf pathway is subdivided into reciprocal ligand-receptor pairs so that the target of one tissue's (e.g., epithelial) Fgf is a different tissue (e.g., mesenchyme). This *asymmetric* mode of morphogen action constitutes a DPM (termed *ASM*) in its own right (Newman and Bhat 2009).

Before discussing additional DPMs that mediate morphological complexity later in embryogenesis it is important to recognize that while the modules that have been discussed so far, and their associated genes, are defining features of phylotypic identity in animal taxonomy (basal vs. eumetazoans, diploblast vs. triploblasts, Figures 9.1 and 9.2), the only requirement for their operation is their presence in the context of a critical number of cells. Although by definition, DPMs are not operative in single cells, including the fertilized egg, they can potentially organize forms and patterns in clusters of animal cells that have arisen by aggregation and not only via early embryogenesis. Organisms that develop from aggregates, however, will necessarily incur cellular competition and potential conflict (Michod and Roze 2001; Grosberg and Strathmann 2007; see also Two-Phased Multicellular Evolution, below). This has led to the proposal that the animal egg was an evolutionary innovation that was selected for by its capacity to originate genetically uniform phylotypic lineages and thus, ensure their phylogenetic stability (Newman 2011a, 2014b).

D Pattern Formation: Segmentation and Skeletogenesis

Apart from the described reshaping of cell masses into isolated structural motifs—layers, lumens, folds, appendages, condensations, and so forth—the excitability of metazoan liquid tissues can impose *patterns* on the respective morphogenetic processes, arrangements of the resulting motifs which are typically periodic or otherwise repetitive. The left-right reflective symmetry of the whole body of vertebrate and many invertebrate animals is a case in point, as are the spacing patterns of feathers, hairs and teeth, the nephrons of the kidney, and the villi of the small intestine (Forgacs and Newman 2005, Salazar-Ciudad et al. 2003). But pattern formation mechanisms can also generate regular arrangements of differentiated cells independently of morphogenesis: the alternative layering of classes of visual relay neurons in the lateral geniculate body of the mammalian brain (Horton and Hocking 1996) is one example, as is the stripe pattern of pigmented cells in a zebrafish's epidermis (Yamaguchi et al. 2007).

The patterning of morphological motifs in animal systems can occur *morphostatically*, that is, with the initial establishment of a molecular and cellular pattern (termed a "prepattern") followed by local morphogenesis at specific sites. Frequently, however, pattern formation and morphogenesis occur in a concomitant fashion, that is, *morphodynamically*. The range of phenotypic variations can differ (e.g., continuous or saltational) when mechanisms employing morphostatic or morphodynamic modes of development are altered, for example, by mutation of one or another component. These mechanisms have implications for the evolution of the respective developmental systems, as discussed by Salazar-Ciudad et al. (2003) and by Salazar-Ciudad and Marín-Riera (2013). Here we summarize two categories of pattern formation in animals for which mechanisms employing morphodynamic DPMs have been proposed. Further details and additional references for these descriptions can be found in (Newman 2016b).

Many kinds of animals, ranging across groups as diverse as arthropods, annelid worms, mollusks, and vertebrates, have segmented or partially segmented bodies. Segmentation is established during development when the primary axis of the embryo becomes subdivided into a series of tissue blocks (Figures 9.1 and 9.2). Typically, these tandemly arranged modules appear similar to one another when they first form, but later they follow distinct developmental routes, often driven by positional modulation of the expression of multiple shared genes by gradients of Hox transcription factors. Depending on the type of animal, the varied segment morphologies in the adult form can obscure the original metameric organization.

Arthropods become segmented by utilizing one of two apparently very different processes (Davis and Patel 2002) (Figure 9.2, upper left quadrant). The first is via a hierarchy of gene expression controls, well studied in the fruit fly *Drosophila mela-nogaster*, but also characteristic of other long-germ band insects, where the nuclei of the early embryo exist in a syncytium. The second, more typical mode occurs in the cellularized embryos of short-germ band insects such as grasshoppers and beetles, and in spiders, where segments bud off in a rhythmic sequential fashion from the posterior region of the embryo by a mechanism that depends on one or more molecular clocks. This mechanism has an obvious biogeneric physical basis in the operation of gene expression oscillators, periodically affecting the individuation of cell masses. (The operation of oscillatory gene expression in a multicellular context is the *OSC* DPM).

In contrast, during *Drosophila* segmentation, maternally deposited factors like bicoid cue the nonuniform expression from the embryo's nuclei of a set of gap genes (e.g., hunchback, knirps). These, in turn, induce interdigitated stripes of expression of pair-rule genes (e.g., even-skipped, hairy) which, in a concentration-dependent fashion, specify segment boundary-defining expression of genes such as engrailed. No clear generic physical process underlies this gene regulatory hierarchy, but one can be inferred to have operated at the time of evolutionary divergence of the long-germ band from the short-germ band insects (Salazar-Ciudad et al. 2001).

This is based on the following observations: when an oscillating chemical reaction operates in an extended spatial domain, if the diffusion rates of its components are sufficiently different, the chemical reaction can generate standing waves instead of temporal periodicities. (This is known as the Turing reaction–diffusion mechanism of chemical pattern formation—the *TUR* DPM—after the mathematician who first described it and proposed an embryological role for it; Turing 1952; Kondo and Miura 2010). In this interpretation, the original long-germ band mode arose when the short-germ band oscillator came to function in a syncytium thus transforming into a Turing mechanism. The gene regulatory hierarchy in *Drosophila* and other present-day long germ band insects is proposed to have been built on the resulting standing wave template by subsequent evolution (Salazar-Ciudad et al. 2001).

Vertebrate segmentation, which generates the somites (paired tissue blocks that give rise to the vertebrae and ribs, as well as muscles and other tissues which later lost their discrete character), occurs by a process that resembles the short-germ band insect and arachnid modes (Fleming et al. 2015) (Figure 9.2, upper right quadrant). Specifically, it is based on an oscillatory process that involves the periodic expression of the Hes1 gene, which is homologous to the *Drosophila* pair-rule gene hairy. The physical basis of vertebrate somitogenesis provides insight into the kinds of variation in this trait that has arisen over the course of evolution; simply tuning the ratios between embryo elongation and clock period can dramatically alter the final number of somites that develop. Zebrafish, for example, have 30 such units, mice have 65, and snakes have as many as 500 (Gomez et al. 2008).

The paired appendages of the jawed vertebrates (gnathostomes) similarly appear to employ a pattern forming mechanism of the oscillatory-Turing family of processes or at least evolved to do so by the time the stereotypical tetrapod limb emerged in this lineage (Newman and Müller 2005; Bhat et al. 2016). The fins or limbs of vertebrates are epithelial–mesenchymal extensions of the body wall shaped by liquid and elastic tissue properties. In addition, they contain various arrangements of endoskeletal nodules, plates and parallel rods that arise from mesenchymal condensations leading to cartilaginous primordia typically replaced by bone later in development (Newman and Bhat 2007). In the lobe-finned fish, which include all extant tetrapods and their direct ancestors, the appendicular skeleton typically exhibits increasing numbers of elements along the proximodistal (body wall-to-digit tip) axis (Figure 9.2, lower right quadrant).

Turing-type mechanisms have been advanced for skeletal patterning networks in different tetrapod classes. In the mouse, a gene regulatory network consisting of the two toolkit morphogens, Bmp2 and Wnt, and the cartilage master transcription factor, Sox9, exhibit reaction-diffusion dynamics to specify the periodic spacing pattern of digits (Raspopovic et al. 2014). In this case, the *TUR* DPM operates independently of cell movement and is thus *morphostatic* in the sense of Salazar-Ciudad et al. (2003). In the chicken, galectin-1a and galectin-8, two members of the galectin family of carbohydrate-binding proteins that may have also mediated the origination of some forms of triploblasty (see above), form a multiscale skeletal pattern formation network that is strictly dependent on cell movement and thus, constitutes an explicitly *morphodynamic TUR* DPM (Bhat et al. 2011, Glimm et al. 2014). Comparative phylogenomics has suggested that the evolution of galectin protein conformation and gene regulation modules around the time of the fin-to-limb transition fine-tuned the network so as to render it capable of producing characteristic tetrapod limb skeletal arrangements (Bhat et al. 2016). Changing combinations of these two self-organizing mechanisms appear to have been employed in various fin- and limb-bearing vertebrates over the course of evolution (Newman et al. 2018).

9.3 "SOLID TISSUES" AND "CELLULAR SOLIDS": PLANTS AND FUNGI

Unlike the morphogenetically active tissues of animals, which exist in a liquid state, the tissues of multicellular plants and fungi are solid, or more properly speaking *cellular solids*, that is, composites consisting of a solid phase (cell walls) and a fluid phase (cytoplasm). Changes in physiological conditions, such as an increase or decrease in turgor, provoke reciprocal changes in the mechanical properties of cell walls by either placing walls in increased or decreased tension. Thus, contrary to the inertness invoked by their description as solids, living plant tissues are highly dynamic (more so, in some ways, than animal tissues) in that they can be modified in response to internal and external physiological signals and mechanical forces. This feature is particularly true for living tissues such as parenchyma and collenchyma, which have thin cell walls that can respond dramatically to changes in turgor. In addition, plant tissues undergo morphogenesis by reshaping effects that are formally consistent with the solid or cellular solid state—local gain and loss of mass, changes in stiffness or viscoelasticity—but accomplished by means unknown in nonliving solids, that is, cell expansion, proliferation, cell death, cell wall lysis, and reconstitution.

A PLANT AND FUNGAL PATTERNING MODULES

The previous sections showed that the developmental mobilization of very dissimilar molecular systems or processes could produce much the same phenotypic effects. This dictum has been formalized for multicellular animals, using DPMs, each of which involves one or more sets of shared gene networks, their products, and physical processes common to all living things. In theory, these DPMs or their analogs can operate in plants and fungi, as well as animals because of fundamental similarities among all eukaryotic cells. Consider for example cell-to-cell adhesives.

All eukaryotic cells have the capacity to secrete polysaccharides and structural glycoproteins that self-assemble to form extracellular matrices around animal and plant cells. Both types of matrices contain interpenetrating polymeric networks that employ hydroxyproline-rich glycoproteins (HRGPs) as major scaffolding components (collagen in animals and the HRGP extensin superfamily in various algae and in the embryophytes). These proteins generally form elongated, flexible, rod-like molecules with marked peptide periodicity (much like the modularity seen in mussel adhesives) with repeat motifs dominated by hydroxyproline in a polyproline II helical formation extensively modified by arabinosyl/galactosyl side chains. It is possible therefore that this superfamily of cell-to-cell adhesives evolved by the co-option of an ancestral gamete–gamete self-recognition or cell-adhesion-to-substratum toolkit. Likewise, the evolutionary expansion of preexisting gene families encoding regulatory proteins in combination with novel physical and regulatory interactions resulting from such expansions may also have played critical roles and may even have driven the evolution of multicellular complexity, as illustrated by the basic helix-loop-helix (bHLH) protein family involved in diverse cellular developmental processes in plants and animals (reviewed by Feller et al. 2011) and a wide array of microtubule-associated proteins in algae, embryophytes, fungi, and metazoans (Gardiner 2013).

Nevertheless, the DPMs identified in animal systems (Newman and Bhat 2008, 2009) cannot be applied directly to plant or fungal development because of substantive differences among these three major eukaryotic clades. For example, during animal development, cells are typically free to migrate and slide past one another in ways that permit differential adhesion, cortical tension, and other processes that can facilitate cell sorting and tissue self-assembly (Maître et al. 2012). In contrast, plant cells have rigid cell walls that are typically firmly fixed to one another. Plant signaling molecules also can act intercellularly, as well as intracellularly as transcriptional modulators and determinants of tissue, as well as cell fate, thereby blurring the functional separation of gene regulatory networks affecting multi- as opposed to single-cell differentiation. Although the intercellular transport of developmental transcription factors is known in animal systems, it is rare (Prochiantz and Joliot 2003). Further, while cell polarity in plants and animals both involve rearrangements of the cytoskeleton, in plants the key molecular components are PIN and PAN1 proteins (Dhonukshe 2009; Zhang et al. 2012), mediating, respectively, auxin transport, and asymmetric cell division, whereas in animals PAR proteins organize the nonuniform placement of cell surface cadherins and integrins. Finally, cell division mechanics and the deposition of cell walls differ even among desmids and in different filamentous ascomycetes (Hall et al. 2008; Seiler and Justa-Schuch 2010).

In light of these and other issues, Hernández-Hernández et al. (2012) proposed a preliminary set of six DPMs associated particularly with critical embryophyte developmental processes:

1. The formation and orientation of a future cell wall (FCW)
2. The production of cell-to-cell adhesives (ADH)
3. The formation of intercellular lines of communication and spatial-dependent patterns of differentiation (DIFF)
4. The establishment of axial and lateral polarity (POL)
5. The creation of lateral protrusions or buds (BUD)
6. The construction of appendicular leaf-like structures (LLS)

For the purposes of this chapter, only the first four of these modules (i.e., FCW, ADH, DIFF, and POL) are relevant because cell-to-cell adhesion and intercellular communication are the *condicio sine qua non* of simple multicellularity across all eukaryotic clades and because these modules operate in a pairwise manner in many multicellular algae and fungi, as well as in the land plants (Figure 9.3).

For example, among embryophytes, the ADH and FCW modules operate in tandem because the presence of adhesive pectin polysaccharides in the middle lamella is associated with the deposition of the future primary cell walls of adjoining cells. The cell wall begins to be formed from cell plates during cytokinesis, such that cell adhesion is the default state (Figure 9.3). Additionally, the proportion and chemical

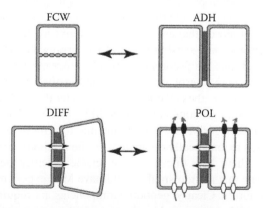

FIGURE 9.3 Paired dynamic patterning modules (indicated by arrows) participate in the evolution of multicellularity in plants and fungi. The acquisition of each of these modules is required for the evolution of multicellularity. These modules operate in pairs for organisms with cell walls because cell-to-cell adhesion is related to the location of a new cell wall and because intercellular communication operates in tandem with cell polarity. *Abbreviations*: ADH, the capacity for cell-to-cell adhesion. DIFF, the establishment of intercellular communication and cellular differentiation; FCW, the future cell wall module (establishes the location and orientation of the new cell wall); POL, the capacity for polar (preferential) intercellular transport. (Adapted from Hernández-Hernández, V. et al., *Int. J. Dev. Biol.*, 56, 661–674, 2012.)

state (e.g., level of esterification) of each of the cell wall components is spatiotemporally regulated over the course of development, locally as well as globally, adjusting the mechanical properties of cells and tissues and contributing to the regulation of cell and organ growth in size, as well as to organogenesis.

A somewhat analogous system operates during the extension of fungal hyphae. The DIFF and POL modules are also functionally interconnected because both are required for cell-type specification and intercellular communication. Thus, among embryophytes, DIFF and POL involve the transport of metabolites, transcription factors, and phytohormones through plasmodesmata. In some developmental systems, plasmodesmata also enable a type of generic physicochemical reaction-diffusion patterning mechanism (Benítez et al. 2011) that includes a lateral inhibitory component. Experimental evidence in *Arabidopsis thaliana* and other model plant systems likewise, shows that auxin flow and cell wall mechanical forces reciprocally interact during the emergence of polarity, whereas auxin promotes polar expansion by localized cell wall loosening, involving the acidification of the apoplast and the concomitant disruption of noncovalent bonds among cell wall polysaccharides. The preferential localization of PINs (or their transporting vesicles) that determines auxin fluxes also targets loci for future cell wall loosening.

As with the animal DPM framework, each of the FCW, ADH, DIFF, and POL modules involves the participation of generic physical mechanisms such as mechanical forces. Consider for example how the FCW module operates in embryophytes in response to mechanical stresses. Centrifugation experiments of both haploid and diploid land plant cells show that the position of the interphase nucleus (which prefigures the preprophase band and the phragmoplast) establishes the location of the future division plane. On the basis of these and other observations, Besson and Dumais (2011) proposed that embryophyte cell division involves a microtubule (MT)-length-dependent force-sensing system that permits the cytoskeleton to position the nucleus (and thus the preprophase band) into a mechanically equilibrated location. If the nucleus in interphase is positioned artificially off-center, the MTs radiating from it, outward to the cell cortex, will recenter the nucleus based on differences in the tensile forces generated among the MTs differing in length. Collectively shorter as opposed to longer MTs would be favored to achieve an equilibrium configuration that would axiomatically coincide with the minimal area plane. Cells that are too large would have MTs that would be unable to tether the nucleus to some cell wall facets; cells that are too small would have MTs experiencing compressive rather than tensile forces. Clearly, genomic components are required for the operation of the FCW module as revealed by the persistent participation of subfamily III leucine-rich repeat-receptor-like kinases in symmetric and asymmetric cell division (Zhang et al. 2012).

Thus, organisms may rely on physical forces to establish a simple default developmental condition, but they must modify their responses to these forces to achieve alternative developmental options. This is illustrated by how cell wall stresses induce the synthesis of different chitin synthase enzymes to rescue alternative septation and cytokinetic patterns in mutated yeast cells (Walker et al. 2013), or how the formation of the structures prefiguring the appearance of villi in the gut of the chick embryo relies on compressive mechanical forces generated by the differentiation of nearby smooth muscle tissue that causes the buckling of endoderm and mesenchyme (Shyer et al. 2013).

B Mapping Modules into Morphospaces

The roles played by the FCW, ADH, DIFF, and POL modules during the evolution of multicellularity are shown when their functionalities are mapped onto a morphospace identifying the major plant body plans and when this map is informed with a series of morphological transformations predicted by a simple multilevel selection model for the evolutionary appearance of multicellularity. In general terms, a morphospace is a representation of all the theoretically possible phenotypes within a specific group of organisms (McGhee 1999). Each axis defining the domains within a morphospace represents a developmental variable or process that describes or obtains a phenotypic character (with one or more character states). Each intersection of two or more axes identifies a hypothetical phenotype with the character states specified by the variables or processes stipulated by the participating axes. A morphospace for plant body plans was constructed previously using four developmental axes, each with two character states (Niklas 2000): (1) whether cytokinesis and karyokinesis are synchronous, (2) whether cells remain aggregated after they divide, (3) whether symplastic continuity or some other form of intercellular communication is maintained among neighboring cells, and (4) whether individual cells continue to grow indefinitely in size (Figure 9.4). The intersections of these axes identify four major body plans, each of which can be theoretically either uninucleate or multinucleate: (1) the unicellular body plan, (2) the siphonous/coenocytic body plan, (3) the colonial body plan, and (4) the multicellular body plan. The addition of a fifth axis—the orientation(s) of cell division—distinguishes among the various tissue constructions of the multicellular plant body plan: (1) the unbranched filament, which results when cell division is confined to one plane of reference, (2) the branched filament (with or without a pseudoparenchymatous tissue construction), which requires two planes of cell division, and (3) the parenchymatous tissue construction, which requires three planes of cell division.

A review of the secondary and primary literature treating the algae shows that all but two of the 14 theoretically possible phenotypes are represented by one or more species. It also reveals considerable homoplasy among various plant lineages. For example, the unicellular multinucleate variant with determinate growth is represented by the chlorophycean alga *Bracteacoccus* and the ulvophycean alga *Chlorochytridium*, the colonial multinucleate body plan by the chlorophycean algae *Pediastrum* and *Hydrodictyon*, the siphonous body plan by the ulvophycean alga *Caulerpa*, the xanthophycean alga *Vaucheria*, the multicellular multinucleate (siphonocladous) branched variant by the rhodophycean alga *Griffithsia*, and the ulvophycean alga *Cladophora*. Among the multinucleate multicellular (siphonocladous) variants differing in tissue construction, the unbranched and branched filamentous variants are represented by the ulvophycean algae *Urospora*, and *Acrosiphonia*, respectively; the siphonocladous body plan with a pseudoparenchymatous tissue construction is represented by species within the ulvophycean genus *Codium*.

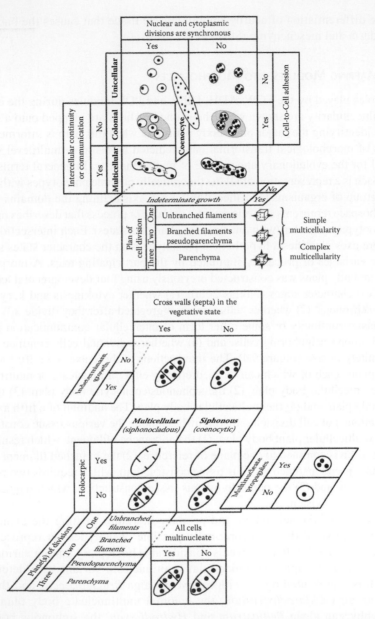

FIGURE 9.4 A morphospace for the four major plant body plans shown in bold (unicellular, siphonous/coenocytic, colonial, and multicellular) resulting from the intersection of five developmental processes: (1) whether cytokinesis and karyokinesis are synchronous, (2) whether cells remain aggregated after they divide, (3) whether symplastic continuity or some other form of intercellular communication is maintained among neighboring cells, and (4) whether individual cells continue to grow indefinitely in size. Note that the siphonous/coenocytic body plan may evolve from a unicellular or a multicellular progenitor. The lower panels dealing with the plane of cell division, localization of cellular division, and symmetry pertain to the evolution of complex multicellular organisms. (Adapted from Niklas, K.J., *Ann. Bot.*, 85, 411–438, 2000, with permission).

The two variants that are not represented by any known species are the uninucleate indeterminate (siphonous) body plan and the parenchymatous siphonocladous variants. The absence of the former may be the result of physiological constraints imposed by the volume of cytoplasm that a single nucleus can sustain, a hypothesis proposed by Julius Sachs (1892) and recently revisited in the context of the midblastula transition in animal ontogeny (Collart et al. 2013). A convincing explanation for the absence of a parenchymatous siphonocladous body plan remains problematic. Nevertheless, the evolutionary significance of the ADH, FCW, DIFF, and POL modules in light of the plant body plan morphospace is obvious: ADH is required for the construction of the colonial and multicellular body plans; FCW participates in the synchronicity of cyto- and karyokinesis and participates in the tissue construction of a multicellular plant body; and DIFF and POL are required for intercellular physiological coordination and cellular specialization.

The ADH, FCW, DIFF, and POL modules also establish character polarities in the context of a multilevel selection theory for the evolution of multicellularity (Folse and Roughgarden 2010; Niklas and Newman 2013); see TWO-PHASED MULTICELLULAR EVOLUTION. This theory identifies the unicellular body plan as ancestral to the colonial body plan that in turn is ancestral to a truly multicellular body plan, that is, it identifies a unicellular colonial multicellular body plan transformation series that requires ADH to establish and maintain a colonial body plan and FCW, DIFF, and POL to coordinate and specify intercellular activities to achieve an integrate multicellular phenotype whose complexity exceeds simple dyatic interactions among individual cells.

The volvocine green algae provide particularly valuable insights into aspects of the unicellular->colonial->multicellular transition series (Bonner 2000). The ancestral state in the volvocines is inferred to be a unicellular organism probably similar to extant species of *Chlamydomonas*. Transformation of the unicellular cell wall into ECM (seen in the Tetrabaenaceae->Goniaceae->Volvocaceae transformation series), incomplete cytokinesis (seen in the Goniaceae->Volvocaceae transformation series), and the appearance of additional derived traits produce forms, ranging from simple cellular aggregates (e.g., *Tetrabaena socialis*) to colonies with complex, asymmetric cell division, to quasi-multicellular organisms with full germ-soma division of labor (e.g., *Volvox carteri*) (Kirk 2005). In the case of the latter, cytoplasmic bridges with multiple functionalities are maintained among neighboring cells, that is, they participate in the mechanics of kinesin-driven inversion, and they serve as conduits to provide nutrients to developing gonidia. In adult plants, these bridges are extensive in number and broader than plasmodesmata (~200 nm in diameter). These bridges are developmentally severed in some volvocine taxa, which provide interesting examples of a multicellular to colonial transformation series.

Evidence for a transition from colonial to multicellular life-forms in animals and in different algal lineages is reviewed by Niklas and Newman (2013).

C TWO-PHASED MULTICELLULAR EVOLUTION

But how does a colonial aggregate of cells achieve individuality? Multilevel selection theory identifies two evolutionary stages—an alignment-of-fitness phase (denoted as MLS1) in which genetic similarity among adjoining cells prevents cell–cell conflict, and an export-of-fitness phase (denoted as MLS2) in which cells become interdependent and collaborate in a sustained physiological and reproductive effort (for a general review, see Folse and Roughgarden 2010). Phyletic analyses of lineages in which obligate multicellularity has evolved are consistent with this MLS1 and MLS2 model. They also show that lineages characterized by species with clonal group formation are more likely to have undergone an evolutionary transition to obligate multicellularity than lineages characterized by species with nonclonal group formation. MLS1 is typically achieved by a unicellular bottleneck, which occurs in every organism's life cycle, for example, a spore, zygote, or uninucleate asexual propagule (Niklas and Newman 2013). This bottleneck establishes genetic homogeneity among subsequently formed cells (or, more precisely, among nuclei) even among asco- and basidiomycete heterokaryotic fungi for which experimental data indicate competition among genetically different nuclei sharing the same cytoplasm. For example, nuclear ratios of heterokaryons in the ascomycetes *Penicillium cyclopodium* and *Neurospora crassa* are reported to change depending on environmental conditions in ways that reflect the underlying fitness of the constituent homokaryons grown in isolation (Jinks 1952; Davis 1960).

James et al. (2008) report similar results for *Heterobasidion parviporum* and conclude that this basidiomycete violates the standard model of what constitutes an individual since genetically different nuclei compete among themselves to form homokaryotic hyphae. However, it must be recognized that an absence of conflict does not mean an absence of competition—and competition can be a good thing. Indeed, there is evidence to suggest that competitive–cooperative interactions have shaped form–function relationships even at the simple molecular level (Foster 2011). Consider that many developmental processes employ lateral inhibition (see discussion of the *LAT* DPM in MULTILAYERING AND LUMEN FORMATION, above) in which neighboring genomes compete to adopt the same cell fate, for example, during gonad development of *Caenorhabditis elegans*, cells compete to develop into either a terminally differentiated cell or a ventral uterine precursor cell, which is determined by the relative amounts of the Lin-12 receptor and its Lag-2 ligand (Greenwald 1998).

It is equally important to recognize that mitosis does not invariably result in genetically identical derivative cells even in the absence of mutation or chromosomal aberrations. Preferential sister chromatid segregation is observed in plants, fungi, and animals. Further, during the early development of female mammals, one of the two X chromosomes is randomly silenced and faithfully perpetuated during subsequence cell proliferation (Chow et al. 2005). Methylation patterns of cytosine in CpG doublets and other epigenetic changes provide additional avenues for establishing genetically different groups of cells in the same organism, each of which required the evolution of stable interdependent cell lineages sharing the same genome but expressing different gene network patterns. Indeed, epigenetic mechanisms may be critical to *maintaining* multicellularity. Consider that the principal limitation to achieving and maintaining cooperation is the appearance of defectors in an evolutionary game setting (i.e., participants that consume resources but fail to confer any benefit to other players). An obvious example of cellular defectors is animal neoplasms, which may have deep genetic roots in terms of the regulation of cell proliferation and de-differentiation (Davies and Lineweaver 2011).

Numerous mechanisms to maintain cooperation and reduce or eliminate defectors have been suggested, among which the effects of group selection, direct and indirect reciprocity, network structure, and tag-based donation schemes are perhaps best known. However, all these mechanisms require players to remember past proceedings or to possess some method of recognizing one another like players in the same game. Epigenetic mechanisms, as well as signaling pathways that connect metabolic status with nutrient availability or other environmental factors (e.g., the TOR signaling pathway; (Huang and Fingar 2014; Rexin et al. 2015) provide one solution to dealing with defectors, while the unicellular bottleneck provides, at least initially, a homogeneous collection of cooperating cells. There are other tactics as well. Game-theoretical models show that resource limitations can cause the rules of a game to change in ways that foster cooperation among players with no memory and no recognition of one another (Requejo and Camacho 2013). Likewise, zero-determinant models show that altruistic and generous strategies can sustain cooperation and reduce negative interactions (Hilbe et al. 2014).

It is worth noting further that cheater mutants in the social amoeba *Dictyostelium discoideum* and the mouse *Mus musculus* are reported to cooperate in ways that conform to normal developmental patterns and that do not disrupt the functionality of the collective organism (Santorelli et al. 2008; Dejosez et al. 2013), which indicates cooperation may be an emergent property of ancient and robust gene networks (see also Nanjundiah and Sathe; 2011). This is consistent with theoretical considerations indicating that "cheaters" can evolve to function as asexual propagules in very ancient proto-life cycles (Komarova 2014). It is also the expectation of the DPM framework for the origination of morphological motifs and body plans (Newman and Bhat 2008, 2009), since the mesoscale effects mobilized by the interaction toolkit genes and their products become active only when a critical number of cells are present. This makes cooperation (aka self-organization) physically inescapable even when the cells (apart from having a common complement of toolkit gene) are genetically heterogeneous. As discussed in Triploblasty and Organogenesis, above, the innovation of the egg-stage of animal development may have been a strategy to promote phyletic stability by suppressing such heterogeneity (Newman 2011a, 2014b).

In summary, cooperation among cells and nuclei can evolve along a number of routes and with different consequences, which may explain why colonial life-forms, multinucleate cells, and multicellularity are not uncommon.

Nevertheless, MLS2 requires that selection shift from the level of individual cells to the level of an emergent entity that reproduces a functionally integrated phenotype with a heritable fitness (typically followed by some degree of cellular specialization). The key difference between MLS1 and MLS2 is that the fitness of the cell-group (aka "the colony") is an additive function of the fitness of individual cells, whereas the fitness of a multicellular organism is nonadditive. Put differently, the evolution of a multicellular organism requires a means to guarantee the heritability of fitness at the emergent level of the multicellular entity. In some, but not all multicellular organisms, this guarantee is accomplished by sequestering a germline, for example, animals, and embryophytes, respectively. It is noteworthy that, with very few exceptions (e.g., *Volvox*), the separation of a germline from the soma does not occur in the land plants nor in any algal lineage, that is, somatic embryogenesis is the norm.

A germ-soma separation may be an indirect consequence of the necessity to compensate for the increasing costs of evolving a progressively larger body size. Body size matters because the probability of compounding a genetic error or mutation increases as a function of the number of cell divisions required to achieve the size of a mature organism. Small multicellular organisms have a lower probability of introducing errors into their reproductive cells, whereas progressively larger organisms escape Muller's ratchet (the inevitable accumulation of deleterious mutations) by ultimately sequestering cells in a germline. It is also worth noting that cellular specialization may evolve more readily in larger organisms than in smaller because unsuccessful attempts at specialization are more easily tolerated in larger organisms.

However, obligate sexual reproduction is not required to override the conflict between a multicellular individual and its constituent cells (Buss 1987). As noted, in the absence of somatic mutations, the presence of a spore or similar reproductive unit assures a unicellular bottleneck regardless of the type of life cycle. Although it can be difficult for asexual organisms to escape the consequences of Muller's ratchet (the irreversible buildup of deleterious mutations) and its consequences on fitness, even an asexual organism experiences an alignment-of-fitness by means of unicellular propagules, which can purge deleterious genomic changes as a consequence of the death of individual propagules.

Likewise, multicellularity is not required for cellular specialization. Unicellular bacteria, algae, yeast, and amoeba exhibit alternative stable states of gene expression patterns and manifest alternative cell morphologies during their life cycles, often as a result of competing processes, for example, motility versus mitosis. This feature is particularly intriguing in light of the studies showing that seemingly random fluctuations in cellular dynamics may provide a simple switch for changing cell fate. For example, *Bacillus subtilis* can exist in two stable forms, called vegetative and competent, under conditions of nutrient deficiency. A simple mathematical model, using a stochastic algorithm can predict how and when these two cellular conditions are decided based on the level of biological noise in the system (Maamar et al. 2007). In addition, mathematical models indicate that cellular differentiation can emerge among genetically identical cells in response to competing for physiological processes (Kaneko and Yomo 1999), or simply because of the metabolic costs of switching the tasks a cell must perform to stay alive or complete its life cycle. In more derived lineages, an alignment-of-fitness can compensate for conflicts of interest among cellular components such that a division of cellular labor becomes possible and even necessary. Even a loose colony of cells can have emergent biological properties that give it a collective edge in which every cell benefits (Solé and Valverde 2013). The origin of the cellular differentiation (DIFF) module, therefore, may reside in the inherent multistability of complex gene regulatory networks with somatic or reproductive functional roles for different cell-types possibly established by natural selection.

Finally, it is important to recognize that once a functionally adequate body plan or organ form arises in a multicellular lineage, it does not stop evolving, even if its architecture remains stable (phyletic stasis). Integration and reconfiguration (rewiring) of physiology and development can lead to extensive changes in the ways a given form is realized (developmental system drift; True and Haag 2001), leading to differences in genotype–phenotype relationships between divergent species and even between individuals of the same species (Narasimhan et al. 2016). Whether occurring by random effects or through stabilizing selection, the physical origins of developmental processes will become obscured by such complexifying transformations, making the analysis of ontogeny intractable without a concomitant consideration of phylogeny.

9.4 CONCLUDING REMARKS

The morphological theme of multicellularity and the confluence of generic and genetic processes by which it was achieved in different clades continue to draw attention to classical but largely unanswered questions in microbiology, botany, zoology, and mycology. Among these is the relationship between the organism and the cell, and the extent to which an organism's external form (morphology) and internal structure (anatomy) are necessarily interrelated. An evolutionary-developmental perspective in tandem with the growth of molecular biological techniques has informed the pursuit of these and other questions, but we are still remarkably ignorant about many fundamental processes. A contributing factor to this ignorance is the assumption that patterning processes and the mechanisms accounting for them are the same in different organisms. The fact that fungal mitotic divisions are intra-nuclear, whereas microtubules invade the nuclear space after the dissolution of the nuclear envelope to form the division spindle in most plant and animals cells is sufficient to caution against canonical discussions about cell division. Another factor is that molecular sequence homology does not necessarily translate into morphogenetic or organographic homology. The various cases presented in this chapter are sufficient to show that this is not invariably true and that detailed analyses are required to determine whether two structures or processes are truly developmentally homologous. A third factor is a paucity of phylogenetically disparate model organisms to answer questions that span vastly different life-forms.

A related problem is that many of the model developmental systems that are currently available are species drawn from late-divergent lineages that manifest many derived character states, which need to be juxtaposed with data drawn from species from early-divergent persistent lineages to assess ancestral character states. A recent example is how the absence of SMG1 in *Arabidopsis thaliana* led some workers to conclude that this core kinase in the nonsense mRNA decay pathway was generically absent in earlier-divergent plant groups (see, however, Lloyd and Davies 2013).

The study of the evolution of multicellularity draws these and other limitations into sharp relief. By so doing, it also provides a venue in which to resolve them by synthesizing information from fields of study as diverse as paleontology and proteomics, and as in our focus in this chapter, the physics of materials. The DPM framework, by superseding the either/or of developmental genetics and biomechanics permits a testable approach to long-standing problems like analogy and homology, structure and function, and the relationship between the inheritance of genes and the inheritance of pattern and form. Moreover, with increasing knowledge of the molecules and mobilized physical effects in animal, plant, and fungal systems, it can account for why these multicellular organisms exhibit the morphological and pattern motifs they do, and why they have exhibited their characteristic intragroup similarities and extra-group differences since their divergence.

These and other fundamental questions puzzled Wilhelm Hofmeister (1824–1877), Léo Errera (1858–1905), William Bateson (1961–1926), and other developmental biologists with an evolutionary perspective on their subject (De Beer, 1938), and they continue to do so today. Their resolution can only be achieved by a concerted interdisciplinary approach in which research questions are carefully but phylogenetically broadly framed. The literature reviewed here shows that this can be (and is being) done, but that much more is needed to marry the *Weltanschauungen* of evolutionary and developmental biology.

REFERENCES

Abascal, F. and Zardoya, R. 2013. Evolutionary analyses of gap junction protein families. *Biochim Biophys Acta* 1828(1): 4–14. doi:10.1016/j.bbamem.2012.02.007.

Abedin, M. and King, N. 2008. The premetazoan ancestry of cadherins. *Science* 319(5865): 946–948.

Adams, J. C., Chiquet-Ehrismann, R., and Tucker, R. P. 2015. The evolution of tenascins and fibronectin. *Cell Adh Migr* 9(1–2): 22–33. doi:10.4161/19336918.2014.970030.

Amack, J. D. and Manning, M. L. 2012. Knowing the boundaries: Extending the differential adhesion hypothesis in embryonic cell sorting. *Science* 338(6104): 212–215. doi:10.1126/science.1223953.

Benítez, M., Monk, N. A., and Alvarez-Buylla, E. R. 2011. Epidermal patterning in *Arabidopsis*: Models make a difference. *J Exp Zool B Mol Dev Evol* 316(4): 241–53. doi:10.1002/jez.b.21398.

Bertrand, S., Iwema, T., and Escriva, H. 2014. FGF signaling emerged concomitantly with the origin of Eumetazoans. *Mol Biol Evol* 31(2): 310–318. doi:10.1093/molbev/mst222.

Besson, S. and Dumais, J. 2011. Universal rule for the symmetric division of plant cells. *Proc Natl Acad Sci U S A* 108(15): 6294–6299. doi:10.1073/pnas.1011866108.

Bhat, R., Chakraborty, M., Glimm, T., Stewart, T. A., and Newman, S. A. 2016. Deep phylogenomics of a tandem-repeat galectin regulating appendicular skeletal pattern formation. *BMC Evol Biol* 16(1):1 62. doi:10.1186/s12862-016-0729-6.

Bhat, R., Lerea, K. M., Peng, H., Kaltner, H., Gabius, H. J., and Newman, S. A. 2011. A regulatory network of two galectins mediates the earliest steps of avian limb skeletal morphogenesis. *BMC Dev Biol* 11: 6. doi:10.1186/1471-213X-11-6.

Bonner, J. T. 2000. *First Signals: the Evolution of Multicellular Development*. Princeton, NJ: Princeton University Press.

Borghi, N., Lowndes, M., Maruthamuthu, M., Gardel, M. L., and Nelson, W. J. 2010. Regulation of cell motile behavior by crosstalk between cadherin- and integrin-mediated adhesions. *Proc Natl Acad Sci U S A* 107(30): 13324–13329. doi:10.1073/pnas.1002662107.

Brunkard, J. O. and Zambryski, P. C. 2016. Plasmodesmata enable multicellularity: New insights into their evolution, biogenesis, and functions in development and immunity. *Curr Opin Plant Biol* 35: 76–83. doi:10.1016/j.pbi.2016.11.007.

Budd, G. E. and Jensen, S. 2015. The origin of the animals and a 'Savannah' hypothesis for early bilaterian evolution. *Biol Rev Camb Philos Soc.* doi:10.1111/brv.12239.

Buss, L. W. 1987. *The Evolution of Individuality*. Princeton, NJ: Princeton University Press.

Chen, J. Y., Bottjer, D. J., Oliveri, P., Dornbos, S. Q., Gao, F., Ruffins, S., Chi, H., Li, C. W., and Davidson, E. H. 2004. Small bilaterian fossils from 40 to 55 million years before the cambrian. *Science* 305(5681): 218–222.

Chow, J. C., Yen, Z., Ziesche, S. M., and Brown, C. J. 2005. Silencing of the mammalian X chromosome. *Annu Rev Genomics Hum Genet* 6: 69–92. doi:10.1146/annurev. genom.6.080604.162350.

Cole, G. T. 1996. Basic biology of fungi. In *Medical Microbiology* (Ed.) Baron, S. University of Texas Medical Branch at Galveston, Galveston, TX.

Collart, C., Allen, G. E., Bradshaw, C. R., Smith, J. C., and Zegerman, P. 2013. Titration of four replication factors is essential for the Xenopus laevis midblastula transition. *Science* 341(6148): 893–896. doi:10.1126/science.1241530.

Croll, A. B., Massa, M. V., Matsen, M. W., and Dalnoki-Veress, K. 2006. Droplet shape of an anisotropic liquid. *Phys Rev Lett* 97: 204502.

Davies, P. C. and Lineweaver, C. H. 2011. Cancer tumors as Metazoa 1.0: Tapping genes of ancient ancestors. *Phys Biol* 8(1): 015001. doi:10.1088/1478-3975/8/1/015001.

Davis, G. K. and Patel, N. H. 2002. Short, long, and beyond: Molecular and embryological approaches to insect segmentation. *Annu Rev Entomol* 47: 669–699. doi:10.1146/ annurev.ento.47.091201.145251.

Davis, R. H. 1960. Adaptation in pantothenate-requiring Neurospora. II. Nuclear competition during adaptation. *Am J Bot* 47: 648–654.

Dayel, M. J., Alegado, R. A., Fairclough, S. R., Levin, T. C., Nichols, S. A., McDonald, K., and King, N. 2011. Cell differentiation and morphogenesis in the colony-forming choanoflagellate Salpingoeca rosetta. *Dev Biol* 357(1): 73–82. doi:10.1016/j.ydbio.2011.06.003.

De Beer, G. R. 1938. *Evolution; Essays on Aspects of Evolutionary Biology, Presented to Professor E. S. Goodrich on His Seventieth Birthday*. Oxford, London: Clarendon Press.

Degnan, B. M., Adamska, M., Richards, G. S., Larroux, C., Leininger, S., Bergrum, B., Calcino, A., Taylor, K., Nakanishi, N., and Degnan, S. M. 2015. In *Evolutionary Developmental Biology of Invertebrates* (Ed) Wanninger, A., pp. 65–106. Vienna, Austria: Springer-Verlag.

Dejosez, M., Ura, H., Brandt, V. L., and Zwaka, T. P. 2013. Safeguards for cell cooperation in mouse embryogenesis shown by genome-wide cheater screen. *Science* 341(6153): 1511–1514. doi:10.1126/science.1241628.

Dhonukshe, P. 2009. Cell polarity in plants: Linking PIN polarity generation mechanisms to morphogenic auxin gradients. *Commun Integr Biol* 2(2): 184–190.

Feller, A., Machemer, K., Braun, E. L., and Grotewold, E. 2011. Evolutionary and comparative analysis of MYB and bHLH plant transcription factors. *Plant J* 66(1): 94–116. doi:10.1111/j.1365-313X.2010.04459.x.

Fidler, A. L., Vanacore, R. M., Chetyrkin, S. V., Pedchenko, V. K., Bhave, G., Yin, V. P., Stothers, C. L. et al. 2014. A unique covalent bond in basement membrane is a primordial innovation for tissue evolution. *Proc Natl Acad Sci U S A* 111(1): 331–336. doi:10.1073/pnas.1318499111.

Fleming, A., Kishida, M. G., Kimmel, C. B., and Keynes, R. J. 2015. Building the backbone: The development and evolution of vertebral patterning. *Development* 142(10): 1733–1744. doi:10.1242/dev.118950.

Folse, H. J. and Roughgarden, J. 2010. What is an individual organism? A multilevel selection perspective. *Q Rev Biol* 85(4): 447–472.

Forgacs, G. and Newman, S. A. 2005. *Biological Physics of the Developing Embryo*. Cambridge, UK: Cambridge University Press.

Foster, K. R. 2011. The sociobiology of molecular systems. *Nat Rev Genet* 12(3): 193–203. doi:10.1038/nrg2903.

Foty, R. A., Forgacs, G., Pfleger, C. M., and Steinberg, M. S. 1994. Liquid properties of embryonic tissues: Measurement of interfacial tensions. *Phys Rev Lett* 72: 2298–2301.

Gardiner, J. 2013. The evolution and diversification of plant microtubule-associated proteins. *Plant J* 75(2): 219–229. doi:10.1111/tpj.12189.

Gilbert, S. F. and Barresi, M. J. F. 2016. *Developmental Biology*. 11th edition. Sunderland, MA: Sinauer Associates.

Glimm, T., Bhat, R., and Newman, S. A. 2014. Modeling the morphodynamic galectin patterning network of the developing avian limb skeleton. *J Theor Biol* 346: 86–108. doi:10.1016/j.jtbi.2013.12.004.

Gomez, C., Ozbudak, E. M., Wunderlich, J., Baumann, D., Lewis, J., and Pourquié, O. 2008. Control of segment number in vertebrate embryos. *Nature* 454 (7202): 335–339.

Greenwald, I. 1998. Notch signaling: Lessons from worms and flies. *Genes & Development* 12: 1751–1762.

Grosberg, R. K. and Strathmann, R. 2007. The evolution of multicellularity: A minor major transition? *Annu Rev Ecol Evol Syst* 38: 621–654.

Hagadorn, J. W., Xiao, S., Donoghue, P. C., Bengtson, S., Gostling, N. J., Pawlowska, M., Raff, E. C. et al. 2006. Cellular and subcellular structure of neoproterozoic animal embryos. *Science* 314(5797): 291–294.

Halbleib, J. M. and Nelson, W. J. 2006. Cadherins in development: Cell adhesion, sorting, and tissue morphogenesis. *Genes Dev* 20(23): 3199–3214. doi:10.1101/gad.1486806.

Hall, B. K. and Miyake, T. 2000. All for one and one for all: Condensations and the initiation of skeletal development. *Bioessays* 22(2): 138–147.

Hall, J. D., McCourt, R. M., and Delwiche, C. F. 2008. Patterns of cell division in the filamentous Desmidiaceae, close green algal relatives of land plants. *Am J Bot* 95(6): 643–654. doi:10.3732/ajb.2007210.

Heisenberg, C. P. and Bellaiche, Y. 2013. Forces in tissue morphogenesis and patterning. *Cell* 153(5): 948–962. doi:10.1016/j.cell.2013.05.008.

Hernández-Hernández, V., Niklas, K. J., Newman, S. A., and Benítez, M. 2012. Dynamical patterning modules in plant development and evolution. *Int J Dev Biol* 56(9): 661–674. doi:10.1387/ijdb.120027mb.

Hilbe, C., Wu, B., Traulsen, A., and Nowak, M. A. 2014. Cooperation and control in multiplayer social dilemmas. *Proc Natl Acad Sci U S A* 111(46): 16425–16430. doi:10.1073/pnas.1407887111.

Horton, J. C. and Hocking, D. R. 1996. An adult-like pattern of ocular dominance columns in striate cortex of newborn monkeys prior to visual experience. *J Neurosci* 16(5): 1791–1807.

Huang, K. and Fingar, D. C. 2014. Growing knowledge of the mTOR signaling network. *Semin Cell Dev Biol* 36: 79–90. doi:10.1016/j.semcdb.2014.09.011.

James, T. Y., Stenlid, J., Olson, A., and Johannesson, H. 2008. Evolutionary significance of imbalanced nuclear ratios within heterokaryons of the basidiomycete fungus Heterobasidion parviporum. *Evolution* 62(9): 2279–2296. doi:10.1111/j.1558-5646.2008.00462.x.

Jinks, J. L. 1952. Heterokaryosis; A system of adaption in wild fungi. *Proc R Soc Lond B Biol Sci* 140(898): 83–99.

Kaltner, H. and Gabius, H. J. 2012. A toolbox of lectins for translating the sugar code: The galectin network in phylogenesis and tumors. *Histol Histopathol* 27(4): 397–416.

Kaneko, K. and Yomo, T. 1999. Isologous diversification for robust development of cell society. *J Theor Biol* 199(3): 243–356.

Karner, C., Wharton, K. A., and Carroll, T. J. 2006a. Apical-basal polarity, Wnt signaling and vertebrate organogenesis. *Semin Cell Dev Biol* 17(2): 214–222.

Karner, C., Wharton, Jr., K. A., and Carroll, T. J. 2006b. Planar cell polarity and vertebrate organogenesis. *Semin Cell Dev Biol* 17(2): 194–203.

Keller, R., Shook, D., and Skoglund, P. 2008. The forces that shape embryos: Physical aspects of convergent extension by cell intercalation. *Phys Biol* 5(1): 15007.

Kirk, D. L. 2005. A twelve-step program for evolving multicellularity and a division of labor. *Bioessays* 27(3): 299–310. doi:10.1002/bies.20197.

Komarova, N. L. 2014. Spatial interactions and cooperation can change the speed of evolution of complex phenotypes. *Proc Natl Acad Sci U S A* 111(Suppl 3): 10789–10795. doi:10.1073/pnas.1400828111.

Kondo, S. and Miura, T. 2010. Reaction-diffusion model as a framework for understanding biological pattern formation. *Science* 329(5999): 1616–1620. doi:10.1126/science.1179047.

Krieg, M., Arboleda-Estudillo, Y., Puech, P. H., Kafer, J., Graner, F., Muller, and Heisenberg, C. P. 2008. Tensile forces govern germ-layer organization in zebrafish. *Nat Cell Biol* 10(4): 429–436. doi:10.1038/ncb1705.

Kutschera, U. and Niklas, K. J. 2016. The evolution of the plant genome-to-morphology auxin circuit. *Theory Biosci* 135(3): 175–186. doi:10.1007/s12064-016-0231-0.

Lander, A. D. 2007. Morpheus unbound: Reimagining the morphogen gradient. *Cell* 128(2): 245–256.

Lanna, E. 2015. Evo-devo of non-bilaterian animals. *Genet Mol Biol.* doi:10.1590/S1415-475738320150005.

Levin, M. 2014. Molecular bioelectricity: How endogenous voltage potentials control cell behavior and instruct pattern regulation in vivo. *Mol Biol Cell* 25(24): 3835–3850. doi:10.1091/mbc.E13-12-0708.

Levine, H. and Ben-Jacob, E. 2004. Physical schemata underlying biological pattern formation-examples, issues and strategies. *Phys Biol* 1(1–2): P14–P22. doi:10.1088/1478-3967/1/2/P01.

Lloyd, J. P. and Davies, B. 2013. SMG1 is an ancient nonsense-mediated mRNA decay effector. *Plant J* 76(5): 800–810. doi:10.1111/tpj.12329.

Loh, K. M., van Amerongen, R., and Nusse, R. 2016. Generating cellular diversity and spatial form: Wnt signaling and the evolution of multicellular animals. *Dev Cell* 38(6): 643–655. doi:10.1016/j.devcel.2016.08.011.

Long, M., VanKuren, N. W., Chen, S., and Vibranovski, M. D. 2013. New gene evolution: Little did we know. *Annu Rev Genet* 47: 307–333. doi:10.1146/annurev-genet-111212-133301.

Maamar, H., Raj, A., and Dubnau, D. 2007. Noise in gene expression determines cell fate in *Bacillus subtilis*. *Science* 317(5837): 526–529. doi:10.1126/science.1140818.

Maître, J. L., Berthoumieux, H., Krens, S. F., Salbreux, G., Julicher, F., Paluch, E., and Heisenberg, C. P. 2012. Adhesion functions in cell sorting by mechanically coupling the cortices of adhering cells. *Science* 338(6104): 253–256. doi:10.1126/science.1225399.

Mayr, E. 1988. *Toward a New Philosophy of Biology*. Cambridge, MA: Harvard University Press.

Mayr, E. 1982. *The Growth of Biological Thought: Diversity, Evolution, and Inheritance*. Cambridge, MA: Belknap Press.

McGhee, G. R. 1999. *Theoretical Morphology: The Concept and Its Applications, Perspectives in Paleobiology and Earth History*. New York: Columbia University Press.

Michod, R. E. and Roze, D. 2001. Cooperation and conflict in the evolution of multicellularity. *Heredity* 86(Pt 1): 1–7.

Miller, D. J. and Ball, E. E. 2005. Animal evolution: The enigmatic phylum placozoa revisited. *Curr Biol* 15(1): R26–R28.

Miller, P. W., Clarke, D. N., Weis, W. I., Lowe, C. J., and Nelson, W. J. 2013. The evolutionary origin of epithelial cell-cell adhesion mechanisms. *Curr Top Membr* 72: 267–311. doi:10.1016/B978-0-12-417027-8.00008-8.

Nance, J. 2014. Getting to know your neighbor: Cell polarization in early embryos. *J Cell Biol* 206(7): 823–832. doi:10.1083/jcb.201407064.

Nance, J. and Zallen, J. A. 2011. Elaborating polarity: PAR proteins and the cytoskeleton. *Development* 138(5): 799–809. doi:10.1242/dev.053538.

Nanjundiah, V. and Sathe, S. 2011. Social selection and the evolution of cooperative groups: The example of the cellular slime moulds. *Integr Biol (Camb)* 3(4): 329–342. doi:10.1039/c0ib00115e.

Narasimhan, V. M., Hunt, K. A., Mason, D., Baker, C. L., Karczewski, K. J., Barnes, M. R., Barnett, A. H. et al. 2016. Health and population effects of rare gene knockouts in adult humans with related parents. *Science* 352(6284): 474–477. doi:10.1126/science.aac8624.

Newman, S. A. 2011a. Animal egg as evolutionary innovation: A solution to the embryonic hourglass puzzle. *J Exp Zool B Mol Dev Evol* 316(7): 467–483. doi:10.1002/jez.b.21417.

Newman, S. A. 2011b. The developmental specificity of physical mechanisms. *Ludus Vitalis* 19(36): 343–351.

Newman, S. A. 2014a. Physico-genetics of morphogenesis: The hybrid nature of developmental mechanisms. In *Towards a Theory of Development* (Eds.) Minelli, A. and Pradeu, T., pp. 95–113. Oxford, London: Oxford University Press.

Newman, S. A. 2014b. Why are there eggs? *Biochem Biophys Res Commun* 450(3): 1225–1230. doi:10.1016/j.bbrc.2014.03.132.

Newman, S. A. 2016a. Biogeneric developmental processes: Drivers of major transitions in animal evolution. *Phil Trans R Soc Lond B Sci* 371(1701): 20150444.

Newman, S. A. 2016b. Origination, variation, and conservation of animal body plan development. *Rev Cell Biol Mol Med* 2(3): 130–162.

Newman, S. A. and Bhat, R. 2007. Activator-inhibitor dynamics of vertebrate limb pattern formation. *Birth Defects Res C Embryo Today* 81(4): 305–319.

Newman, S. A. and Bhat, R. 2008. Dynamical patterning modules: Physico-genetic determinants of morphological development and evolution. *Phys Biol* 5(1): 15008.

Newman, S. A. and Bhat, R. 2009. Dynamical patterning modules: A pattern language for development and evolution of multicellular form. *Int J Dev Biol* 53(5–6): 693–705. doi:10.1387/ijdb.072481sn.

Newman, S. A., Bhat, R., and Mezentseva, N. V. 2009. Cell state switching factors and dynamical patterning modules: Complementary mediators of plasticity in development and evolution. *J Biosci* 34(4): 553–572.

Newman, S. A., Glimm, T., and Bhat, R. 2018. The vertebrate limb: An evolving complex of self-organizing systems. *Prog Biophys Mol Biol.* doi: 10.1016/j.pbiomolbio.2018.01.002.

Newman, S. A. and Linde-Medina, M. 2013. Physical determinants in the emergence and inheritance of multicellular form. *Biol Theory* 8: 274–285.

Newman, S. A. and Müller, G. B. 2005. Origination and innovation in the vertebrate limb skeleton: an epigenetic perspective. *J Exp Zoolog B Mol Dev Evol* 304(6): 593–609.

Nichols, S. A., Roberts, B. W., Richter, D. J., Fairclough, S. R., and King, N. 2012. Origin of metazoan cadherin diversity and the antiquity of the classical cadherin/beta-catenin complex. *Proc Natl Acad Sci USA* 109(32): 13046–13051. doi:10.1073/pnas.1120685109.

Nielsen, C. 2012. *Animal Evolution: Interrelationships of the Living Phyla.* 3rd edition. Oxford, UK: Oxford University Press.

Niklas, K. J. 2000. The evolution of plant body plans—A biomechanical perspective. *Ann Bot* 85: 411–438.

Niklas, K. J. and Newman, S. A. 2013. The origins of multicellular organisms. *Evol Dev* 15(1): 41–52. doi:10.1111/ede.12013.

Niklas, K. J. 2016. *Plant Evolution: An Introduction to the History of Life.* Chicago, IL: The University of Chicago Press.

Niklas, K. J. and H. C. Spatz. 2012. *Plant physics.* Chicago, IL: The University of Chicago Press.

Odling-Smee, J., Laland, K. N., and Feldman, M. W. 2003. *Niche Construction: the Neglected Process in Evolution.* Princeton, NJ: Princeton University Press.

Prochiantz, A. and Joliot, A. 2003. Can transcription factors function as cell-cell signalling molecules? *Nat Rev Mol Cell Biol* 4(10): 814–819.

Raspopovic, J., Marcon, L., Russo, L., and Sharpe, J. 2014. Modeling digits. Digit patterning is controlled by a Bmp-Sox9-Wnt Turing network modulated by morphogen gradients. *Science* 345(6196): 566–570. doi:10.1126/science.1252960.

Rebscher, N., Deichmann, C., Sudhop, S., Fritzenwanker, J. H., Green, S., and Hassel, M. 2009. Conserved intron positions in FGFR genes reflect the modular structure of FGFR and reveal stepwise addition of domains to an already complex ancestral FGFR. *Dev Genes Evol* 219(9–10): 455–468. doi:10.1007/s00427-009-0309-5.

Requejo, R. J. and Camacho, J. 2013. Scarcity may promote cooperation in populations of simple agents. *Phys Rev E Stat Nonlin Soft Matter Phys* 87(2): 022819. doi:10.1103/PhysRevE.87.022819.

Rexin, D., Meyer, C., Robaglia, C., and Veit, B. 2015. TOR signalling in plants. *Biochem J* 470(1): 1–14. doi:10.1042/BJ20150505.

Rieseberg, L. H., Archer, M. A., and Wayne, R. K. 1999. Transgressive segregation, adaptation and speciation. *Heredity* 83(Pt 4): 363–372.

Rieseberg, L. H., Widmer, A., Arntz, A. M., and Burke, J. M. 2003. The genetic architecture necessary for transgressive segregation is common in both natural and domesticated populations. *Philos Trans R Soc Lond B Biol Sci* 358(1434): 1141–1147.

Rosenberg, L. E. 2008. Legacies of Garrod's brilliance. One hundred years–and counting. *J Inherit Metab Dis* 31(5): 574–579. doi:10.1007/s10545-008-0985-8.

Sachs, J. 1892. Beiträge zur Zellentheorie. Energiden und Zellen. *Flora* 75: 57–67.

Salazar-Ciudad, I., Jernvall, J., and Newman, S. A. 2003. Mechanisms of pattern formation in development and evolution. *Development* 130(10): 2027–2037.

Salazar-Ciudad, I. and Marín-Riera, M. 2013. Adaptive dynamics under development-based genotype-phenotype maps. *Nature* 497(7449): 361–364. doi:10.1038/nature12142.

Salazar-Ciudad, I., Solé, R. V., and Newman, S. A. 2001. Phenotypic and dynamical transitions in model genetic networks. II. Application to the evolution of segmentation mechanisms. *Evol Dev* 3: 95–103.

Santorelli, L. A., Thompson, C. R., Villegas, E., Svetz, J., Dinh, C., Parikh, A., Sucgang, R. et al. 2008. Facultative cheater mutants reveal the genetic complexity of cooperation in social amoebae. *Nature* 451(7182): 1107–1110. doi:10.1038/nature06558.

Savagner, P. 2015. Epithelial-mesenchymal transitions: From cell plasticity to concept elasticity. *Curr Top Dev Biol* 112: 273–300. doi:10.1016/bs.ctdb.2014.11.021.

Sax, D. F., Early, R., and Bellemare, J. 2013. Niche syndromes, species extinction risks, and management under climate change. *Trends Ecol Evol* 28(9): 517–523. doi:10.1016/j.tree.2013.05.010.

Sebé-Pedrós, A., Irimia, M., Del Campo, J., Parra-Acero, H., Russ, C., Nusbaum, C., Blencowe, B. J., and Ruiz-Trillo, I. 2013. Regulated aggregative multicellularity in a close unicellular relative of metazoa. *Elife* 2: e01287. doi:10.7554/eLife.01287.

Seiler, S. and Justa-Schuch, D. 2010. Conserved components, but distinct mechanisms for the placement and assembly of the cell division machinery in unicellular and filamentous ascomycetes. *Mol Microbiol* 78(5): 1058–1076. doi:10.1111/j.1365-2958.2010.07392.x.

Shyer, A. E., Tallinen, T., Nerurkar, N. L., Wei, Z., Gil, E. S., Kaplan, D. L., Tabin, C. J., and Mahadevan, L. 2013. Villification: How the gut gets its villi. *Science* 342(6155): 212–218. doi:10.1126/science.1238842.

Sinha, S. and Sridhar, S. 2015. *Patterns in Excitable Media: Genesis, Dynamics, and Control.* Boca Raton, FL: CRC Press.

Solé, R. V. and Valverde, S. 2013. Before the endless forms: Embodied model of transition from single cells to aggregates to ecosystem engineering. *PLoS One* 8(4): e59664. doi:10.1371/journal.pone.0059664.

Steinberg, M. S. 2007. Differential adhesion in morphogenesis: A modern view. *Curr Opin Genet Dev* 17(4): 281–286. doi:10.1016/j.gde.2007.05.002.

True, J. R. and Haag, E. S. 2001. Developmental system drift and flexibility in evolutionary trajectories. *Evol Dev* 3(2): 109–119.

Tsarfaty, I., Resau, J. H., Rulong, S., Keydar, I., Faletto, D. L., and Vande Woude, G. F. 1992. The met proto-oncogene receptor and lumen formation. *Science* 257(5074): 1258–1261.

Tsarfaty, I., Rong, S., Resau, J. H., Rulong, S., da Silva, P. P., and Vande Woude, G. F. 1994. The Met proto-oncogene mesenchymal to epithelial cell conversion. *Science* 263(5143): 98–101.

Tucker, R. P. 2004. The thrombospondin type 1 repeat superfamily. *Int J Biochem Cell Biol* 36(6): 969–974. doi:10.1016/j.biocel.2003.12.011.

Tucker, R. P. 2013. Horizontal gene transfer in choanoflagellates. *J Exp Zool B Mol Dev Evol* 320(1): 1–9. doi:10.1002/jez.b.22480.

Turing, A. M. 1952. The chemical basis of morphogenesis. *Phil Trans Roy Soc Lond B* 237: 37–72.

Walker, L. A., Lenardon, M. D., Preechasuth, K., Munro, C. A., and Gow, N. A. 2013. Cell wall stress induces alternative fungal cytokinesis and septation strategies. *J Cell Sci* 126(Pt 12): 2668–2677. doi:10.1242/jcs.118885.

Webster, G. and Goodwin, B. C. 1996. *Form and Transformation: Generative and Relational Principles in Biology.* Cambridge, UK: Cambridge University Press.

Widelitz, R. B. and Chuong, C. M. 1999. Early events in skin appendage formation: Induction of epithelial placodes and condensation of dermal mesenchyme. *J Invest Dermatol Symp Proc* 4(3): 302–306.

Winfree, A. T. 1994. Persistent tangled vortex rings in generic excitable media. *Nature* 371(6494): 233–236.

Yamaguchi, M., Yoshimoto, E., and Kondo, S. 2007. Pattern regulation in the stripe of zebrafish suggests an underlying dynamic and autonomous mechanism. *Proc Natl Acad Sci U S A* 104(12): 4790–4793.

Yang, Y. 2009. Skeletal morphogenesis during embryonic development. *Crit Rev Eukaryot Gene Expr* 19(3): 197–218.

Yang, Z., Huck, W. T., Clarke, S. M., Tajbakhsh, A. R., and Terentjev, E. M. 2005. Shape-memory nanoparticles from inherently non-spherical polymer colloids. *Nat Mater* 4(6): 486–490.

Zhang, X., Facette, M., Humphries, J. A., Shen, Z., Park, Y., Sutimantanapi, D., Sylvester, A. W., Briggs, S. P., and Smith, L. G. 2012. Identification of PAN2 by quantitative proteomics as a leucine-rich repeat-receptor-like kinase acting upstream of PAN1 to polarize cell division in maize. *Plant Cell* 24(11): 4577–4589. doi:10.1105/tpc.112.104125.

Index

Note: Page numbers followed by f, t, and n refer to figures, tables and notes, respectively.

Printed and bound by CPI Group (UK) Ltd, Croydon, CR0 4YY

24/10/2024

01778308-0009